Second Edition

Geology
of the
Pacific Northwest

S0-AVN-044

Second Edition

Geology
of the
Pacific Northwest

William N. Orr
University of Oregon
Elizabeth L. Orr
University of Oregon

WAVELAND
PRESS, INC.
Long Grove, Illinois

For information about this book, contact:
Waveland Press, Inc.
4180 IL Route 83, Suite 101
Long Grove, IL 60047-9580
(847) 634-0081
info@waveland.com
www.waveland.com

Copyright © 2002, 1996 by William N. Orr and Elizabeth L. Orr
Reissued 2006 by Waveland Press, Inc.

10-digit ISBN 1-57766-480-9
13-digit ISBN 978-1-57766-480-2

All rights reserved. No part of this book may be reproduced, stored in a retrieval system, or transmitted in any form or by any means without permission in writing from the publisher.

Printed in the United States of America

7   6   5   4   3

Dedicated to the Wood Sprite and Earth Mother

CONTENTS

CHAPTER 11

Coast Province of Oregon and Washington   260

PREFACE TO THE SECOND EDITION

Since publication of the first edition, public appreciation of geologic processes and hazards has improved substantially.

Although the eruption of Mount St. Helens is two decades in the past, we continue to learn from that episode. Recovery of the landscape from the devastation has been more rapid than imagined possible. Additionally, the close monitoring of activity just prior to that event has helped in the prognosis of several volcanic eruptions elsewhere that resulted in lives saved.

Continued oceanographic exploration of the fabric of offshore microplates has revealed far more undersea activity than expected. Revisiting several sites examined during the 1970s and 1980s shows profound changes to the sea floor in the vicinity of spreading ridges as well as along major fault traces.

The role of plate subduction and ensuing earthquakes in northwest geologic history is becoming clearer. New evidence on the age of the landslide at Bonneville along the Columbia River suggests it was triggered by the last great earthquake 300 years ago. Seismic studies have even succeeded in tying specific quakes in the Pacific Northwest to catastrophic tsunamis in Japan. Perhaps most important, an awareness of subduction earthquakes and resulting tsunamis has im-proved to the point that the media and politicians have begun to talk about preparedness. If even some of the modest predictions are realized, a significant quake event could have a critical impact on the Northwest.

An important trend is the increasing focus on environmental concerns as they relate to geology and mining. In the face of a healthy economy, old mines have been reopened, while those under operation have continued and expanded. Under mandatory environmental regulations, new mining strategies as well as cleanup are having a salutary effect. Running counter to these concerns, mining laws dating back to the 1800s remain unchanged.

Geology is playing an expanding role in urban development because of the dispersal of human populations throughout the region. Efforts to fill in and populate areas previously regarded as unbuildable due to implicit geologic hazards have placed many homeowners in harm's way. Minimal appreciation of earth processes would have curbed such decisions.

Thanks go to the following people for reviewing the second edition manuscript: Anthony J. Irving, University of Washington, Paul Karl Link, Idaho State University, and Steve Macias, Olympic College.

Eugene, Oregon, 2001

PREFACE TO THE FIRST EDITION

One of the pleasures of the past 28 years in the Pacific Northwest has been to participate in and watch the unfolding of geologic thought and theory as it is applied to this unique area. In that regard our motive was to produce a book which explains the role of tectonic plate movements and particularly accretionary events as fundamental processes of northwest geologic history. Only after the Pacific Northwest had been fabricated from exotic pieces did sedimentary layers, ash, lavas, and glacial debris cover the original terranes.

It is hoped our audience will appreciate that the events which took place here beginning millions of years in the past are part of an ongoing progression. Even though such a vast area as the Pacific Northwest is treated in some detail, the casual reader as well as those with geologic training should be able to make use of this volume.

This manuscript was written during the years 1994 and 1995, and in keeping with the peregrinations of continental plates that permeate the text, we wrote the book while living at Friday Harbor, Washington, Christchurch, New Zealand, Hood River, and Charleston, Oregon.

We owe a great deal to others for research assistance, reviews, comments, and photographs. Particularly beneficial chapter reviews were by John Armentrout, Mobil Research, Robert Carson, Whitman College, Darrel Cowan, University of Washington, Paul Hammond, Portland State University, Greg Harper, SUNY, Buffalo, Fred Miller, U.S. Geological Survey, Spokane, Ray Price, Queens University, Jack Rice, University of Oregon, and Tracy Vallier, U.S. Geological Survey, Menlo Park. We appreciated assistance by Reed Hollinshead at the Idaho Department of Transportation, Carol McKillip, Coquille, Elice Mattison, California Division of Mines and Geology, Norton Orr, Beaverton, Cindy Poon at the Canadian Geological Survey Photo Library, John Rensberger, University of Washington, Weldon Rau, Washington Division of Geology and Earth Resources, Jim Walker and staff at the Washington Department of Transportation, personnel and facilities of the University of Oregon Map Library, the Biology Laboratory of the University of Washington, and Klaus Neuendorf, Oregon Department of Geology and Mineral Industries. The staffs at the British Columbia Archives and Record Service and the Ministry of Environment Lands and Parks, the Special Collections at the University of Oregon, the Craters of the Moon National Monument and Hagerman Fossil Beds in Idaho, as well as the Idaho and Siskiyou County Historical Societies were most pleasant and accommodating. Finally, we relied on the facilities of the public library at Hood River.

We especially would like to thank Chris Suczek at Western Washington University, Bellingham, and Greg Wheeler of California State University, Sacramento, for their assistance and comments on the entire manuscript.

Eugene, Oregon, 1995

Geologic time scale chart — *Geologic events and environments of the northwest.*

ERA	PERIOD	EPOCH	M.Y.A	PLATE MOVEMENTS	VOLCANICS & INTRUSIVES	CLIMATIC CHANGE & ENVIRONMENT
Cenozoic	Quaternary	Holocene	.01	Continued subduction of Juan de Fuca Plate with major earthquakes in N.W. on a 300/500 yr. cycle	Eastern Snake River basalt flows	Lake Bonneville flood
Cenozoic	Quaternary	Pleistocene	1.8		Lavas of Modoc Plateau and Medicine Lake Highlands begin	Multiple Missoula glacial floods / Continental glaciers cover much of British Columbia
Cenozoic	Tertiary	Pliocene	5.3		High Cascade Volcanism	Temperate climate
Cenozoic	Tertiary	Miocene	23.7	Uplift of Coast Range in Oregon/Washington	Multiple ash flow tuffs east of Cascades / Columbia River/Chilcotin back arc spreading volcanism — Snake River plain volcanism begins / Strawberry Volcanics / Steens basalt	Dry cooler climate east of the Cascades / Development of Cascade rainshadow
Cenozoic	Tertiary	Oligocene	36.6	Collision of multiple small terranes in N. Washington Coast Range and S.E. British Columbia	John Day volcanism in E. Oregon	
Cenozoic	Tertiary	Eocene	55	Kula plate completely subducted beneath N. America / Decrease in rate of Farallon plate subduction / Steeper angle to subducting plate	Western Cascade volcanism begins / Clarno-Challis volcanism in E. Oregon, Idaho / Volcanic origin of Oregon/Washington Coast Range as Crescent/Siletz terrane	Warm, wet tropical climate
Cenozoic	Tertiary	Paleocene	66	Increase in rate of Farallon plate subduction / Shallower angle of subduction	Mechosin volcanics in B.C.	
Mesozoic	Cretaceous		144	Folding of B.C. Coast Range / Insular superterrane collides with N. America	Idaho batholith / Batholith intrusions throughout northwest	Major Late Mesozoic transgressive seas cover N.W.
Mesozoic	Jurassic		208	Folding of Omineca belt / Intermontane superterrane collides with N. America	Separation of North American continental plate with opening of the Atlantic Ocean	
Mesozoic	Triassic		245	Breakup of Pangea supercontinent begins		
Paleozoic	Permian		286	Final assembly of Pangaea supercontinent		Phosphoria Fm. in inland sea of E. Idaho
Paleozoic	Pennsylvanian		320		Intermontane, Insular terranes develop as island arcs in S.W. Pacific	
Paleozoic	Mississippian		360	Ocean shoreline across Washington and Idaho		
Paleozoic	Devonian		408	Collision of Klamath arc and development of Antler highlands from Montana to southern California		Blue Mtn. limestones and coral reefs form in tropical Pacific Ocean
Paleozoic	Silurian		440	Assembly of Pangea begins		
Paleozoic	Ordovician		510	E. Klamath volcanic archipelago begins to form		
Paleozoic	Cambrian		540			
Proterozoic			2500	Breakup of early supercontinent / Formation of early supercontinent	Oldest rocks in Pacific Northwest — Purcell-Belt supergroup (1.3 to .8 billion yrs.)	
Archean			4000		Oldest known rocks on earth	
			4500		Formation of the earth	

E.W.0.

Geologic events and environments of the northwest.

CORNERSTONES OF PACIFIC NORTHWEST GEOLOGY

Along the Kitsumkalum River, east of Prince Rupert on the British Columbia coast, unmistakable U-shaped valleys, silt-laden rivers, and mountain peaks dusted with snow are all signs of ongoing geologic activity in this region (B.C. Ministry of Environment, Lands, and Parks).

Over three decades ago the earth sciences experienced a major revolution when the old idea of continental drift was reexamined in light of more recent discoveries and restructured as the theory of global or plate tectonics. During plate-tectonic processes, as enormous crustal pieces move, continents on the surface break up and separate. If continental fragments are pulled away from each other, they must collide and merge elsewhere. Clearly to change from a system of rigid fixed continents to one of multiple crustal plates splitting and rejoining required significant adjustments throughout the realm of geologic thought.

As the implications of plate tectonics were more widely recognized, it became obvious that tectonism or mountain building played a critical role in most of the natural processes of the earth. Throughout the millennia, a wide variety of tectonic activities altered the physical architecture of ancient continents. These included growth by the intermittent addition of terranes—fragments of crustal rocks—to the continental margins as well as by severe and ongoing deformation by folding and faulting, and alteration by heat and pressure of metamorphism. One of the most encompassing consequences of tectonics is volcanism that can modify landscapes and even climates. Tremendous changes in erosion, sedimentation, temperature, and rainfall are brought on by volcanic activity, which normally destroys local plants and animals, even as it opens new environments. These ongoing periodic events are reflected by the quality, quantity, and distribution of fossil remains.

Although today there is general agreement as to the speed and direction of moving tectonic plates, details of the mechanisms by which these great slabs move elude geologists. Similarly, the process of delineating and mapping smaller terranes embedded in larger continents is well underway, but the tracing of these fragments back to their points of origin is just beginning.

OVER THE PAST 400 MILLION YEARS, the movement of tectonic plates has affected the Pacific Northwest as the eastern Pacific margin from California to Alaska experienced multiple collisions of small and large pieces of the crust. Also called terranes, tectonic slabs separate, move, and collide in predictable patterns. Plates split apart and drift away from each other, allowing cracks to open and fresh lava to erupt onto the ocean floor. On a large scale, extensive volcanic episodes may even cause global sea levels to rise.

Often fragments of the oceanic crust first amalgamated or clustered together before being swept up by the western edge of the North American continent. Since oceanic plates are thinner and heavier than the continental plate, they are typically carried beneath the larger continent to be melted and recycled in a process called subduction. Other pieces are scraped off the upper surface of the descending slab, piling up in a broken mixture or melange, and wedged against the edge of the larger continent in the accretion process. At least 100 identifiable terrane fragments, slivers of the earth's crust, have been annexed to the eastern Pacific Rim, each collision event effectively increasing the coastal margin by tens of miles.

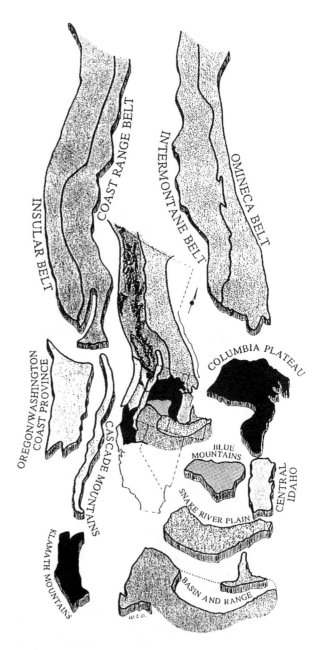

Figure 1.1 The Pacific Northwest can be subdivided into a dozen physiographic areas and almost 50 smaller terranes and geologic provinces.

Terrane rocks reflect a variety of previous oceanic settings from tropical island chains with fringing coral reefs to high latitude cool waters. These environments can be deciphered by examining the package of fossils and rocks unique to each terrane. Marine animals are useful in correlating time and events as well as determining relationships between individual terrane fragments.

Near the base of the terrane slabs, ophiolites, or sequences of deep ocean crust containing fossiliferous marine clays, pillow lavas, and intrusive rocks, have been embedded in the continent by plate collisions. Because they have been impregnated with minerals dissolved in hot springs on the ocean floor, ophiolites play an important economic role in the Pacific Northwest.

The process of terrane annexation by collision and accretion triggers both deep intrusive batholiths as well as extrusive or surface volcanic activity. Once the leading edge of the subducting plate sinks from 50 to 75 miles below the continent, the crust thickens at the same time that intense pressure and elevated temperatures partially liquify the rocks. These fluids invade the surrounding strata as small plutons or larger batholiths that bind the descending slab to the overriding mass upon cooling. As terranes adhere, the subduction process stops, and a new collision zone opens on the oceanic side of the freshly accreted fragments.

Accompanying collision and accretion, the hot liquid rock or magma from the melting process makes its way through the upper slab to appear on the surface as lava and volcanic debris. When this happens offshore, the resulting volcanoes may project above the water in an island chain or archipelago. Volcanic islands and ocean plateaus, which are carried toward the subduction

Figure 1.2 Lava that erupts underwater assumes a distinctive pillow shape (courtesy M.F. Meier, U.S. Geological Survey).

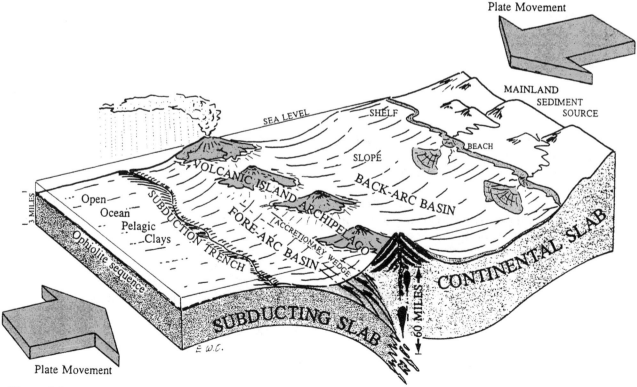

Figure 1.3 A variety of environments are associated with colliding tectonic plates.

Figure 1.4 Lying only a few miles offshore from the Golden Gate Bridge at San Francisco, the Farallon Islands gave their name to the once splendid Farallon plate, now largely subducted beneath North America (California Division of Mines and Geology).

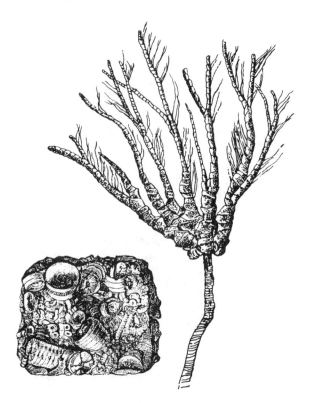

Figure 1.5 Criniods or sea lilies were common marine fossils in terrane rocks of the Pacific Northwest.

trench by continued motion of the plates, eventually merge and accrete to the mainland.

Between the trench that marks the subduction point and the island arc, a shallow fore-arc basin may form, whereas a back-arc basin separates the volcanic islands and the continent. Over long spans of time, these basins are filled with sediments eroded from both the archipelago and mainland, preserving a record of ongoing environmental changes and collisions.

Magma, which emerges on land as lava and ash, erupts through conduits, cracks, and fissures in the crust to affect the topography and environment in a number of ways. Stratovolcanoes, huge calderas, incandescent ash clouds, and widespread lava flows wreck havoc on the local erosion and deposition cycle. Ash dispersing high in the atmosphere can even cause dramatic temperature fluctuations on a global scale. Volcanic flows or heavy ash falls, which kill and bury plants and animals, create new habitats by damming streams to develop lakes. Soils that form on the surface of lavas provide fertile sites for further growth. These cyclic processes have interacted throughout western North America over the past millions of years.

FOR SUCH A SMALL AREA, the northeast Pacific Rim has a complex structural makeup dominated by four large crustal plates, the Farallon, Kula, Pacific, and North American. Toward the end of the Cretaceous period and beginning of the Tertiary, these plates merged with one another along the west coast. The fast-moving northbound Kula plate, originally derived from the slower east-moving Farallon in the Late Mesozoic, was subducted beneath Alaska by Late Eocene time. The

Farallon plate is about to suffer a similar fate beneath the North American plate, and its tiny trailing edges, renamed the Gorda, Juan de Fuca, and Explorer, are found today as microplates off the coast of northern California, Oregon, Washington, and British Columbia.

Faults and ocean ridge systems, which define the boundaries between plates and plate margins, are the focus of continuous earthquake and volcanic activity. The Cascadia subduction zone, which runs parallel to the coast northward from Cape Mendocino, California, to British Columbia, marks the western-most edge of the North American plate. The Mendocino fault is the boundary between the Pacific, Gorda, and North American plates; and the Cape Mendocino junction, where the south end of Cascadia subduction zone meets the Mendocino and San Andreas faults, is the site of numerous recent earthquakes.

TECTONIC PLATE MOVEMENTS, EROSION, GLACIATION, and similar geologic processes are almost painfully slow, but looking at these activities throughout the enormous span of time covered in the history of the earth imparts a perspective not found elsewhere. Geologic time measured in millions or even billions of years is compressed into a remarkably thin rock record. Despite the critical importance of geologic time to ancient events and environments, the tendency to trivialize its effect is almost irresistable. For example, over

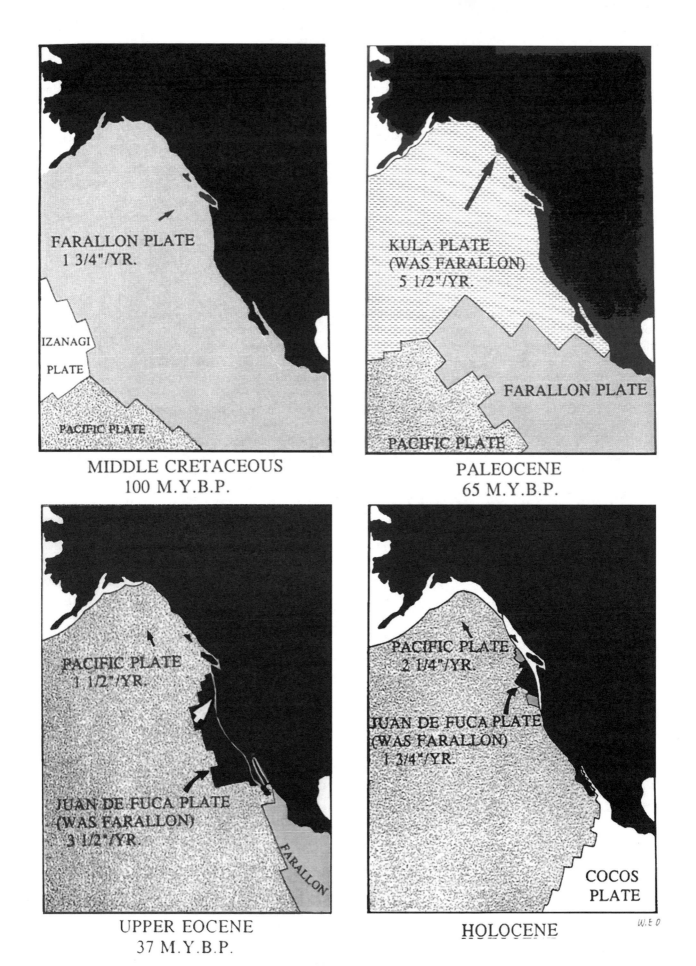

Figure 1.6 From the Late Mesozoic to the present, four major plates have governed the tectonic evolution of the Pacific Northwest. Arrows show plate direction and speed (after Engebretson, Cox, and Gordon, 1985).

an 11 million year interval the Columbia River basalts flooded across the Pacific Northwest to extrude more than 42,000 cubic miles of lava. What at first glance appears as a major volcanic episode is diminished somewhat when it is noted that individual flows occurred at intervals averaging 35,000 years apart. Like a movie projector run at high speed or in slow motion, geologic time can be made to emphasize or minimize any natural event.

IN THE LATE PRECAMBRIAN AND EARLY PALEOZOIC, 1 billion to 400 million years ago, Pacific Northwest geology involved the opening of a great rift across a large supercontinent. North America, Europe, and Asia, as a single vast continent, broke free from an immense landmass that lay to the west. As the crustal slabs separated, submarine volcanic eruptions along cracks or ocean spreading ridges between the plates displaced a sufficient amount of water to elevate global sea levels.

On a wide continental shelf across British Columbia, Washington, Oregon, southwest Idaho, Nevada, and northern California up to 30,000 feet of marine sediments accumulated. Beneath the fluctuating ocean waters a thin wedge of chert, mudstone, siltstone, volcanic rocks, and limestone along the shoreline thickened westward in the deepening seaway from a thin edge. Some of the oldest rocks in the Pacific Northwest are Late Pre-

cambrian Belt-Purcell sediments deposited in marine embayments of the ocean in western Montana, northern Idaho, and southeast British Columbia.

Fossil corals, brachiopods, trilobites, and intermittent algal reefs are locally abundant in these sediments across the ancient Paleozoic shelf. Near Field, British Columbia, the middle Cambrian Burgess shale, with beautifully preserved carbon film impressions of even the fleshy tissue of arthropods and other invertebrates, is one of the most famous fossil localities known. Far from representing simple primitive marine life, this early Paleozoic fauna was diverse and highly evolved.

FROM THE LATE PALEOZOIC through Mesozoic, 300 to 70 million years ago, the shallow marine seas that dominated the northwest were gradually replaced by smaller marginal oceans produced when volcanic island archipelagos collided with the mainland. Late in this interval, the subduction of oceanic crust and the related development, movement, and accretion of island arcs and adjacent basins were the major tectonic processes. These volcanic island systems originated some distance offshore before being transported eastward where several amalgamated or combined with others, all merging with North America. It was during this time span that most of the terranes were affixed to the continent between southern Alaska and central California.

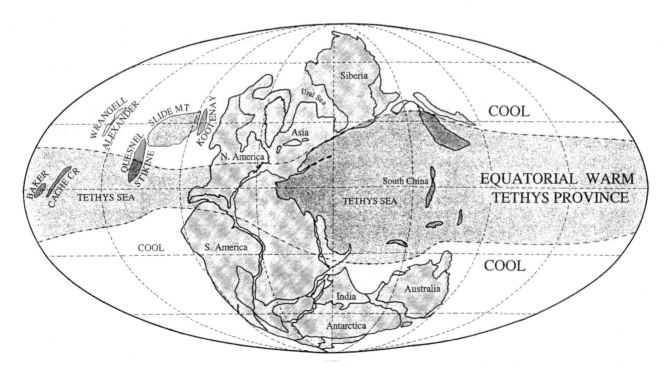

Figure 1.7 Middle Permian global geography shows incoming terranes west of North America (after Ziegler, et al., 1979).

The accretion of terrane rocks took place in parallel belts, which reflect a variety of tectonic and depositional environments. Sediments eroded from the North American mainland and deposited on the continental shelf provide a record of local conditions. In contrast, sediments that accumulated around more distant volcanic island archipelagos to the south reflect tropical marine environments completely unlike those near North America.

Paleozoic and Mesozoic terrane fragments are known from the Blue Mountains of Oregon, Idaho, and Washington, the Klamath Mountains of California and Oregon, the Insular and Intermontane regions of British Columbia and northern Washington, and from the North Cascades of Washington and British Columbia. Smaller terranes are scattered across the northern Basin and Range area of southeast Oregon, Idaho, and northcentral Nevada.

Figure 1.8 As the Fraser River turns east toward Hope, British Columbia, it divides the Skagit terrane that makes up the Skagit Mountains on the right from the Hozameen Range, which is composed of the Hozameen terrane, on the left (B.C. Ministry of Environment, Lands, and Parks).

The diverse Blue Mountains, composed of five accreted terranes, provide a link between the sheet-like exotic rocks of the Klamath Mountains and those in northern Washington and British Columbia. In British Columbia, numerous slabs of the Insular and Intermontane superterranes first merged or amalgamated offshore to be transported and eventually annexed to North America. The Insular had migrated from a subtropical to temperate realm before accreting between Vancouver Island and southeast Alaska. In southern British Columbia and northern Washington the oldest rocks in the North Cascades are a composite platform of terranes dating back over 500 million years. Smaller terranes on southern Vancouver Island, the Strait of Juan de Fuca, and in the Olympic Peninsula were latecomers that were annexed with ongoing subduction during the Tertiary period.

THE IMPACT OF REPEATED terrane collisions severely wrinkled and folded the western margin of North America into the mountainous welts that make up the physiography of the Omineca and Coast Range. At the same time the San Juan Islands, North Cascades, Klamaths, and Blue Mountains were dissected by a network of faults. Simultaneously, fore-arc depressions were forming between subduction trenches and volcanic island arcs. The large Bowser basin in northcentral British Columbia, the Queen Charlotte basin offshore, the Nanaimo on Vancouver Island, and the Chuckanut trough in Washington appeared along the continental margin. Following intensive erosion from surrounding highlands, debris was flushed into these depressions and interlarded with fossil shells and organic material such as plants, which were later converted to coal.

BELTS OF GRANITIC BATHOLITHS mark major accretionary sutures in the Pacific Northwest. The Idaho batholith glued terranes of the Blue Mountains to the continent, just as the Sierra Nevada batholith annealed the Western Sierra-Klamath belt, and the Coast Range plutons in British Columbia bound the Intermontane and Insular superterranes.

Of these plutons, the most immense are the Idaho batholith, which underlies most of central Idaho today, and the Coast Range batholith in British Columbia. Emplaced while terranes were being annexed, plutons were injected as pockets of liquid magma and partial melts into the surrounding strata before cooling deep within the earth's crust. Averaging over 5 miles in thickness, the bulbous subterranean mass of the Idaho batholith was emplaced about 100 million years ago, while the Coast Range granites, composing over 85% of the province, were intruded during a long time interval from the Jurassic to Miocene.

BY THE EARLY CENOZOIC, 55 million years ago, the Pacific Northwest had begun to approximate its present geographic configuration. The coastline, which

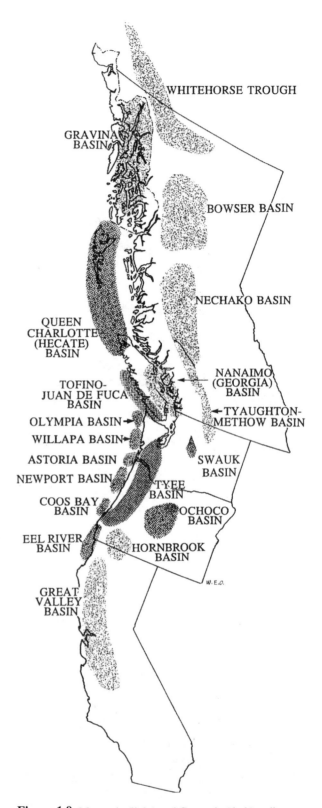

Figure 1.9 Mesozoic (light) and Cenozoic (dark) sediment basins of the northwest (after Haggard, 1993; Snavely, 1987).

Figure 1.10 Smaller islands in the Hecate lowland in British Columbia are granitic batholith rocks, while Vancouver Island, in the distance across Johnston Strait, is composed of accreted terranes (B.C. Ministry of Environment, Lands, and Parks).

ran from California inland through Oregon and Washington to exit near Vancouver Island in Canada, was established by withdrawal of the Cretaceous seaway and uplift of mountain ranges. Warm humid Eocene climates were gradually replaced by cool dry conditions of the Miocene and Pliocene and ultimately by ice age temperatures of the Pleistocene.

The action of shifting crustal plates brought significant changes to the geologic record of the Eocene. These included episodes of volcanism, intrusion, faulting, erosion, and rapid sedimentation. The previous region-wide pattern of deposition into fore-arc basins, along the continental margin was altered to one of local volcanic eruptions and accumulations of sediment. Older rocks were shattered by faults and deeply eroded before being covered.

Ruptures in the crust developed offshore between the Kula and Farallon plates. The newly formed Kula plate was pulled northward, and the Farallon was subducted beneath the continent as an ocean spreading ridge developed between the two slabs. Eocene pillow basalts from submarine eruptive centers along rifts parallel to the coast piled up in a linear volcanic platform. Atop this oceanic plateau, volcanic ash, silt, and sand derived from the rapidly eroding Klamath and Western Cascade mountains accumulated as the foundation of the incipient Coast Range of Oregon and Washington.

In the Middle Eocene another major change in tectonic style occurred when the convergence of the Farallon plate beneath North America slowed, and the angle of the subducting slab steepened. This initiated local volcanics as well as basins throughout a wide interior

Figure 1.11 Early Cenozoic shoreline and environments.

area. Across Montana, Wyoming, Idaho, and northeast Washington, explosive andesitic and rhyolitic activity during the Challis episode overlapped in time with ash and lava of the Sanpoil Formation in northeast Washington and that of the Clarno Formation in northcentral Oregon. Large-leafed tropical plants, seeds, fruits, and an unusual assortment of mammals were repeatedly overwhelmed and buried by blankets of volcanic debris in central Oregon when the loose Clarno ash mixed with water to form thick, heavy mudflows and nearly perfect conditions for fossil preservation.

A variety of smaller sedimentary basins the length of the continent reflect changing plate tectonic patterns. In the Omineca and Intermontane provinces of British Columbia and northern Washington, numerous discontinuous north-south depressions were produced as the brittle crust split and opened when the north-moving Kula plate pulled against the edge of the North Ameri-

can slab. In these depositional centers near Kamloops and Princeton in southcentral British Columbia and Republic, Washington, a varied flora and fauna was entombed in layers of ash and lava mixed with freshwater sediments. West of Wenatchee, Washington, similar depressions, delineated by faults, were filled with sand, mud, and conglomerate. An abundant flora in the Swauk and Chumstick basins here, indicative of stream and lake environments, later yielded coal. Similar basins occurred on the broad swampy coast from the Great Valley of California and Tyee basin of southern Oregon to Puget Sound and the Strait of Juan de Fuca, Washington. Tropical conditions around these basins are reflected by the plant fossils.

A discontinuous chain of metamorphic rock exposures from British Columbia to Mexico may have formed when the crust was distorted by opposing plate motions. The same tensional stresses that produced the sedimen-

Figure 1.12 Late Cenozoic volcanic centers in the northwest (after Wood and Kienle, 1990; Smith and Luedke, 1984).

tary troughs opened sections of the crust to reveal old deeply buried rocks called core complexes. Domed upward, core complexes underlie many smaller mountain ranges in the northwest to reveal much older crustal rocks beneath the surface.

The end of the Eocene epoch brought renewed volcanism immediately east of the coast. This marked the beginning of the older Western Cascade volcanic arc that would ultimately dominate the region through the Middle and Late Tertiary. A significant change in plate movements offshore, as the angle of the subducting Farallon plate flattened, triggered a broad band of Western Cascade volcanic activity. About 40 million years ago volcanoes of the Western Cascades in southern Washington, Oregon, and northern California began to erupt ash and lava, building a plateau of volcanic debris. Miocene elevation and tilting of this platform created an effective atmospheric barrier removing moisture from east-moving oceanic air masses as they rose along the front of the range. Even today this rainshadow persists, creating the dry eastern desert behind a rainy, heavily vegetated western slope.

VOLCANIC EPISODES that began in the Late Eocene continued through the Oligocene and into the Early Miocene. Between 35 and 20 million years ago, low topographic relief and spacious savannahs characterized the inland basins of central Oregon, Washington, and northern California. The upward arching of the Blue Mountains in Oregon depressed the John Day basin along its western flank. Enormous clouds of glowing rhyolitic ash called ignimbrites poured down the sides of local volcanic cones smothering and entombing entire temperate populations of animals and plants for which the John Day Formation is famous.

THE MIOCENE EPOCH, FROM 20 TO 5 MILLION years ago, was distinguished by the slow ascent of the Coast Range and Olympic Peninsula west of the Cascades as well as by voluminous outpourings of lava to the east. Uplifting of coastal mountains in British Columbia, Washington, and Oregon, which began in the Early Tertiary and continues to the present, is proceeding in concert with the ongoing subduction of the Farallon Plate beneath the continent.

In British Columbia episodes of uplift and subsidence alternated from north to south. While elevation of the coast in southeast Alaska and northern British Columbia coincided with subsidence of the Queen Charlotte basin, the final stage, beginning 5 million years ago, raised Queen Charlotte Islands and depressed Hecate Strait.

Similarly the Eocene volcanic and sedimentary platform, which makes up the foundation of the Oregon and Washington coast mountains, was elevated from the Early to Late Miocene so that by the Pliocene it was well above sea level. The western edge of the coast was

Figure 1.13 The Columbia River at Biggs Junction, Oregon, cuts through layered Miocene flood basalts (Oregon Highway Dept.).

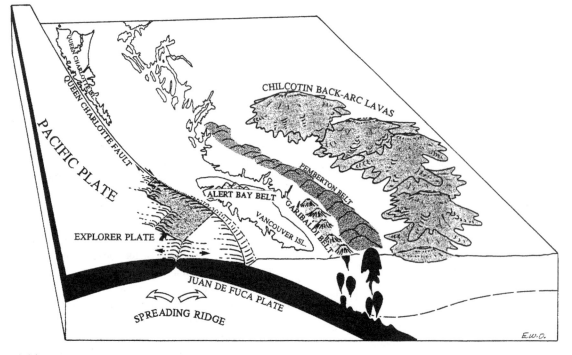

Figure 1.14 Volcanic and structural components of southwest British Columbia can be related directly to tectonic plate configurations (after Souther, 1991).

Figure 1.15 Immediately east of the Coast Range in British Columbia, the Stikine Plateau features three volcanic centers, Level Mountain, Mt. Edziza, and Hoodoo Mountain, all part of the Stikine volcanic belt (B.C. Ministry of Environment, Lands, and Parks).

marked by embayments and a broad shelf. In southwest Washington and northwest Oregon thick Tertiary sediments, known for a rich fossil deposit of marine fauna that includes invertebrates, fish, and mammals, accumulated in this setting. As the Coast Range rose, the Willamette and Puget basins simultaneously subsided, and the ocean retreated westward.

The array of volcanic extrusions that took place during this interval was unmatched elsewhere in the world. Lavas of Steens Mountain in Oregon, the Warner Range in California, and Snake River Plain of Idaho, the basalts of the Columbia River plateau, as well as five separate volcanic provinces across British Columbia, the Yukon, and southeast Alaska severely changed the topography of the northwest. Although extremely copious, eruptions of basaltic lavas in Steens Mountain lasted only 50,000 years and were confined to a small area of southeast Oregon. The Warner basalt, which spanned the Miocene to Pliocene epochs, was also confined to a limited area of northern California.

In contrast, widespread rhyolitic ash and lava eruptions within the Owyhee highlands mark the starting point of the Yellowstone hot spot. Beginning 16 million years ago, movement of the North American Plate over an active, deeply buried, volcanic source or mantle plume generated the eastward propagating Owyhee-McDermitt, Bruneau-Jarbidge, Magic Reservoir, and Heise volcanic fields, before culminating in the easternmost episode, the Island Park-Yellowstone center in Wyoming. This volcanic track reflects the speed and direction of the North American Plate since Middle Miocene time.

Figure 1.16 The Snake River winds across the flat lava plateau near Shoshone Falls, Idaho (Idaho Historical Society).

The most prodigious volcanic activity of the Pacific Northwest was the enormous outpourings of Columbia River basalts that constructed a broad lava platform across northeast Oregon and southeast Washington with embayments into Idaho. Related to crustal stretching between the volcanic archipelago and mainland, this episode would result in one of the largest sequences of flood basalts in the world. In central British Columbia, Late Miocene Chilcotin basalt flows, from a multitude of volcanic centers, are strikingly similar in timing, tectonic setting, and origin to those of the Columbia plateau.

As with earlier volcanic events, lavas of the Miocene Columbia River Group were paramount in their effect on the biota and local topography. Profound environmental alterations brought about by these lava flows were the periodic diversions of stream drainages. While major rivers were able to reestablish their courses or alter direction around lavas that plugged their valleys, smaller streams ponded into vast lakes. Between erup-

tions, soil layers, which developed atop basalts, provide a remarkably complete record of grasslands, forests, and animals from this interval.

Five volcanic chains across British Columbia and Alaska, beginning in the Late Oligocene and continuing into the Pleistocene, were triggered by the multiple effects of plate subduction, mantle hot spots, and crustal spreading. Both parallel and at angles to the coast, the Wrangell, Stikine, Anahim, Pemberton, and Garibaldi volcanic zones were responsible for intrusives and extensive volcanic calderas as well as widespread lava and ash.

VOLCANIC EVENTS CONTINUED TO DOMINATE northwest geologic history during the Pilocene and Pleistocene epochs at a time when glaciers and ice sheets frequently interacted with erupting lavas. Changes in Late Cenozoic plate motions contributed to violent eruptions of the emerging andesitic High Cascade peaks as well as to oozing basalts of the Snake River Plain and Basin and Range. A reconfiguration of

Figure 1.17 Layers of Quadra Sand on Vancouver Island were deposited around the apron of Pleistocene ice masses (courtesy J.J. Clague, Geological Survey of Canada).

the Juan de Fuca slab 9 million years ago, which was being thrust beneath North America, pushed the entire Cascade volcanic arc progressively to the east. This motion produced a narrow linear chain of High Cascade stratocones and lava flows from northern California to southern British Columbia. Eruptions from mounts Lassen and Shasta in northern California, Mount Hood in northern Oregon, and Mount St. Helens in southern Washington within the last two centuries are indications of an active underlying subduction system.

The most recent volcanic covering across the Snake River Plain as well as in the wide-flung Basin and Range of southeast Idaho, Oregon, and northern California relates to extensional tectonics. An Eocene phase of distortion and stretching of the crust continued into the Late Tertiary when reorientation of the subducting Farallon plate produced fracturing and faulting. As continental crust stretches and thins from tension, tears appear, opening conduits that allow lavas to make their way to the surface.

In contrast to earlier explosive rhyolitic eruptions, basalts of this later period issued slowly from elongate fissures spreading out into flat rough irregular fields. Among the largest of these is the Great Rift that crosses the Snake River Plain in Idaho from northwest to southeast. At the northern end of the Rift, Craters of the Moon displays billowy pahoehoe lava, spatter cones, fissures, lava caves, and pits in one of the most extensive volcanic fields in the United States. Within the Basin and Range of central Oregon and northern California, a series of fractures culminated in volcanic events at the pic-

turesque Newberry caldera, Mt. Hoffman, and Medicine Lake Highland as recently as 1000 years ago.

The appearance of the characteristic topography of elongated mountain ranges interspersed with valleys accompanied volcanism in the Basin and Range. Faults, which broke the surface into uplifted block mountains, can be reconstructed to show that this province has been stretched to almost twice its original width. Even today strong earthquakes in southeast Idaho suggest that the Basin and Range continues to be distorted by tensional forces.

THE PLEISTOCENE ICE AGES, far from marking the end of the geologic story, demonstrate that dynamic processes are a continual part of the Pacific Northwest evolution. The onset of the Pleistocene was initiated by the build-up of glacial ice masses brought on by dramatic fluctuations in the climate. Most of British Columbia was covered by ice at some time during the Quaternary period when several major cycles of growth and decay of ice sheets took place. Beginning about 2 million years ago, continental glaciers in British Columbia moved southward. These immense masses of ice reached their maximum extent 15,000 years ago when great lobes occupied the Puget Sound, Okanagan, and Pend Oreille areas of Washington and Idaho. Today continental glaciers are restricted to Antarctica and Greenland. The impact of Pleistocene glaciation was more limited south of the continental ice sheet. This area was characterized by ice caps formed on high summits, as well as valley glaciers, which deepened preexisting stream channels.

Figure 1.18 The vast ice field of Columbia glacier in the Chugach Mountains north of Valdez, Alaska, displays spectacular crevasses and cracks (courtesy M.F. Meier, U.S. Geological Survey).

No geologic agent is more effective in carving up landscapes than ice masses. The entire geological cycle of erosion and deposition is greatly accelerated during a glacial phase. Enormous layers of gravel, sand, silt, and clay, some measuring up to 70 feet thick, cover the surface of some regions today as evidence of past glaciation. Frequently mounds of debris blocked rivers and impounded lakes that are still present throughout the northwest. Symmetrical U-shaped valleys, serrated ridges, and deep fiords are also the work of moving ice, whose effects, although perhaps not as dramatic as those of an erupting volcano, are equally permanent.

One of the most spectacular events of the Ice Ages was a series of stupendous floods that swept across northern Idaho, Washington, and into Oregon from lakes impounded by ice dams. These periodic deluges left their mark by picking up and moving enormous amounts of sediment, recontouring plateaus, and enlarging valleys. Even though dozens of innundations issued from a glacial lake near Missoula, Montana, only a single large overflow originated from Ice Age Lake Bonneville in Utah. This flood left unmistakable deposits along the Snake River Plain as it made its way to the Columbia River.

ALTHOUGH TODAY THE PACIFIC NORTHWEST is regarded as volcanically and seismically benign in comparison to southern California or other locations around the Pacific Rim, the geologic record does not support this. Only 7000 years ago Mt. Mazama (Crater Lake) and Newberry volcano erupted catastropically, spreading ash and rock over thousands of square miles. Devastating Pleistocene floods took place at irregular 50 to 200-year intervals. Earthquakes of incredible power have shaken the Pacific Northwest as the subduction zone between the descending Juan de Fuca slab and overlying continental plate released and slipped on a 300- to 500-year cycle. Along with volcanic eruptions and earthquakes, raised marine terraces on the coast are one of the fastest acting Holocene geologic occurrences. As the Juan de Fuca Plate is being jammed beneath North America, terraces, such as those at Cape Blanco, are rising at the phenomenal rate of 1 inch in 3 years. The Pacific Northwest seems poised for a natural catastrophe from any one of these geologic agents.

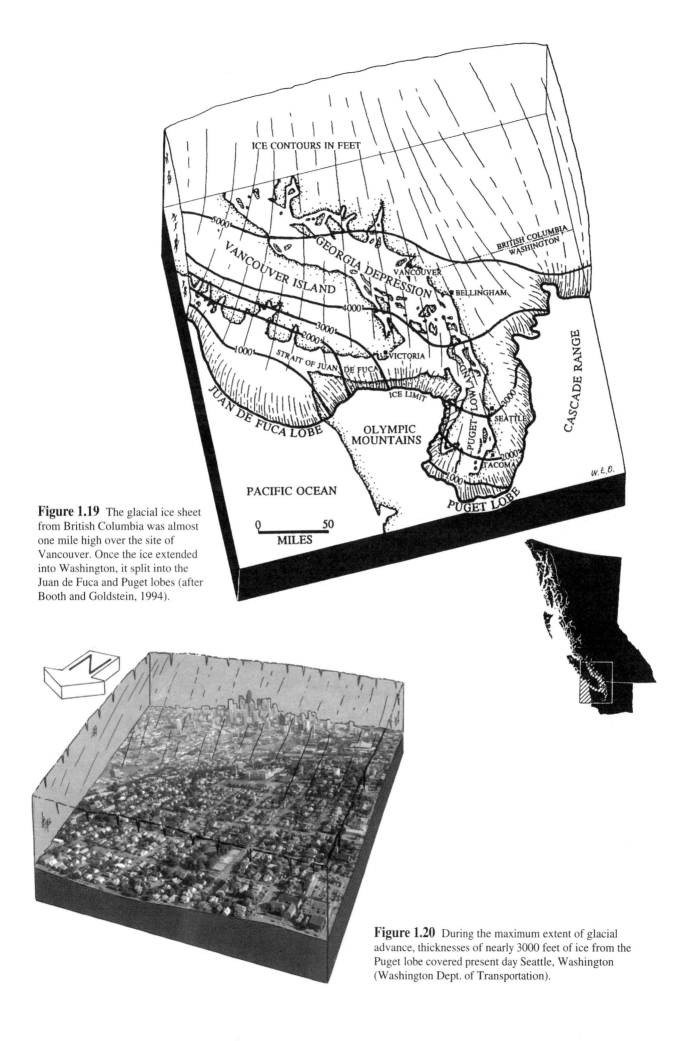

Figure 1.19 The glacial ice sheet from British Columbia was almost one mile high over the site of Vancouver. Once the ice extended into Washington, it split into the Juan de Fuca and Puget lobes (after Booth and Goldstein, 1994).

Figure 1.20 During the maximum extent of glacial advance, thicknesses of nearly 3000 feet of ice from the Puget lobe covered present day Seattle, Washington (Washington Dept. of Transportation).

Additional Readings

Harris, A.G., 1990

McKee B., 1972

McPhee, J., 1981

McPhee, J., 1993

Norris, R.M., and Webb, R.W., 1990

Orr, E.L., and Orr, W.N., 1999

Yorath, C.J., 1990

OMINECA-INTERMONTANE PROVINCE

The Selkirk Mountains in the southeastern Omineca region reach elevations of 11,000 feet (B.C. Ministry of Environment, Lands, and Parks).

The Omineca-Intermontane province is a wide strip of dissected high plateaus and scattered mountains that begins in Alaska and extends southeast through British Columbia and the northeast corner of Washington, the Idaho panhandle, and Montana. Along its southern boundary the province disappears beneath the Columbia River basalts between Brewster, Washington, and Idaho. The Northern Cascades of Washington and the Coast Mountains of British Columbia form the western margin, and the Tintina-Rocky Mountain trench marks the eastern margin. A long series of interweaving faults, the Tintina-Rocky Mountain structure runs for 1600 miles from Alaska to Montana between the Rocky Mountains and the Cassiar, Omineca, Cariboo, Selkirk, and Cabinet ranges. In British Columbia, summits reach 8000 feet in the Cassiar and Omineca ranges, more than 11,000 feet in the Selkirks, and 7000 feet in the Cabinet Mountains of Idaho.

Across the center of the region a succession of uplands, the Yukon, Stikine, and Interior plateaus in British Columbia and the Okanagan highland in Washington, have deeply incised drainage systems and elevations of 3000 to 4000 feet. Scattered throughout the interior, thousands of sinuous lakes are glacially scoured and dammed stream valleys.

The extensive Columbia and Fraser rivers drain the southern end of British Columbia. Both river systems originate in the Rocky Mountain trench and flow northwest in a parallel course for about 150 miles, before turning abruptly south. The Columbia passes Revelstoke before reversing itself south to enter Washington, but the Fraser moves northward a considerable distance past Prince George before curving southward. Originating in the same trench, the smaller Kootenay River follows a meandering pathway south into Montana, then northwest back into British Columbia, before joining the Columbia.

INTRODUCTION

Central British Columbia, lying between the Coast Range and Rocky Mountains, has been divided into two great north-south belts, the Omineca on the east and the Intermontane on the west. Composed of smaller terranes, the wide Intermontane superterrane merged with and was accreted to North America in the Middle Mesozoic. The force of collision stacked great thicknesses of sedimentary rocks upon each other, folding, faulting, uplifting, and metamorphically altering the existing edge of the continent into the mountainous Omineca welt. The intrusion of plutons, throughout much of the Omineca, was accompanied by emplacement of metallic

Figure 2.1 Deeply incised in Quaternary fill, the Fraser River near Clinton, British Columbia, cut its channel at the end of the Ice Ages (B.C. Ministry of Environment, Lands, and Parks).

ores that instilled considerable mineral wealth into this region.

Profound charges in tectonic style from compressional in the Mesozoic to stretching or tensional in the Early Tertiary were brought about by adjustments to the direction in which the crustal plates were moving. Ancient core complex rocks, which had been metamorphically altered deep in the earth, were thrust close to the surface, while basins, bordered by faults, were dispersed across the province as products of this tensional phase. Throughout the Eocene epoch dozens of long narrow trenches collected volcanic debris and organically rich fluvial sediments that subsequently altered to coal seams of unusual thickness.

The next chapter to the geologic history here was written during the Ice Ages when multiple climatic fluctuations buried the area beneath vast sheets of moving glacial ice. Today scenic glacial lakes as well as thick deposits of clay, silt, and gravel are reminders of the ice masses that melted just a few thousand years ago.

GEOLOGY OF THE OMINECA-INTERMONTANE PROVINCE

The geology of southeast Alaska, the Yukon, District of Mackenzie, British Columbia, Alberta, northeast Wash-

Figure 2.2 In this view from the north, the Fraser River in the Rocky Mountain trench marks the boundary between the Cariboo Mountains in the distance and the Rocky Mountains in the foreground (B.C. Ministry of Environment, Lands, and Parks).

ington, and the Idaho panhandle is arranged in five parallel belts that follow a curving northwest by southeast trend. The structure and origin of the continuous Foreland, Omineca, Intermontane, Coast, and Insular belts are the essence of the geologic history of this region during the Paleozoic and into the Tertiary.

Great sheet-like terranes as well as slivers and fragments of crustal rocks are the building blocks that make up the large belts. A terrane designates a rock package having a history which is substantially different from that of the adjacent rocks. Within each terrane, rocks

and fossils delineate specific depositional or volcanic environments. Although many terranes may be exotic or imported from long distances, others originate at their present location. The meaning of the word terrane has evolved over the past decade so a rock sequence that was designated as a group or even a formation in the past might be interpreted as a terrane today. The progressive growth by attachment of terrane rocks along the leading edge of a moving crustal slab takes place when a large landmass such as a continent collides with and sweeps up smaller exotic pieces. This

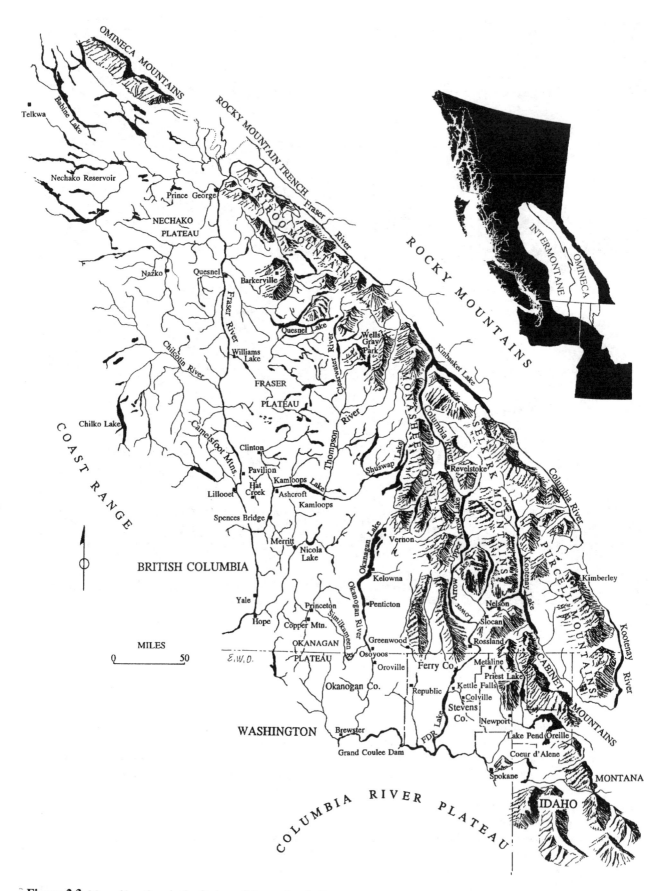

Figure 2.3 Map of locations in the Omineca-Intermontane belt.

Figure 2.4 Gold played a significant role in bringing inhabitants and civilization to the Omineca region. Near Boundary Falls the trail that ran along the west side of the creek became a meeting place for prospectors. Today the concrete supports and wooden trestle for the woodstave pipe can still be seen (B.C. Archives and Records Service).

process called accretionary tectonics dominates Pacific Northwest geology.

The accretion process involves a complex chain of events. Prior to collision, terranes were often clustered or amalgamated offshore into larger superterranes. As crustal plates move against one another, usually the thinner heavier oceanic slabs plunge beneath the lighter continental mass. Accretion takes place when the lower plate is welded to the edge of the overriding continent. Following the impact, terranes are distorted and intruded by plutons. A belt of altered and intruded strata that was wrinkled into large-scale continuous ridges or welts by the impact of collision marks the suture between the newly arrived terrane and the larger continent. The most profound damage from collision is evident adjacent to the area of attachment but diminishes away from this zone so that low subdued folds are the only evidence of the impact further up on the continent. Boundaries between crumpled overlapping terranes are often difficult to draw on a map, and it is only in cross-section that the orientation of slabs with respect to each other is clear.

THE OMINECA AND INTERMONTANE BELTS were fabricated during an early phase of accretion. In British Columbia and the northwestern United States the first collision event took place during Middle Jurassic time between 190 and 160 million years ago as the Intermontane superterrane was accreted with North America. This event severely folded and altered the border of the continent in a wide zone called the Omineca crystalline belt. A second major collision between 130 and 80 million years ago in the Cretaceous period occurred when the Insular superterrane merged with the western face of the Intermontane producing another deformed welt designated as the Coastal belt.

The exact timing of these two major accretionary events is not clear, and there is evidence that the Intermontane and Insular superterranes may have even been joined as an immense terrane prior to collision with North America. This huge hypothetical terrane has been called "Baja, British Columbia" for its similarity to the continental peninsula of Baja, California, currently tearing loose from Mexico and moving northward.

THE EARLIEST PRECAMBRIAN history of the Omineca Intermontane province dates back just under a billion years when a supercontinent broke into two pieces along what now is the western region of North America. One piece was the ancestral Pacific Northwest that lay beneath a wide ocean with a shoreline across eastern British Columbia, Idaho, Nevada, and southern

Figure 2.5 Distribution of Intermontane and Omineca belts in eastern Canada.

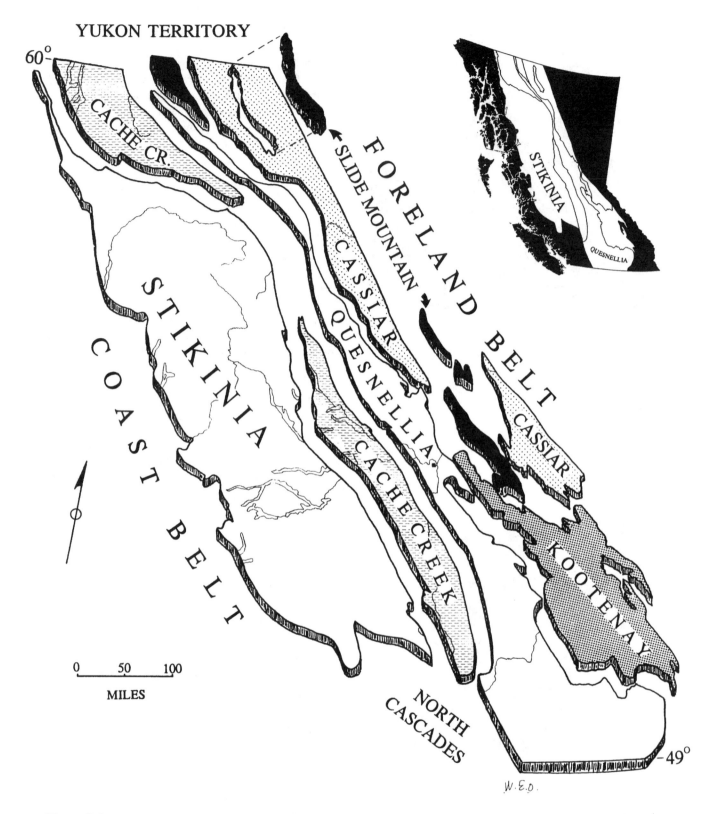

Figure 2.6 Distribution of terranes in the Omineca-Intermontane belts of British Columbia (after Gabrielse and Yorath, 1991).

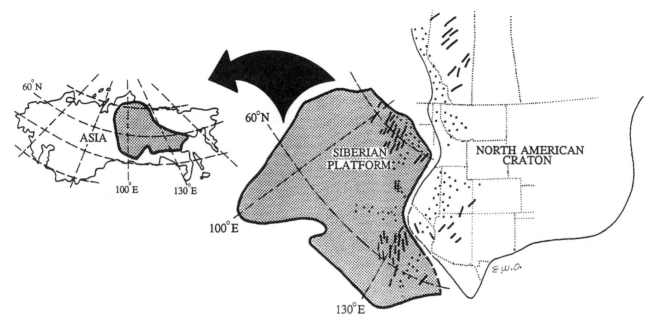

Figure 2.7 Reconstruction of the breakup of an immense supercontinent into two pieces, the North American landmass and the Siberian platform (after Sears & Price, 1978).

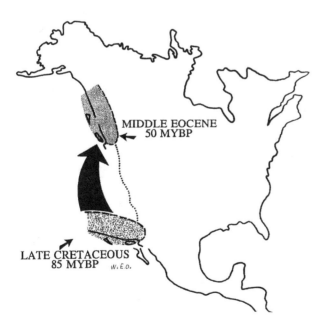

Figure 2.8 Proposed model for the northward movement of "Baja, British Columbia," from south to north along the coast of North America (after Gabrielse and Yorath, 1991).

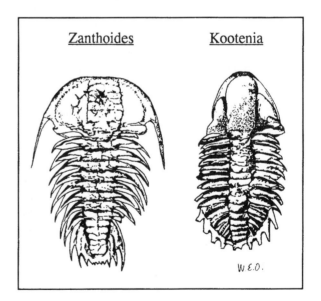

Figure 2.9 Trilobites, living in the Cambrian ocean 550 million years ago, were segmented crab-like scavengers.

Figure 2.10 Displaying the serrated edge of the colony, graptolites were attached to balloon-like structures which kept them afloat (Wash. Division of Geology and Earth Resources).

Figure 2.11 The distribution of sediments of the Proterozoic Windermere, Purcell, and Belt supergroups defines the western margin of ancestral North America (after Burchfiel, Cowan, and Davis, 1992).

California. The identity and fate of the other continental piece still remains an unanswered question. The missing continental mass may be a large chunk of eastern Siberia that matches nicely with Precambrian structures in Montana and southern California if given a quarter turn clockwise. A second possibility for the missing piece would be an enormous plate that later split up to yield Antarctica, Australia, India, and China.

A volcanic rift system gradually separated the two gigantic blocks, causing prodigious eruptions of lava onto the sea floor and raising global sea levels. A significant biologic event which accompanied the expanding seaways was the first appearance of shelled invertebrate animals that flourished from the beginning of the Cambrian period about 550 million years ago. Dominated by trilobites, the Paleozoic fauna included molluscs, corals, and brachiopods, as well as tiny fish-like vertebrates represented today only by tooth-like conodont microfossils. Blue-green algae or cyanobacteria, common in the Precambrian seas, continued to construct thick limestone reefs into the early Paleozoic.

Ocean shelf, slope, and deep water environments are represented by marine deposits throughout the province. The middle Cambrian Metaline Limestone of northeast Washington characterizes shelf conditions with trilobite fragments and even intact carapaces. In British Columbia, outer shelf carbonates of the Nelway Formation, from the same interval, are curiously devoid of fossils. On top of the Metaline, the dark black Ledbetter slates of Ordovician to Silurian age, with graptolite fossils preserved as a pyrite [fool's gold] film, reflect open ocean waters of slope and greater depths. Small colonial graptolite fossils, resembling a saw blade, inhabited surface wa-

ters where they were suspended from balloon-like floats. In the Idaho panhandle the Gold Creek, Rennie Shale, and Middle to Upper Cambrian Lakeview strata illustrate similar widespread shallow continental shelf conditions. Layers of the Lakeview are dominated by algae shoals and are rich with a trilobite hash.

Fossils are an excellent means of determining the age of sediments. Trilobites evolve rapidly in shallow shelf limestones and shales. Floating near the surface of an open ocean setting, graptolites are utilized to decipher deep water muds, while the swimming conodonts represent a broad spectrum of marine environments.

THE FORELAND BELT, stretching from Alaska across eastern British Columbia, the Yukon, District of Mackenzie, western Alberta, and into Idaho and Montana, encompasses all of the Mackenzie and Rocky Mountains and defines the western continental edge of ancestral North America. The oldest sediments here were deposited along the seaway that opened during Late Precambrian time.

Forming a wedge-shaped package within this belt, sedimentary rocks, known as the Belt Supergroup in the United States and the Purcell Supergroup in Canada, range from 1.7 to .8 billion years old. Mud cracks, salt crystals, and ripple marks within these rocks are all distinct signs of shallow water conditions indicating the Belt-Purcell group filled a former basin. When continents split, the fracture is not a clean line but is broken by cracks running deep into the separate pieces. These

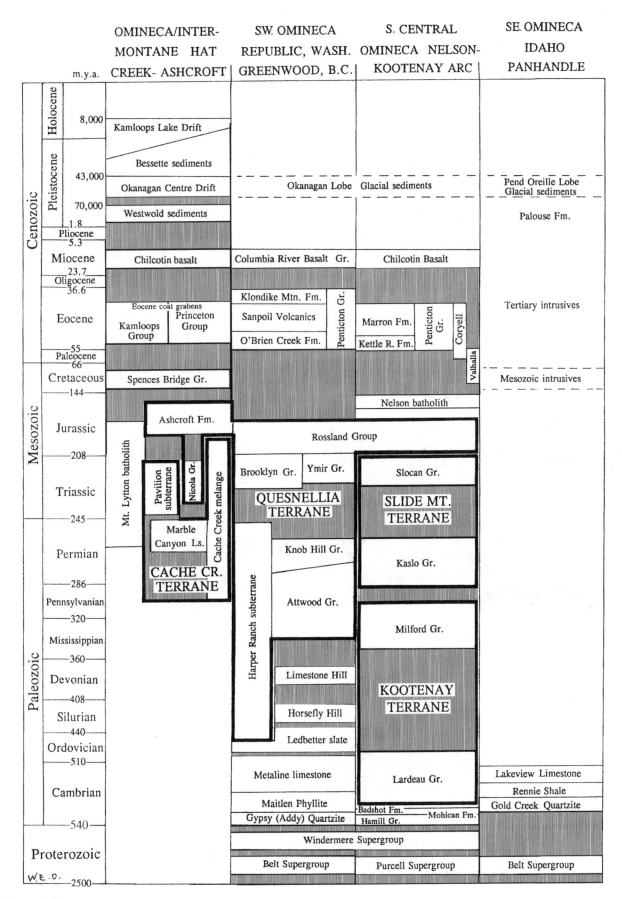

Figure 2.12 Omineca-Intermontane stratigraphy (after Gabrielse and Brookfield, 1988; Wheeler and McFeely, 1991).

cracks or rifts immediately act to trap sediments along the newly formed continental margins and often become the mouths of large rivers.

The thin eastern edge of the Belt is algae-rich limestone, reddish oxidized shale, and mudstone, while toward its western edge the Belt-Purcell inflates to 50,000 feet of finely laminated mudstone deposited as deep ocean turbidites. Stromatolite algae of the genus *Collenia* grew from cabbage-sized colonies to dense masses hundreds of feet in diameter forming much of the limestones. Significantly no evidence of burrowing animals or "bioturbation," which characterizes modern deep sea settings, is found in this strata. Belt-Purcell rocks and fossils are nearly identical to those in a large tract east of the Urals in Siberia, which may be the missing Precambrian fragment of the North American western margin.

Above the Belt-Purcell sediments, Late Proterozoic sand and limestone of the Windermere Supergroup, exposed in southern British Columbia, were also deposited

along the newly formed continental margin. Lower in the Windermere sequence there is evidence of deposition within local rift basins.

Thrust faults that stacked rocks of the Foreland belt in sheets leaning to the east have shortened this region by as much as 100 miles. This thrust zone is separated from southcentral Idaho and western Montana by the Lewis and Clark lineament that trends west-northwest across the Idaho panhandle and into Montana. An enormous linear feature visible on the earth's surface from space, the lineament marks a shear zone up to 20 miles wide of overlapping faults that have been active from the Late Precambrian through the Recent.

Similar to the rocks of ancestral North America, shelf limestones of the Cassiar terrane suggest it has been displaced by faults several hundred miles northward from a southerly origin.

THE OMINECA BELT, just west of the Foreland, is a mountainous strip that runs from the Yukon plateau southeastward through the Cassiar, Omineca, and Purcell mountains of British Columbia before disappearing beneath the basalt plateau of Washington. The Jurassic collision of the Intermontane superterrane with North America produced the extensive raised folds of the Omineca welt. Distorted strata within this strip are predominantly sedimentary layers deposited as part of the Precambrian and Early Paleozoic seaway.

Granitic intrusives, distributed throughout the Omineca, are yet another result of the tectonic accretion. Because of collisions the oceanic slab descended, the crust compressed and thickened while heat from below caused the rocks to melt at depth, injecting liquid magma in the form of plutons into the overlying strata. Plutonic rocks in the Omineca belt were emplaced during four intervals between the Late Triassic and Eocene. Younger plutons are widespread south of Prince George, but Middle Cretaceous granites are the most extensive and voluminous throughout the belt.

Figure 2.13 Precambrian stromatolite algae display typical banding in cross-section.

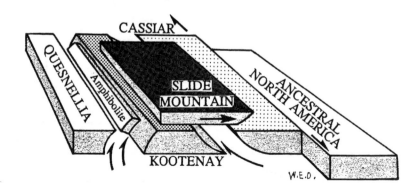

Figure 2.14 Structural relationships of terranes in the vicinity of Prince George, British Columbia (after Gabrielse and Yorath, 1991).

The Kootenay arc, Purcell trench and anticlinorium, and similar structures in the Omineca belt are all produced by collision processes. Lying between North America and accreted terranes of the Intermontane belt, the 250-mile long Kootenay arc is an immense upfold of intensely deformed rocks from Revelstoke, British Columbia, through the Idaho panhandle and eastern Washington. Highly distorted and sheared rocks within the arc are Precambrian and Paleozoic shelf, slope, and ocean basin sediments that delineate the westernmost edge of the ancient North American continent and Kootenay terrane.

Figure 2.15 Upper Moyie Falls at the top of the Idaho panhandle in Boundary County are made up of rocks of the Purcell trench (Idaho Dept. of Transportation).

Within the Kootenay arc, rocks of the Kootenay terrane, folded and thrust back over the North American continent, represent the oldest of the sheets of accreted terranes. Although rocks of this terrane are similar in age to those of the Foreland belt along the margin of the continent, the Kootenay sequence was deposited in deep water in the gap between oncoming terranes and the mainland.

After accretion of terranes in the Jurassic, sediments of the Kootenay arc were invaded by granitic plutons of the Kuskanax and Nelson batholiths west of Kootenay Lake followed by the Bayonne batholith east of the lake in British Columbia. Also a product of collision processes, the 94 million-year old Spirit pluton beneath northeast Washington is significantly older than adjacent rocks of the Bitteroot lobe of the Idaho batholith.

Extending from Coeur d'Alene, Idaho, to southern British Columbia, the Purcell trench is a gigantic structural trough lined with Quaternary alluvium. This depression separates high grade metamorphic rocks on its western margin from the relatively unaltered Belt-Purcell sediments on the east. Several periods of advance and retreat of the Pend Oreille ice lobe, between 25,000 and 13,000 years ago, left a thick covering of glacial fill within the trench, largely obscuring earlier formations. Paralleling the Purcell trench, the huge up-folded Purcell anticlinorium is mainly older Precambrian Belt-Purcell rocks that have been lightly altered, faulted, and sometimes overturned. Bounded by the Kootenay arc on its western border, the anticlinorium terminates against the Lewis and Clark lineament and forms the base of the Purcell Mountains.

THE INTERMONTANE BELT is a level plateau in contrast to the mountainous Omineca to the east and the Coast Range to the west. This belt extends south from the Yukon to expand in central British Columbia before narrowing and disappearing in Washington beneath the Columbia River basalts.

This plateau consists of rocks of the Intermontane superterrane which is, in turn, a collage of smaller terranes. Thrust against the Omineca, scraps and fragments of these Intermontane terranes are scattered for 1000 miles parallel to the Tintina-Rocky Mountain trench in British Columbia. Major terranes include the Slide Mountain on the east, the Quesnellia, Cache Creek, and most westerly Stikinia. Rocks that make up these terranes were deposited in different oceanic and volcanic island environments from the late Paleozoic and Mesozoic.

The Slide Mountain terrane, which is oceanic in composition, was heavily compressed or telescoped during accretion of the Intermontane superterrane. Permian microfossils in this fragmented strip are similar to those found in Quesnellia as well as in Texas and New Mexico suggesting the terrane travelled some distance before attaching to the continent.

Extending over 1200 miles, the fault-bounded Quesnellia terrane is exposed in a broad band in Washington and British Columbia but tapers to a few thin slivers in Alaska. Primarily of Late Triassic through Middle Jurassic age, Quesnellia sediments originated as a volcanic island archipelago that was later invaded by granitic plutons.

Rocks of the Middle Paleozoic to Jurassic Cache Creek terrane form two tapering belts up to 80 miles wide from the Yukon to southern British Columbia. These rocks are mainly oceanic in complexion, and Permian microfossils of the Cache Creek are tethyan or tropical in nature, implying the slab developed in much warmer southerly latitudes before moving north to accrete to North America. Within this terrane the Mississippian through Permian coral atoll reefs found in northern British Columbia near Whitehorse are remarkable for their longevity. A thick reef was continuously constructed over an interval of 100 million years in an area where the sea floor was slowly subsiding at a rate of about 60 feet per one million years. Finally when sea levels dropped rapidly at the end of the Permian, the reef was exposed, killing the coral.

Constructed as a volcanic island chain from the Middle Paleozoic to Jurassic, Stikinia is the largest terrane in British Columbia with areal estimates as high as 150,000 square miles. Stretching from Alaska to southern British Columbia, the terrane balloons to almost 200 miles wide in the upper Stikine River watershed. Located in the Telkwa Range of central British Columbia, a reef within this terrane suggests warm highly productive waters. Clams and other reef dwellers that inhabited the steep-sided coral framework were covered with ash from nearby volcanic vents to bury and preserve them in pristine condition.

MESOZOIC COLLISIONS AND UPLIFT in the Omineca and Coast belts brought about local folding to create the Bowser, Sustut, and Nechako basins as part of the Intermontane belt. From the Jurassic to Tertiary the uplifted eroding Omineca Mountains shed debris eastward into a broad inland continental seaway of the Foreland belt and westward into these local marine depressions.

Formation of the large Bowser basin accompanied accretion of the Stikinia to the Cache Creek terrane in the Middle Jurassic. Enclosing an area of 2000 square miles, the basin was bordered by the Omineca belt on the east, the Coast Belt on the west, the Stikine arch to the north, and the Skeena arch to the south. Chert, basalt, and limestone of the uplifted Cache Creek terrane and Skeena arch were eroded and carried into the depression to accumulate over 10,000 feet of Middle Jurassic to Early Cretaceous sediments of the Bowser Lake Group. Fine-grained marine and nonmarine sediments, containing layers of organic matter, spread out into deltas toward the south and west to fill the basin and develop thin coal beds.

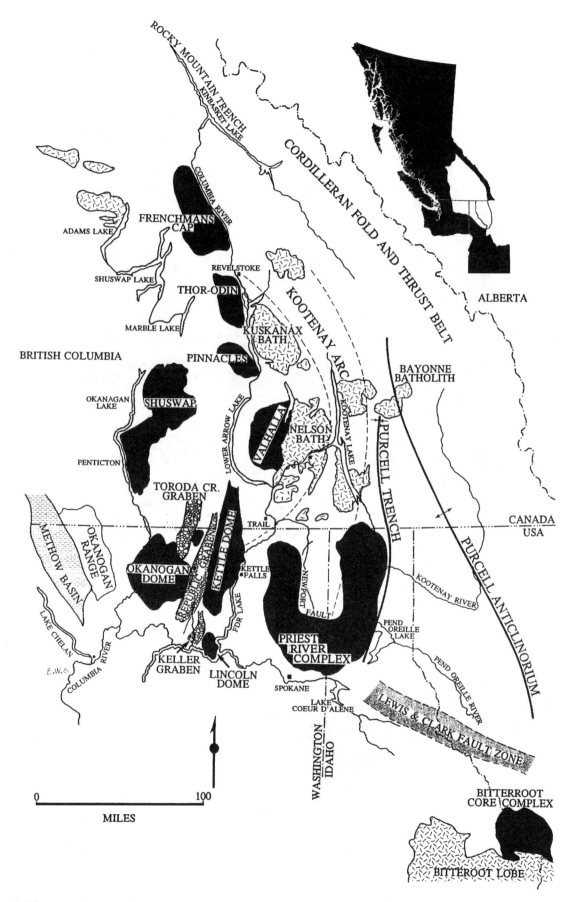

Figure 2.16 Domed core complexes, granitic intrusions, grabens or troughs, and other structures of the southern Omineca-Intermontane province (after Fyles, 1970; Gabrielse and Yorath, 1991).

Figure 2.17 The Quesnel River flows through terrane rocks of the Intermontane belt (B.C. Archives and Records Service).

Figure 2.18 Sequence in the life of the Late Triassic Telkwa reef in the Intermontane belt. Oyster-like clams and corals built a reef mound that was eventually suffocated by a covering of ash and lower sea level (after Stanley and McRoberts, 1993).

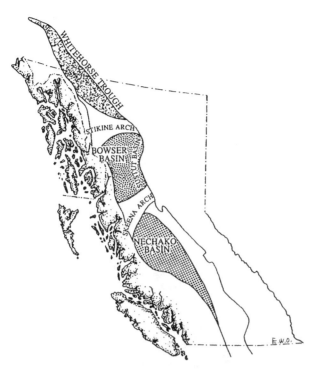

Figure 2.19 Mesozoic basins and other structures of the Intermontane belt (after Gabrielse and Yorath, 1991).

The later Cretaceous to Early Tertiary Sustut-Skeena trough, which formed a narrow belt along the eastern margin of the Bowser basin, received copious amounts of mica-rich sediments of the Sustut Group from the Omineca belt as well as volcanic debris from the uplifted Coast landmass to the west. South of the Skeena arch, marine and nonmarine sediments of the Cretaceous Nechako basin can be traced from the Bowser basin.

THE OPENING OF THE CENOZOIC in the Omineca-Intermontane province saw the uplift and exposure of core complex rocks, the appearance of a mosaic of tensional fault-bounded troughs or grabens, and accompanying volcanism brought on by stretching and thinning of the crust. These events were produced by the opposing movements of large tectonic plates. A shift from a compressional to shearing motion resulted when the northward-moving Kula plate ground against the west-bound margin of North America, yielding elongated north-south faults.

WITHIN THE EASTERN PORTION OF THE OMINECA BELT, a discontinuous chain of isolated metamorphic and plutonic rock exposures designated as

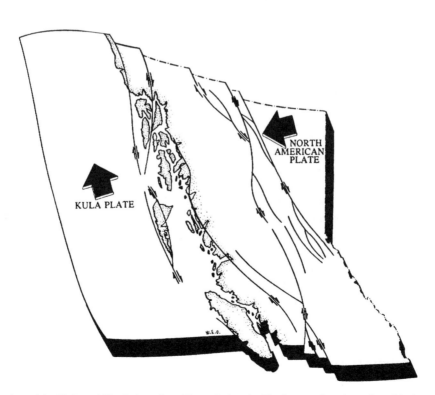

Figure 2.20 Interaction of the Kula and North American Plates during the Tertiary produced a series of faults and tensional basins in this province.

core complexes, extends over 250 miles from southeast British Columbia into Washington and Idaho. This sinuous belt, composed of about 25 complexes, continues through eastern Nevada, southwest Arizona, and into Mexico. Although their origin is unclear, the formation of core complexes in Mesozoic time relates to the effects of collisions by superterranes that severely compressed and altered rocks in the margin of the ancestral North American continent. Deep within the crust, these rocks were deformed and recrystallized, then later lifted close to the surface by the same tensional stresses that produced Eocene faulting.

Located at roughly 50-mile intervals in British Columbia, metamorphic rocks of the Shuswap core complex include Frenchmans Cap at the center of the Monashee Mountains near Revelstoke, Mt. Thor and Mt. Odin, Pinnacles, and Valhalla and Passmore beneath the peaks at Valhalla Park. In northeast Washington this string of core complexes continues and expands to include the Okanogan and Kettle domes, separated by the Republic graben, as well as the Lincoln gneiss and Spokane domes. In Idaho the Priest River complex lies beneath the panhandle while the northcentral Bitterroot, central Pioneer, and southern Albion mountains are also underlain by core complex rocks.

Typically these metamorphic or plutonic complexes are bordered by faults that separate them from a lid or veneer of younger strata. The shallow spoon-shaped Newport fault that loops south across the Washington-Idaho border near Newport, Washington, is an example of such an extension fault. Dated as Middle Eocene at 48 to 50 million years, this fault separates the Priest River core complex from Belt Supergroup and younger rocks.

APPROXIMATELY 40 SMALL STRUCTURAL BASINS, scattered along the Omineca and Intermontane belts from the Yukon to Washington, are yet another by-product of tensional stretching of the crust. Elongated to the north or northeast, the basins or grabens filled with Eocene volcanic debris and lavas from local eruptive centers as well as fluvial sediments from surrounding highlands. In these basins fine-grained claystone is interlayered with plant fragments and low-grade coal beds that sometimes exceed several hundreds of feet in thickness.

In the Omineca-Intermontane typical basins at Hat Creek, Merritt, Princeton, and Quesnel were formed by regional stresses between the Rocky Mountain trench and the Fraser River-Straight Creek fault systems. At the Hat Creek graben west of Kamloops, lake, swamp, and delta environments of the Eocene Kamloops Group persisted for over a quarter of a million years. Four sub-bituminous coal seams at Hat Creek reached a remarkable 1200 feet in combined thickness. At least 7 coal seams in the basins near Princeton, on the other hand, average 100 feet in thickness. Associated with the Princeton Forma-

tion, plant and woody debris within the coal suggest high rainfall in a setting that included open grasslands, lakes, and alpine forests 50 million years ago. There is evidence that flooding and destruction of the vegetation took place repeatedly. The Oligocene-Miocene Australian Creek Formation near Quesnel contains coal layers averaging over 60 feet thick. Coal from these fields has been mined periodically from the late 1800s.

Straddling the international boundary from Christina Lake in British Columbia to the Columbia River in Washington, the Republic graben is the most southerly of the extended belt of Eocene tensional basins. The 60-mile long narrow depression is bordered by northeast-

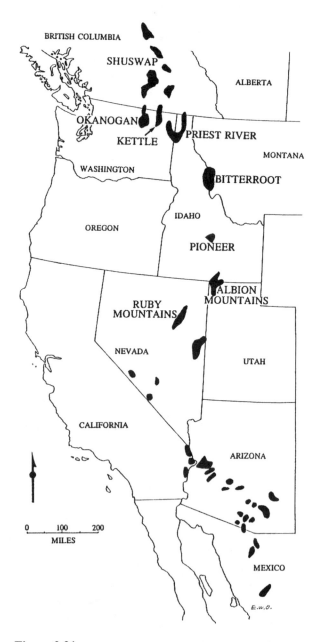

Figure 2.21 Distribution of metamorphic core complexes across western North America reflects widespread tensional plate movements (after Burchfiel, Lipman, and Zoback, 1992).

southwest trending faults that separate the graben from core complexes of the Okanogan dome on the west and the Kettle dome to the east.

Within the Republic depression a large lake ponded atop tuff and andesite layers of the Eocene Sanpoil volcanics. Here lacustrine sediments of the Klondike Mountain Formation contain an outstanding flora of

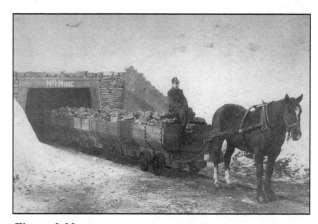

Figure 2.22 The Nicola Coke and Coal Company Mine in the Nicola basin near Princeton extracted Eocene coals from structural troughs (B.C. Archives and Records Service).

fossil dicotyledonous and coniferous leaves of ginkgos, *Metasequoia,* pine, sassafras, katsura, and alder, along with fish and insects. An unusual number of insects such as aphids, leafhoppers, flies, and spittlebugs have been preserved by carbonization, a process where a thin carbon film remains after the insect slowly decays under water. Many species of the flora and fauna are displayed at the Stonerose Interpretive Center in Republic, Washington.

NUMEROUS PLUTON INTRUSIONS, contemporaneous with formation of the grabens, are collectively called the Colville batholith. Lying between the Columbia and Okanogan rivers, this batholith, together with the Republic and Keller grabens and core complex rocks, make up the Okanogan highland of northcentral Washington.

A CENOZOIC PHASE of crustal stretching in the Omineca-Intermontane region occurred simultaneously with volcanic activity of the Pemberton-Garibaldi episode. Both relate to eastward subduction of the Explorer and Juan de Fuca plates beneath the North American continent.

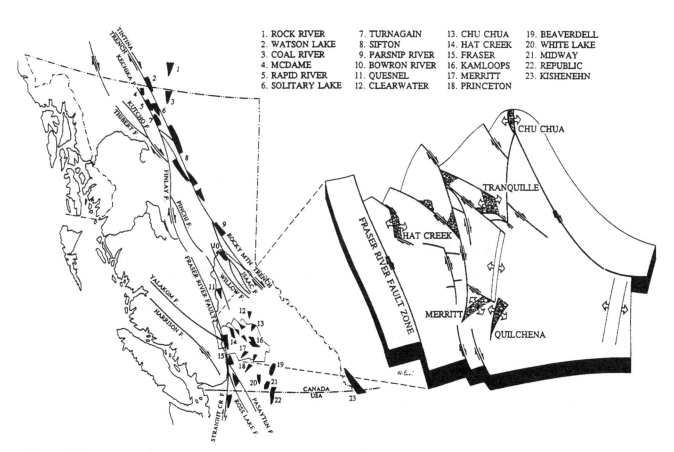

1. ROCK RIVER
2. WATSON LAKE
3. COAL RIVER
4. MCDAME
5. RAPID RIVER
6. SOLITARY LAKE
7. TURNAGAIN
8. SIFTON
9. PARSNIP RIVER
10. BOWRON RIVER
11. QUESNEL
12. CLEARWATER
13. CHU CHUA
14. HAT CREEK
15. FRASER
16. KAMLOOPS
17. MERRITT
18. PRINCETON
19. BEAVERDELL
20. WHITE LAKE
21. MIDWAY
22. REPUBLIC
23. KISHENEHN

Figure 2.23 The Early Tertiary movement of large crustal plates created dozens of tensional basins in the Omineca-Intermontane belt (after Gabrielse and Yorath, 1991).

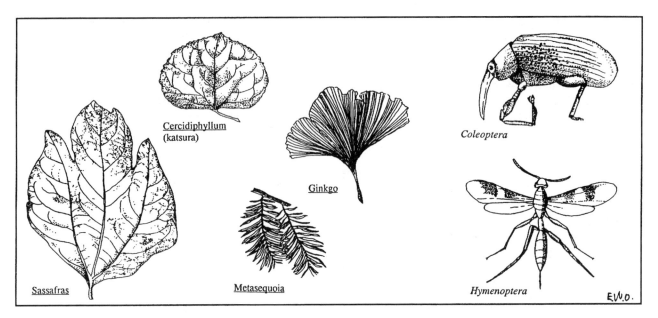

Figure 2.24 Insect wings and bodies as well as plant leaves occur in Eocene Klondike Mountain sediments which filled the Republic graben in Washington around 45 million years ago.

During this episode Chilcotin lava flows were extruded in several stages between 10 to 6 and 3 to 2 million years ago. Building a platform of 20,000 square miles in central British Columbia along the Chilcotin River watershed, a thick mantle of flood basalts covered earlier delta and stream sediments, filled broad valleys, and smoothed over the irregular eroded Eocene-Oligocene landscape.

Even though Chilcotin basalt flows are close to the Columbia River Group in age, volcanic style, and duration, they were not as voluminous. As many as 20 individual thin basalt flows, exuded from a multitude of centers extending northwesterly through the Intermontane belt, coalesced into a layered volcanic platform up to 400 feet thick. Many localities display only a single basalt layer, but at Deadman River north of Kamloops and along the Chilcotin River there are up to seventeen separate flows. Rough pahoehoe lava, broken by coarse jointing, is common as are pillow basalts that resulted when lavas erupted beneath lake waters. Pleistocene glacial action later stripped off almost half of the lava covering.

Outside of the main eruptive belts, Tertiary and Quaternary volcanic activity at Wells Gray Park and in the adjoining Clearwater River drainage southeast of Quesnel took place on the suture between the Quesnellia and Slide Mountain terranes. A number of separate extrusive centers evolved during both glacial and ice-free intervals, and smaller cone-shaped flows were apparently confined beneath thick masses of ice. The well-preserved volcanoes in Wells Gray Park are the youngest. Slightly to the south, flows up to 30 feet thick

from vents along the Clearwater Valley are interlayered with pillow basalts, stream sediments, and even plant material where lavas overwhelmed a forest. These thin flat basalt flows, as well as those in McConnell Creek to the north in the Omineca Mountains, have been deeply dissected by streams.

DURING THE PLEISTOCENE EPOCH virtually all of southwestern Canada was repeatedly glaciated by continental ice sheets that covered much of Alaska, northern Washington, Idaho, and Montana as well. The thickest ice buildup was confined to a broad area between the Rocky Mountains and coastal range of British Columbia. During glacial periods, increased precipitation and cold temperatures brought accumulations of ice that moved down from the high elevations to bury the lowlands. With climatic fluctuations and warming trends during interglacials, the ice was restricted to higher mountain peaks. As the ice sheets thinned and melted, they released huge amounts of sand, mud, and gravel over the surface.

Two major glaciations and three warm interglacials took place in southcentral British Columbia during the Quaternary. Seventy-foot thick Bessette gravel, sand, and silt interspersed with fossil plant and animal remains were deposited during the Olympia interglacial that began 43,000 years ago. Pollen of spruce, pine, birch, and grasses show that the climate during part of this period was similar to that of today.

In colder more northerly localities at Babine Lake, spruce and shrub tundra persisted longer. Wooly mammoths frequented the lakes, grasslands, and forests of

Figure 2.25 Coeur d'Alene Lake in Idaho, one of the many bodies of water throughout the province, was deepened by glaciers and dammed by gravel and sand during the Pleistocene (Idaho Dept. of Transportation).

the province, and the bones of a partially articulated mammoth were recovered during a mining operation on the lake. Wood and other plant fossils buried with the skeleton suggest the beast sank into a muddy quagmire as it died.

The Bessette interglacial was followed by the Fraser glaciation that records the last interval of multiple ice sheet advances. This was a time when mountain glaciers expanded and flowed onto the interior plateaus and lowlands. At the height of the final Fraser glacial advance between 20,000 to 10,000 years ago, rounded lobes of the ice sheet probed southward. North of the Spokane River, the Okanagan lobe merged with the ice mass of the Pend Oreille lobe across Washington, Idaho, and Montana.

Thousands of glacially deepened valleys were dammed by silt and clay to form lakes as the ice melted. Although many of the bodies of water have drained, old shorelines, lake strata, and outlet channels are still visible. Kamloops and Okanagan lakes today occupy the glacially scoured valleys of Pleistocene Lake Deadman and Lake Penticton. The small Nicola and Stump lakes near Merritt are all that remain of larger Ice Age Lake Hamilton. Shuswap Lake, which formed late in the Pleistocene, has also diminished considerably. Within the Rocky Mountain and Purcell trenches, finely laminated silt and clay chronicle numerous lakes that occu-

pied the valley floor during glacial recession. Stories about serpents, dragon-like prehistoric animals, surround many of these deep interior lakes. The Pend Oreille Paddler and the legendary Ogopogo living in Lake Okanagan persist to draw in and consume visitors.

In Idaho the 1225-foot deep U-shaped Pend Oreille lake basin resulted when glacial ice cut into soft underlying sediments. Once the ice began to melt and diminish, it left a residue of gravel banks that blocked the St. Joe River and impounded the beautiful Pend Oreille Lake. Lake Coeur d'Alene, the second largest in Idaho, owes its existance to a high dam of gravel remaining from late Pleistocene Lake Missoula floods.

Both Pend Oreille and Coeur d'Alene lakes provided channels for water released from the Clark Fork River, a branch of the Columbia River, in multiple gigantic Missoula floods across northern Idaho and northwest Washington. These waters coursed through the south end of Pend Oreille Lake before splitting into two routes. One torrent crossed Rathdrum Prairie and went on to innundate the basin at Coeur d'Alene and out to the Spokane River, while a second cascade cut a route through Hoodoo Valley to enter the Little Spokane River near present-day Newport. Blocks of ice, boulders, and sand carried in the maelstrom were eventually swept down the Columbia River to the Pacific Ocean.

IN REGIONS OF STEEP SLOPES AND HEAVY RAINFALL, cities, roads, and rail lines built on and near loose glacial silt and clay are frequently subject to disastrous landslides. Rain storms trigger large mass movements down steep ravines and out onto floodplains, which can have tragic consequences. Although smaller than rock slides, debris slides are much more common and are responsible for over 1/3 of all landslide deaths in western Canada.

In central British Columbia, Kamloops and Prince George are located on Pleistocene debris that are conducive to sliding. From the mid 1800s earth movements repeatedly blocked the Thompson River canyon, forming temporary lakes, and in 1905 a massive slide generated a

Figure 2.26 Remnants of a mammoth were recovered during mining operations at Babine Lake in northcentral British Columbia (skeleton from Flower and Lydekker, 1811).

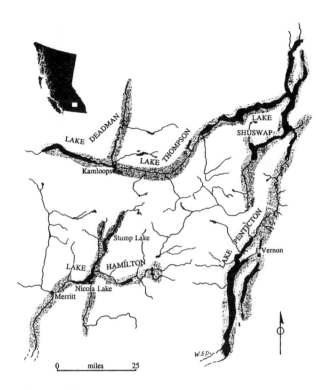

Figure 2.27 Maximum extent of some Pleistocene lakes in southcentral British Columbia (shaded area). Modern lakes in black (after Ryder and Clague, 1989).

15-foot high wave that entered the river at Spences Bridge north of Hope. The wave swept up the valley engulfing a Salish village and killing Chief Lillooet and 15 others. The lake behind the dam lasted for 5 hours. Downriver a few miles, the Drynoch earthflow, which took place several thousand years ago, involved a volume of over 17 million cubic yards of debris. As recently as 1954 a rock avalanche, set off by a small earthquake on the Hope-Princeton Highway, buried four travelers.

MINERAL PRODUCTION

A variety of mineral wealth in the Omineca-Intermontane province includes deposits of gold, silver, lead, zinc, copper, uranium, and molybdenum. After the spectacular gold strikes in California waned, miners searched northward for similar legendary fortunes. Initial examination of gold-laden river gravels in the mid 1800s led to lode discoveries and ultimately to the exploitation of other valuable ores. Mines were located primarily in the Cariboo Mountains, the southeast corner of British Columbia and adjoining Washington, and in the Idaho panhandle.

It is not possible to follow exactly the progress of the army of gold miners, but in 1852 placer gold at Kamloops was first brought to the attention of an

Figure 2.28 Spectacularly thick well-laminated accumulations of Pleistocene lake deposits form cliffs near Kamloops in the Thompson River valley (Canada Geological Survey) (figure at the base of the cliff shows the scale).

Figure 2.31 Hydraulic sluices in the Cariboo mining region of British Columbia separated heavier gold from sand and gravel as it washed down the long troughs (B.C. Archives and Records Service).

Figure 2.29 Glacial Pend Oreille Lake in northern Idaho, with an area of 180 square miles, is the largest lake in the state (Idaho Division of Tourism and Industrial Development).

Figure 2.30 The rumpled surface identifies a landslide near Pavilion, British Columbia. This slow-moving mass of rock and soil occurs in loosely consolidated Cretaceous and Quaternary rocks (Geological Survey of Canada).

agent of the Hudson Bay Company who bought some small nuggets. Within a few years gravels of the Thompson and Fraser rivers were being worked. By 1858 prospectors had pushed as far as Lillooet on the Fraser River, and in spite of the rough inhospitable country and distinctly unfriendly inhabitants over 600 prospectors followed trails to Quesnel. Flowing down from the high mountains, the Fraser River acted as a giant sluice where gold had been trapped for eons in bars and benches on the river bottom. Gravels proved to be very lucrative, and at places such as Hills Bar near Yale less than a square mile of the riverbed yielded $2,000,000 in gold.

THE RUGGED AND REMOTE CARIBOO MOUNTAINS of east-central British Columbia produced more placer gold than any other district in the province. On the celebrated Williams and Lightning creeks, the first $2,000,000 in gold was extracted by less than 1500 people before news and excitment of the finds emptied other mining camps in the northwest. Output was $4,000,000 during the top year in 1863 shortly before the supply of gold was exhausted.

Placer gold in the Cariboo Mountains was recovered from Pleistocene glacial banks strewn across valley floors and slopes. The richest strikes were along the ancient buried stream channels where the gold was situated directly on bedrock. Mining in the loose water-saturated sediments presented some unique problems. Downstream from the gold deposits, shafts were sunk 50 to 100 feet through gravel to pump out excess water. After this, large boulders had to be removed before the tunnels could be lined with closely spaced timbers to prevent

Figure 2.32 Some of the earliest mining efforts in British Columbia were placers along Williams Creek in the Cariboo district that left vast cutover forests and polluted streams (B.C. Archives and Records Service).

collapse. Conditions were still hazardous as water continually showered from overhead, and spring floods frequently filled the workings.

Barkerville, the center of the Cariboo district, was well on its way to becoming a ghost town when the government funded the restoration of many old buildings and mining machinery as an historic site.

BY THE TURN OF THE CENTURY there was a thriving lode gold industry in the Rossland, Hedley, Boundary, and Sheep Creek districts of southeast British Columbia where mineralization occurred in rocks of the Quesnellia terrane. The second largest mining operation in British Columbia was located near Rossland and Nelson where mining began in 1894 when gold ore was sent by rail to Vancouver for processing. Subsequently a dozen smelters were constructed, and locally production attained a peak in 1902 before it slowed and the smelters abandoned around 1928. This district surpassed all other lode areas in western Canada with an astonishing 85 tons of gold removed from an area only a third of a square mile.

In this area metallic ores occur in veins that cut through the Early Jurassic Rossland Group adjacent to the Nelson batholith. Fault zones provided a route for ore-laden hydrothermal fluids to penetrate and precipitate in veins. Mineral-rich veins range from less than 1 inch to several yards wide and extend to great depths in some localities.

THE YMIR-SHEEP CREEK DISTRICT, from Nelson south to the international border, is a narrow band that contains over 100 quartz veins. This region produced around 29 tons of gold from 1899 to 1942. Strata

of the Precambrian Windermere Supergroup was intruded by the Nelson batholith to emplace the ores.

IN THE HEDLEY DISTRICT southwest of Penticton, the Nickel Plate Mine was once one of Canada's largest lode gold producers. Mining began in 1903 but was shut down one year later when the ore bodies were thought to be worked out. However, exploration in 1936 revealed new resources, and today the mines are still in operation. At least 53 tons of gold were recovered from this district along with silver and copper. Ore-rich veins are found within Triassic limestones of the Nickel Plate Formation intruded by Jurassic dikes and granodiorite plutons that form the base of Nickel Plate Mountains.

By 1989 total gold production in western Canada reached 1230 tons, 690 tons from lode and 540 from placers. Most of the lode output was from British Columbia, but the 375 tons of placer gold from the Klondike fields of the Yukon were more than twice that from the Cariboo district. The Yukon placers were among the richest in the world because the gold-bearing river gravels here were not dispersed by glaciation.

MOST OF BRITISH COLUMBIA'S EARLY COPPER output was from the Boundary district near Greenwood, British Columbia, in conjuction with that from Copper Mountain about 70 miles to the northeast. Opening in 1900, the Boundary region yielded 33 tons of gold and over 200 tons of copper ore from mineral veins accompanying the emplacement of Jurassic plutons. Copper Mountain has been open sporadically since first discovered in 1892, and ore bodies here occur in fractured volcanic rocks of the Upper Triassic Nicola Group lying adjacent to Jurassic plutons. Although primarily mined for copper, copper mountain has also yielded 17 tons of gold.

Once it became economically feasible to exploit low-grade copper ores in the 1960s and 1970s, mining began in Highland Valley about 9 miles southeast of Ashcroft. Ores were extracted from granitic intrusives in the Quesnellia terrane. Tighter economic conditions in the 1980s allowed operations to continue only at extremely efficient plants. Between 1962 and 1984 Highland Valley yielded 386 million tons of ore.

IN NORTHEAST WASHINGTON, early mining efforts were focused on the Republic district of Ferry County as well as on the smaller regions of Stevens, Okanogan, and Pend Oreille counties where the opening of the Colville Indian Reservation in 1896 was the stimulus to mineral exploration. For a time at the turn of the century these mining areas were among the most isolated in the United States. Far from railroad lines, they were reached only after long journeys by boat and stage coach over rough dirt roads. Once spurs of the rail line were completed around 1910, removal of ores here was significantly improved, but the output of both placer and

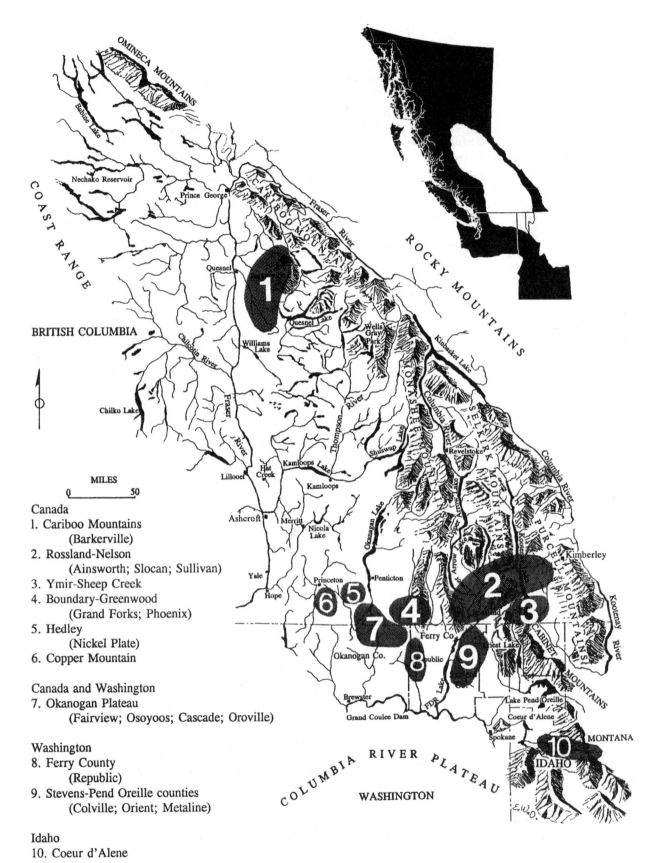

Figure 2.33 Mineral districts of the Omineca-Intermontane province.

Figure 2.34 Barkerville in the Cariboo district was a thriving gold town before the fire of September 16, 1868. A new town rose from the ashes almost immediately, and in recent years the buildings have been preserved as an historic site (B.C. Archives and Records Service).

Figure 2.36 Approximately 300-feet thick and one-mile long, the Sullivan deposit of lead, zinc, and silver at Kimberley, British Columbia, is one of the largest massive sulfides in the world. Over 130 million tons of ore were extracted from rocks of the Proterozoic Purcell Supergroup. The mine here was serviced by a small train and terminal (B.C. Archives and Records Service).

Figure 2.35 A zinc and iron processing operation near Slocan, in the Rossland district of British Columbia (B.C. Archives and Records Service).

Figure 2.37 Built on a mountain of copper in the Boundary-Greenwood district of southern British Columbia, the town of Phoenix was subject to frequent cave-ins (B.C. Archives and Records Service).

Figure 2.38 Loaded ore cars at Old Ironsides, a copper mine at Phoenix, British Columbia (B.C. Archives and Records Service).

Figure 2.39 The long-lived Knob Hill Mine in Ferry County, Washington, produced gold from 1937 onward (Washington Dept. Natural Resources).

Figure 2.40 The Star-Morning Mine in Idaho primarily exploits a single vein. On the west side of the ridge the Morning Mine is rich in lead and silver, while on the east side the Star Mine is zinc-rich (Idaho Dept. of Transportation).

vein gold as well as copper fluctuated considerably over the years.

THE RICHEST SOURCE OF GOLD ORE IN FERRY COUNTY is the Republic graben where veins in the Eocene Sanpoil volcanics have yielded over 2.5 million ounces of gold and 14 million ounces of silver. After opening in the early 1900s, mining as well as crushing operations in the Republic district were periodically closed until production began at the Knob Hill mine. Operating continuously since 1937, the Knob Hill today recovers 95% of the gold and 80% of the silver from processed ore. The 2 millionth ounce of gold was brought up from the Knob Hill shaft in 1989.

Mining in the Republic district received a boost in 1984 with discovery of the Golden Promise ore body. Currently estimated at more than 10 million tons of ore, the Golden Promise had been ignored for years by geologists convinced that these pyroclastic volcanic rocks could not be impregnated with economic deposits. More than 10 years after their discovery, the vein is nearing depletion.

MINES IN STEVENS COUNTY, active until 1959, contributed 32,000 ounces of gold. During most of those years, however, gold was only a by-product from the mining of other ores. Significant amounts of lead and zinc have not been recovered in Washington since the

1980s, but renewed efforts began with a brief open pit mining operation in Stevens County near Metaline from 1992 to 1993.

TAKING IN MINES NEAR OSOYOOS AND OLIVER in British Columbia as well as adjacent Washington, the Okanogan area was initially exploited when gravels along Similkameen River were worked in 1859. However, Oroville, Washington, proved to be the most lucrative at close to 70,000 ounces of gold extracted from both placers and lodes. Output near Oroville was temporarily interrupted when the land was set aside as a Wenatchee tribal reservation. Veins of gold, silver, lead, and copper here are distributed through sedimentary rocks in contact with the Late Cretaceous or Early Tertiary Similkameen batholith.

Approximately 20 miles east of Oroville on Buckhorn Mountain, the Crown Jewel deposit was drilled after environmental assessments were conducted in 1994. Almost 9 million tons of ore were found in zones through Permian to Triassic volcanic rocks intruded by plutons.

IDAHO IS ONE OF THE TOP PRODUCERS of silver, lead, and zinc in the United States. Near Coeur d'Alene in Shoshone County, the rich source of these metallic ores is sediments of Precambrian Belt Supergroup lying within the Lewis and Clark fault zone. Since

Figure 2.41 The extensive metallurgical processing complex at Bunker Hill, Idaho, closed in 1988 (Idaho State Historical Society).

Figure 2.42 The miner and his faithful donkey (Univ. of Oregon Special Collections).

opening in 1884, the famous Coeur d'Alene mining district has yielded over 1 billion ounces of silver, the highest in the world. In addition to silver, 8.5 million tons of lead and 3 million tons of zinc along with antimony, copper, and gold bring the total value of minerals extracted to $4.8 billion.

Over 90 separate mines are sprinkled along the South Fork of the Coeur d'Alene River and its tributaries in the Coeur d'Alene district. Of these, the Bunker Hill silver vein is one of the larger and richest anywhere. In the story often told of its discovery, prospector Noah Kellogg stumbled on the vein while searching for his jackass along the South Fork. Kellogg found the intelligent animal who appeared to be gazing at an outcrop of galena ore shining in the sun. When word of discovery spread, claims were filed, argued, and re-staked in and out of court. Once these disagreements were settled, the Bunker Hill Mine emerged. Most of the immense metallurgical complex closed in 1981, but the mine continues to yield silver, lead, and zinc from two separate veins in tunnels that extend over 150 miles. At present this site is undergoing prolonged environmental cleanup.

The Star-Morning mine near Mullan is comparable to the Bunker Hill in output. In July, 1884, two prospectors happened on the Morning vein which was subsequently developed and operated separately until 1966 when it was combined with the Star Mine. Both exploited the same ore body rich in silver, lead, and zinc. Access for the Star is on the west side of the ridge, and

the Morning entrance is on the east. One of the deepest mining operations known, the Star is accessed by a 7900-foot long shaft which is reached from the 2000 level in an adjoining mine. Because of low prices, the mine closed in 1982.

Along Big Creek, the Sunshine Mine was worked for 25 years at a marginal profit before a 20-foot wide high-grade silver vein was discovered in 1931. In May, 1972, a fire that spread through the mine killed 91 persons in what was one of the worst mining disasters in the history of the U.S. Today the Sunshine is the biggest producer of silver in the world with over 11 thousand tons to its credit.

HISTORICALLY, THE NATURE OF MINING has been to extract maximum profit for minimal expense in equipment and labor. Certainly restoration was seldom a consideration in the past. This outlook has altered today because environmental regulations and laws involve the removal and replacement of contaminated soils as well as those containing toxic materials. In spite of this, new techniques such as heap-leaching with cyanide continue to provide challenges to restoration plans.

ADDITIONAL READINGS

Baird, D.M., 1964

Baird, D. M., 1965

Bennett, E.H., Siems, P.L., and Constantopoulis, J.T., 1989

Breckenridge, R.M., 1989

Cheney, E.S., Rasmussen, M.G., and Miller, M.G., 1994

Church, B.N., 1985

Clague, J.J., 1989

Clague, J.J., et al., 1987

Coney, P.J., 1979

Dawson, G.M., 1888

Dawson, K.M., et al., 1991

Derkey, R.E., 1994

Easterbrook, D.J., 1979

Fox, K.F., and Wilson, J.R., 1989

Fulton, R.J., ed., 1989

Gabrielse, H., and Yorath, C.J., eds., 1991

Geology and Economic Minerals of Canada, 1947

Harms, T.A., and Price, R.A., 1992

Holder, R.W., Gaylord, D.R., and Holder, G.A., 1989

Hoy, T., 1989

Hutchinson, R.W., and Albers, J.P., 1992

Leclair, A.D., Parrish, R.R., and Archibald, D.A., 1993

Lewis, S.E., 1993

Lindgren, W., and Bancroft, H., 1914

McGroder, M.F., and Miller, R.B., 1989

Monger, J.W.H., Price, R.A., and Templeman-Kluit, D.J., 1982

Moye, F.J., 1987

Orr, K.E., and Cheney, E.S., 1987

Price, R.A., Monger, J.W.H., and Roddick, J.A., 1985

Price, R.A., Monger, J.W.H., and Roddick, J.A., 1994

Rehrig, W.A., Reynolds, S.J., and Armstrong, R.L., 1987

Rhodes, B.P., Harms, T.A., and Hyndman, D.W., 1989

Richmond, G.M., 1986

Ryder, J.M., and Clague, J.J., 1989

Stanley, G.D., and McRoberts, C.A., 1993

Templeman-Kluit, D.J., ed., 1985

U.S. Bureau of Mines.[annual]

Watkinson, A.J., and Ellis, M.A., 1987

Wolfe, J.A., and Wehr, W.C., 1991

Insular and Coast Range of British Columbia and Alaska

Looking southwest up Ferris glacier in the St. Elias Range of Alaska, Mt. Fairweather, the highest peak in British Columbia, reaches 15,300 feet. The mountains on the left are in Alaska, and those to the right are in British Columbia (B.C. Ministry of Environments, Lands, and Parks).

High mountain crags, a ragged coastline punctuated by fiords, and hundreds of offshore islands make up the Insular and Coast Range geologic province that extends over 1000 miles from Alaska to Vancouver, British Columbia. While the southwestern margin abuts the Fraser and Puget lowlands, the southeastern boundary merges with the Cascade Mountain range. The Intermontane belt lies on the east with the Pacific Ocean to the west.

Ice-capped peaks within the Coastal belt, many over 9000 feet, drop precipitously into glacially deepened canyons. Approximately 160 miles northeast of Vancouver, Mount Waddington at 13,104 feet above sea level is the highest point along the coast. The west coast of Vancouver Island and the mainland are tattered by scenic fiords projecting as much as 90 miles inland to depths of 2000 feet. Alberni Inlet on southern Vancouver Island and Quatsino Sound on the northern part extend almost completely across the island. Just as Howe Sound stretches toward Mt. Garibaldi, Jervis, Toba, Bute, and Knight inlets reach from the Pacific Ocean far into the coastal mountains.

Offshore islands line the coast. Vancouver Island, 280 miles long and 78 miles at its widest, is the largest in the eastern Pacific. Located about 100 miles from the mainland, the Queen Charlottes are a cluster of small islands of which the larger are Graham, Moresby, Louise, Lyell, and Kunghit. The topography of the offshore islands is subdued in comparison to that of the mainland. At 7000 feet in elevation, the range that forms the back-

bone of Vancouver Island is nearly twice the height of the mountains on Queen Charlotte Islands but still less than the Coast Range. From southern Alaska to Vancouver Island, coastal islands are separated from the mainland by narrow channels, the Clarence Strait, Hecate Strait, Queen Charlotte Strait, and the Strait of Georgia.

Figure 3.2 Even though they are 7000 feet in elevation, the mountains of Vancouver Island are considerably lower than those on the mainland (B.C. Ministry of Environment, Lands, and Parks).

Figure 3.1 Displaying steep headwalls and ice falls, Mt. Tiedemann in the foreground and Mt. Waddington at top center are at the junction of two glaciers. The dirty streaks in the glacier attest to the rapid erosion and movement of the ice (B.C. Ministry of Environment, Lands and Parks).

Figure 3.3 The Insular and Coast Range belts across western British Columbia.

At the southwest corner of the province, the Fraser Lowland is an elongated triangular area of approximately 1500 square miles in British Columbia and the United States. Low rounded hills and wide flat valleys are interposed by bedrock knolls like Burnaby, Grant Hill, and Sumas Mountain. The lowland is dominated by the Fraser River, which discharges its sediment into the Strait of Georgia. Since the Pleistocene, the river has constructed an advancing delta that is now over 20 miles long, 15 miles wide, and growing by as much as 30 feet per year. Presently four channels of the braided river meander through a wetlands of bogs, sandbanks, and small islands.

INTRODUCTION

Western British Columbia is a region long dominated by the movement of enormous crustal plates. The province has been largely built and shaped by changing tectonic patterns, ranging from compression between major colliding terranes to tension from oblique plate interactions. The Coast Mountains and offshore islands of southern Alaska and British Columbia are a patchwork of smaller fragments of exotic rocks and terranes annealed to the western margin of North America. The accretion of the outermost Insular superterrane during the Cretaceous was accompanied by intense metamorphic changes caused by elevated temperatures and pressure, faulting, and uplift along the welt that forms the Coast Range.

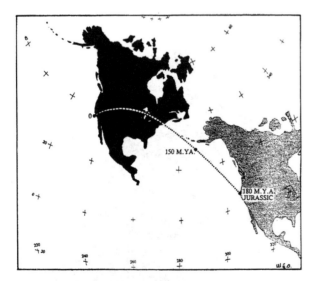

Figure 3.4 Riding atop the North American plate, the North American landmass has changed direction since the Mesozoic. In the Jurassic to Cretaceous the continent moved northwesterly, but today it moves in a west-southwest direction (after Engebretson, Cox, and Gordon, 1985).

The core of the Coast Range belt is predominantly metamorphic rocks and crystalline granitic plutons intruded during accretion.

Events of the Cenozoic can be related to crustal plate interaction as well. Much of the Cenozoic volcanic activity across central and southern British Columbia is a manifestation of hot-spot tracks, faulting along the margins of adjacent crustal plates, and continental volcanic archipelagos above subducting slabs.

No geologic agent has been more important than glaciation in carving up the landscape of this province. Along the edge of the continent as well as on Vancouver and Queen Charlotte islands, valley glaciers cut deep straight fiords far inland to expose details of the Coast Range. While advancing glaciers molded the topography, swollen meltwater lakes and streams left a vast covering throughout British Columbia that today accounts for many of the surface features.

Finally, the offshore submarine geology of the continental shelf, slope, and abyssal deep of the northeast Pacific Ocean provides a glimpse of the crustal plate movements in action. While the present tectonic scenario is not an exact model of the forces that created the Pacific Northwest, it provides a clear view of the relationship between major slabs and the accompanying faulting and volcanic activity that mark the sites and intensity of plate interactions.

GEOLOGY OF THE INSULAR AND COAST RANGE PROVINCE

Of the five belts that reflect ongoing geologic processes across western Canada, the Foreland, Omineca, and Intermontane lie to the east while the Insular and Coast Range are the most westerly. In this province, the Insular and Coast Range belts, welded together by plate collisions, are the two dominant tectonic and physiographic features.

THE INSULAR BELT, taking in the Pacific continental margin and offshore islands, is a complex superterrane of exotic rock fragments that have annealed into a composite crustal mass. Smaller terranes within the belt are unique assemblages of strata that reflect specific depositional environments or volcanic settings. Many were formed offshore as volcanic island archipelagos, ocean basin sediments, or shelf deposits where they were carried atop large moving plates to collide along the western North American coast. Faults, forming the margins of these terranes, attest to their tectonic emplacement into the larger belts. In contrast to the terranes of the Omineca-Intermontane, those of the Insular superterrane have been subjected to much less heat and pressure as a result of metamorphism and plutonic intrusions.

Figure 3.5 Map of locations in the Insular and Coast Mountains of British Columbia and southeast Alaska.

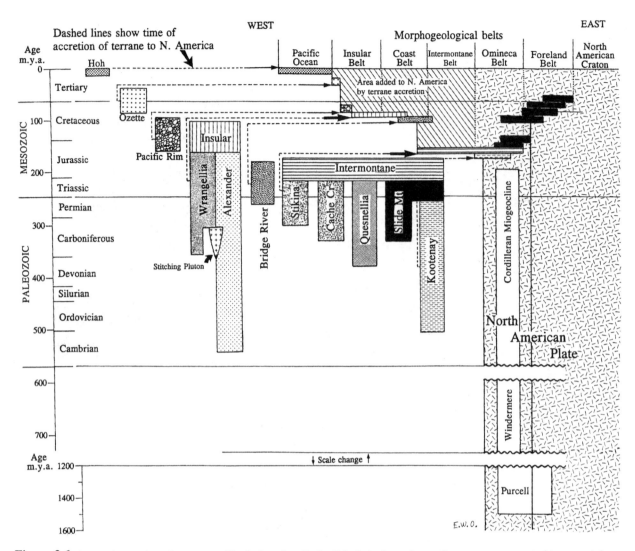

Figure 3.6 Age and accretion of terranes to North America: Each of the belts is made up of two or more accreted terranes (after Price, 1994; Gabrielse and Yorath, 1991).

Two major terranes, Wrangellia and Alexander, each with distinctive rocks and fossils of varying ages and environments, make up the Insular superterrane. On the outboard side of the superterrane, the younger Pacific Rim, Crescent, Yakutat, and Chugach blocks have been progressively shoved beneath the western margin of the Wrangellia and Alexander sheets to create an imbricated stack of terranes. Vancouver and Queen Charlotte islands are portions of the Wrangellia block, while Prince Rupert Island to the north is part of the Alexander slab.

Fossils, paleomagnetic measurements, metamorphism, igneous intrusions, and sediments are among the tools and methods employed to trace the complex history of formation, transportation, amalgamation, and accretion of terranes. Because several lines of evidence are available, it is inevitable that there will be different interpretations as to how the terranes were assembled.

Today, for example, there seems to be little agreement about when the Wrangellia and Alexander terranes amalgamated to form the Insular superterrane or even when collision with the continent took place.

THE COAST RANGE BELT owes its origin to one of the last docking events of major exotic terranes in the Pacific Northwest. This event began in the Early Cretaceous, approximately 100 million years ago. At that time, the Insular superterrane collided with and was welded to the western margin of the Intermontane belt, producing the folds of the coastal belt. In the complex process of accretion, rocks on both sides of the collision boundary were subjected to intensive compression resulting in distortion, shearing, metamorphism, and plutonic intrusions. As the oncoming terrane was squeezed beneath the western margin of

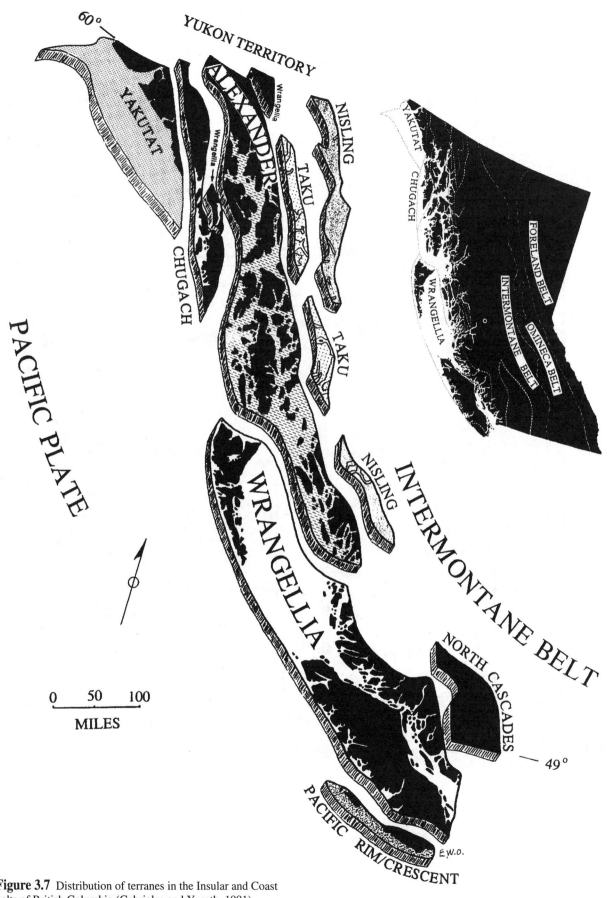

Figure 3.7 Distribution of terranes in the Insular and Coast belts of British Columbia (Gabrielse and Yorath, 1991).

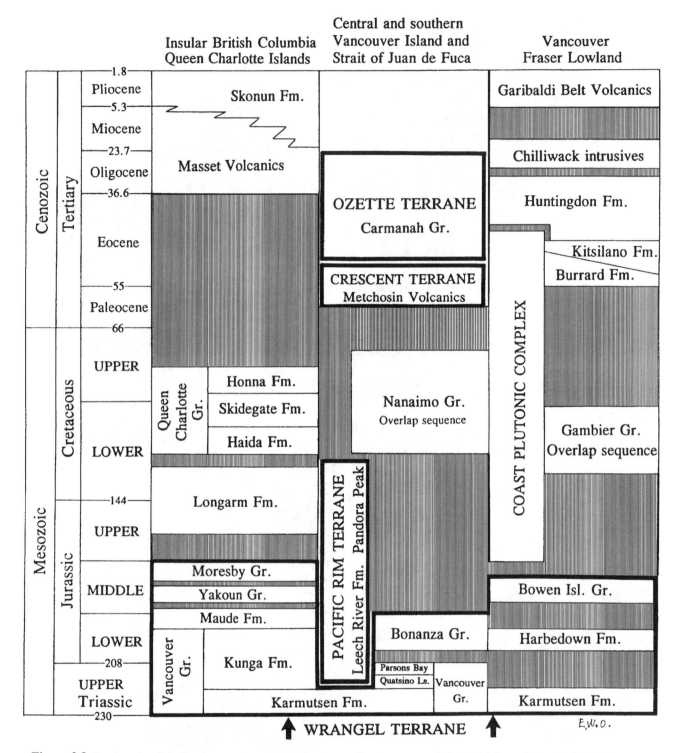

Figure 3.8 Stratigraphy of the Insular superterrane. Dark borders outline the terranes (after Gabrielse and Brookfield, 1988; Wheeler and McFeely, 1991).

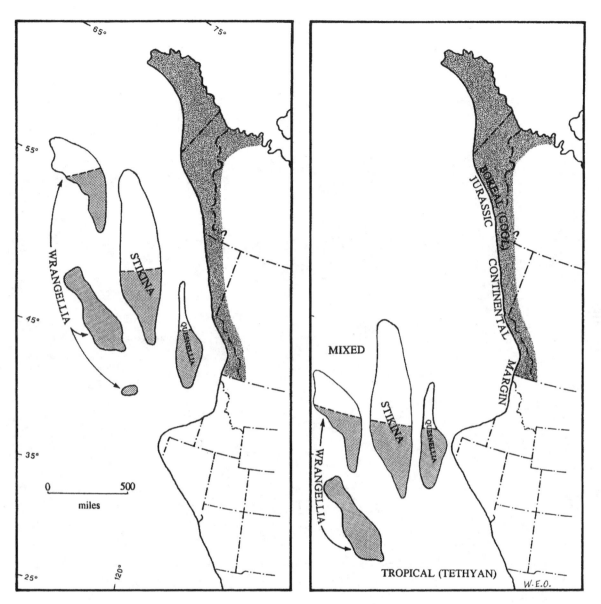

Figure 3.9 Fossils, entombed in exotic terranes, are used to reconstruct ancient environments and trace the wandering continental fragments back to their points of origin. By studying ammonite molluscs in Wrangellia, it has been estimated that this terrane originated at least 1200 miles to the south (after Taylor, et al., 1984).

North America, low gentle folds became more pronounced until the rocks shattered along fault zones. Under this extreme pressure, the rocks failed and telescoped over each other locally, resulting in a shortening and thickening of the crust.

As a result of the collision, plutons, which were insinuated into the Coast Range from the Jurassic to Miocene, account for over 85% of the rocks. This crystalline mass is one of the most extensive in the world. In the intrusion process, plutons invaded older strata at

great depths within the earth to cool as immense batholiths. To melt such large volumes of rock implies tremendous heat over an extended period of time. These conditions could only be met deep in the earth's crust during active subduction. As the moving slab descends 60 to 75 miles beneath the continental plate, it begins to melt. The hot liquid material partially rises in the crust to mobilize and incorporate surrounding rocks. As the magma slowly hardens, large mineral crystals, typical of plutonic bodies, are developed.

Figure 3.10 The granite peaks of the coastal mountains are in the foreground while Mt. Baker of the Washington Cascades is visible in the distance on the upper right. The canyon in the center shows the trademark U-shape of a glacial valley (B.C. Ministry of Environment, Lands, and Parks).

The most significant interval of pluton implacement throughout this province was during the Late Jurassic to Early Cretaceous, and by the Middle Cretaceous intrusive bodies were concentrated only in the Coast Range. During the Early and Middle Eocene extensive intrusion had ended so that plutons from the later Tertiary are much smaller and more shallow.

THE CRETACEOUS PERIOD WORLDWIDE was marked by rising sea levels and transgressive shorelines. In the Pacific Northwest, ocean waters advanced across the continent from the west during the Early Cretaceous, reaching a maximum later during the same interval. Be-

cause of the expanding nature of these oceans, most of the preserved sediments reflect shallow marine environments inhabited by a varied fauna of bottom-dwelling and swimming invertebrates.

The collision of the Insular superterrane during the Cretaceous produced a number of well-defined basins on the west coast of North America from California to Alaska. The largest of these were the Georgia trough, adjacent to Vancouver Island, the Queen Charlotte basin along Hecate Strait, and the Gravina basin east of Prince of Wales Island in southeast Alaska. As the accretion process began, sediments were eroded from the adjacent highlands and deposited into the subsiding troughs. With

Figure 3.11 The rocky west coast of Vancouver Island has few beaches. The distant projection of land is Estevan Point (B.C. Ministry of Environment, Lands, and Parks).

continued collision, the basins filled then shrank in size as they were trapped between merging terranes.

The Georgia or Nanaimo basin, which subsided along the suture between the oncoming Insular superterrane and the coast, lies beneath the Strait of Georgia. This elongate sedimentary trough, which includes the Strait of Georgia, the Nanaimo basin, and the Fraser lowland, was severely compressed and deformed by the collision event.

On the east side of Vancouver Island, the Nanaimo depression was divided into the northern Comox and southern Nanaimo basins by a central upland. Older sediments in the basin were derived locally from the west and southeast, but later deposits were eroded from up-

lifted coastal mountains and the Cascades. Fossil snails, clams, coiled ammonites, and microfossils, preserved in the 10,000-foot thick strata, have been particularly useful in interpreting not only ages but the various environment of the sea locally.

Three environmental settings, nonmarine, shallow coast marine shelf, and deep water channels, have been recognized in the 100 million-year old Nanaimo sediments. Nonmarine intervals with plant fossils and coal occur on the southern end of Vancouver Island, while shell-rich beach deposits are found in the southeast and northwest. Strata representative of deep water channels are scattered throughout Vancouver Island, the Gulf Islands, and San Juans.

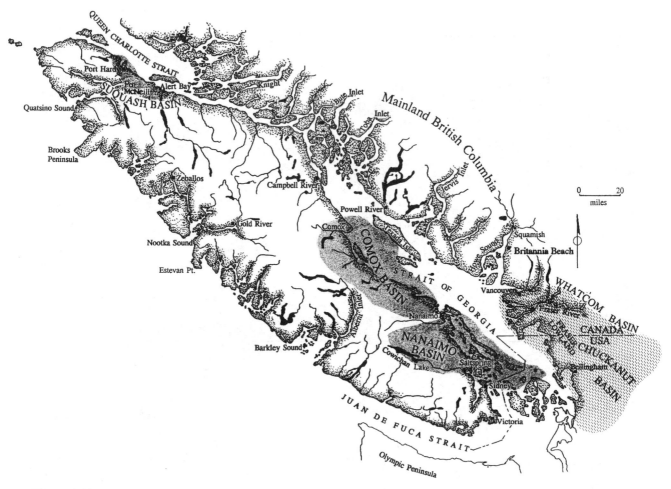

Figure 3.12 During the Cretaceous, sedimentary basins along the shoreline of Vancouver Island received sediments eroded from the uplifted coast (after Pacht, 1984).

During the Eocene and Oligocene epochs, sediments deposited in the smaller Whatcom and Chuckanut basins reflect shallow embayments along the southeastern portion of the larger Georgia trough. Lying along the Fraser and Puget lowlands, sand, mud, and conglomerate of the Eocene Chuckanut Formation near Bellingham, Washington, the Eocene-Oligocene Huntingdon Formation near Abbotsford, British Columbia, and the Middle Eocene Burrard and late Eocene Kitsilano formations near Vancouver, British Columbia, were spread over floodplains by sediment-laden streams.

Near the margins of these Early Tertiary basins, swamps accumulated leaves of palms, cypress, and tree ferns. Even bird tracks from a member of the Heron family are preserved in a shallow water area of the Chuckanut. The pattern of large prints impressed into a rock slab several feet long shows the bird moved with short hops as it preyed on fish and invertebrates around the coastal wetlands.

Marine and nonmarine sedimentary rocks of the Queen Charlotte basin lie beneath the Queen Charlotte Islands, Hecate Strait, and Queen Charlotte Sound, while the boundary between the Wrangellia and Alexan-

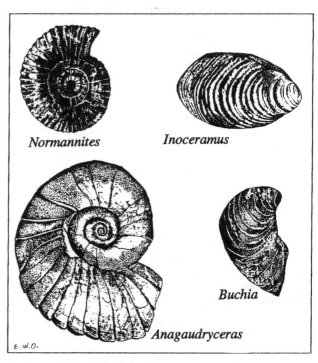

Figure 3.13 Ammonites and clams inhabiting the Jurassic and Cretaceous oceans across Vancouver and the Queen Charlotte Islands.

der terranes is buried under central Graham Island and Hecate Strait. Within the Queen Charlotte basin, over 10,000 feet of sediments span the Cretaceous to Late Tertiary interval. Sand, silt, and conglomerate of the Late Jurassic to Early Cretaceous Longarm Formation and the Middle to Late Cretaceous Queen Charlotte Group were derived from the Coast Range. The Longarm, with abundant molds and broken shells of the flat clam *Inoceramus* was deposited in a deep marine trough, while a protected estuary of the Queen Charlotte Group has ammonites and other shelled invertebrates. Fossil remains of wood and leaves in these rocks reflect the proximity of surrounding lowlands.

In southern Alaska a narrow band of sedimentary and volcanic rocks of the Gravina trough are a Late Jurassic to Middle Cretaceous succession that overlaps the boundary between the Wrangell and Alexandria terranes.

THE CENOZOIC HISTORY of the Insular and Coast Range province was one of tectonic uplift and faulting followed by volcanic activity and intense glacia-

tion. Elevation of the land took place in three stages beginning in the Paleocene, over 55 million years ago, and continuing into the present. The regional focus of uplift alternated between the north and the south. An intensive period of Early Tertiary crustal thickening and mountain-building was followed by Miocene uplift in southern Alaska and the northern British Columbia coast regions that coincided with subsidence of the Queen Charlotte basin. As subduction of the Pacific Plate beneath North America proceeded, the crust was lowered along the western margin of the overriding continental slab, producing a flexural basin inboard of Queen Charlotte Islands. Simultaneously, a broad area of the southern coastal mountains was pushed up, resulting in the present-day topography. The final phase beginning in the Pliocene, about 5 million years ago, and continuing to the present has raised Queen Charlotte Islands and depressed the Hecate Strait.

CENOZOIC VOLCANISM in southwest British Columbia began with Early Eocene Metchosin flows on

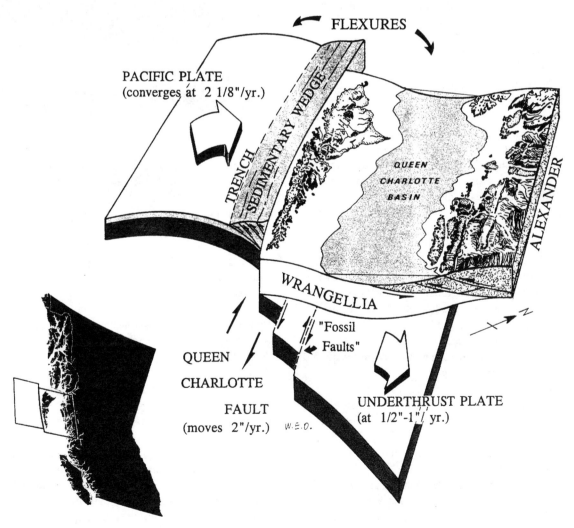

Figure 3.14 As the Pacific Plate converges with North America, it is sheared up by the Queen Charlotte fault, while the Queen Charlotte basin is being flexed downward in the suture between the Wrangellia and Alexander terranes (after Yorath and Hyndman, 1983).

the southern tip of Vancouver Island. These lavas are an extension of the Paleocene to Eocene Coast Range basalts that reach from southwest Oregon to British Columbia. The well-layered succession of the Metchosin includes plutons, dikes, basalt pillows, and lava that represent a partial ophiolite, a vertical sequence of ocean crust.

A later period of eruptions and intrusions lasting from the Miocene to the present is reflected in over 100 volcanic centers in Alaska, British Columbia, and the Yukon. Five volcanic belts, which run parallel or at right angles to the coastline, were generated by diverse tectonic processes that took place here beginning 20 million years ago. The oldest of these are the Pemberton and Garibaldi belts extending north-south along the coast from Queen Charlotte Islands to merge with the Cascade Mountains. The Anahim belt runs from Bella Bella across the Coast Mountains inland to the eastern margin of the Omineca-Intermontane province. On Vancouver Island, Alert Bay volcanism is found from Brooks Peninsula northeast to Port McNeill, while the youngest Wrangell and broad Stikine belts reach across Alaska, into southwest Yukon, and northwest British Columbia.

Active during the Late Miocene, the Pemberton volcanic belt was fed by granitic plutons that intruded much of the Coast Range and Northern Cascades in British Columbia. Volcanic activity within this region was triggered by eastward subduction of the Juan de Fuca and Explorer plates. Early phases of intrusion include the 21 to 16 million-year old Mt. Barr and 29 million-year old Chilliwack batholiths in the Northern Cascades that lie beneath the International Border. In the central Coast Range caldera complexes along with the Masset lavas on Queen Charlotte Islands mark the northern limit of the belt.

Granites of the Franklin Glacier volcanic complex, which underlie an area of 50 square miles near Mt. Silverthrone, were intruded in two phases between 8 and 2 million years ago. Approximately 30 miles northwest, lava flows forming the jagged crests of the Mt. Silverthrone complex date back 750,000 to 80,000 years. The surface of these lavas is deeply eroded, and cones, constructed during the final surge of activity, projected through glacial ice.

Basalt and explosive rhyolite lava of the Masset Formation erupted 20 to 25 million years ago along rifts and faults through older Wrangellia terrane strata on Queen Charlotte Islands. Fed by small granitic plutons aligned northwest throughout the islands, Masset igneous activity is separated by a wide gap of 250 miles from other intrusions of the Pemberton belt. During the final Late Miocene stage of Pemberton volcanism near Mt. Silverthrone, activity shifted westward to the Mt.

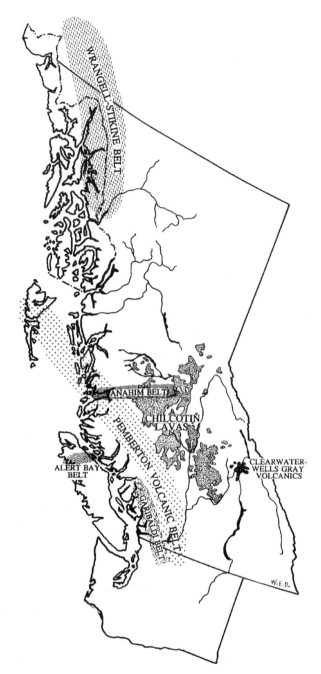

Figure 3.15 Seven Miocene and younger volcanic zones occurred across British Columbia, Alaska, and the Yukon (after Clague, 1981).

Garibaldi volcanic centers at the northern extreme of the Cascade Range.

The Anahim volcanic belt is delineated by a chain of volcanoes and underlying plutons that run from the coast inland to the flatlying Chilcotin lavas of the Omineca-Intermontane area. In fiords and around islands at Bella Bella, granites are exposed, whereas in-

Figure 3.16 Smoke and steam from a volcanic peak within the Wrangell Mountains of southwest Alaska (W.C. Mendenhall, U.S. Geological Survey).

land, shield volcanoes form the Rainbow, Ilgachuz, and Itcha ranges on the southern Nechako plateau. Lavas of the Anahim are distinctly bimodal, erupting in both basaltic and silicic flows that become progressively younger toward the east. Eruptive centers along the coastal islands are 15 million years old in contrast to cones and flows dated at only 7200 years old near Nazko in the interior plateau. This westward progression traces movement of the continent over a hot spot in the mantle.

Volcanic events in the Late Miocene and Pliocene Alert Bay field from 8 to 3.5 million years ago coincide with the shift of plate motion and shift in the focus of activity from the Pemberton to the Garibaldi belt. By Late Pliocene time most of the Farallon tectonic slab had been subducted beneath North America, and the trailing fragments, reorganized and redirected, are designated the Explorer, Juan de Fuca, and Gorda plates. The linear trend of the Alert Bay volcanic belt lines up well with the margin between the Explorer and Juan de Fuca plates. Basaltic lava of the Alert Bay field spread west to east across Vancouver Island from Brooks peninsula to Port McNeill. At Twin peaks, in the center of the island, a large remnant of tuff and basalt covering an area of 10 square miles suggests a previous cone of immense size.

In northern British Columbia and the Yukon, the Late Miocene Wrangell-Stikine volcanic regimes are represented by a variety of cones, plutons, lava fields, and widespread pyroclastic debris that shifted to Alaska where activity still takes place today. At the northern end of the Wrangell belt sequences of basalt flows over 1500 feet thick cover vast areas along the coast where they are preserved on the eastern flank of the Saint Elias Mountains. At the southern end of the zone, the Stikine belt includes Mt. Edziza, Hoodoo Mountain, and a number of smaller craters from Revillagigedo Island, immediately north of Ketchikan in southeast Alaska, to Dease Lake on the Stikine River.

Located about 200 miles north of Prince Rupert, Mt. Edziza covers almost 400 square miles and encompasses one of the largest volcanic complexes in the northwest. Rising 9000 feet above the surrounding plateau, Mt. Edziza began as broad Miocene edifice more than 7 million years ago before thicker lavas and cinders constructed overlapping cones and the high peak. Eruptions of this long-lived center continued into the Pleistocene when many of the craters emitted lava flows beneath ice and snow. North-trending volcanic peaks within this zone trace older Neogene faults. More recent events from this volcanic source took place near the Alaska-Yukon border when eruptions from a vent scattered ash only 1900 to 1200 years ago.

PLEISTOCENE GLACIATION of British Columbia represents the single most profound change to the topography of this region. Throughout the Ice Ages between 1.8 million and 10,000 years ago, British Columbia was besieged by a succession of glacial intervals characterized by merging valley glaciers and an extensive ice sheet that enveloped lowlands and mountains. During glacial advance most of the continental glaciers were confined between the mountain barriers of the Coast Range and Rocky Mountains, while valley glaciers were abundant west of the coastal divide. Warmer interglacials were characterized by melting, thinning, and retreat of the ice along with valley glaciers at high elevations. After melting of the ice mass, the mainland, which had been depressed by the weight of glaciers, slowly rebounded, and ocean waters withdrew. Although these events left prodigious amounts of sediment throughout the valleys and lowlands of British Columbia, deposits on the south coast are among the thickest and most extensive.

In the Quaternary period, the Fraser Lowland was subjected to three major glacial phases and three interglacials. Close to 60,000 years ago the last notable Olympia Interglacial commenced, spreading a thin layer of gravel, sand, and silt of the Cowichan Head Formation in channels, marshes, and on floodplains as the sea level dropped and shorelines moved westward.

Since only high valley glaciers persisted, forests and grassy meadows grew at the lower fringes supporting herds of mammoths, mastodon, musk ox, and horses. Skull, tusk, tooth, and bone fragments of these large mammals, recovered from Late Pleistocene gravels along the Strait of Georgia, Fraser and Puget lowlands, and

Figure 3.17 In the British Columbia Coast Range, the westward view toward Mt. Silverthrone shows evidence of extensive glaciation (B.C. Ministry of Environment, Lands, and Parks).

Vancouver Island, show migration routes to Vancouver Island ahead of the advancing ice sheet. In contrast to scattered mammal remains, fossil marine invertebrate shells are abundant along the Fraser Lowland. Similar to living species, snails and clams are found with occasional crabs, sea urchins, barnacles, and plant fragments.

Over the Cowichan Head Formation, 150 feet of glacial Quadra Sand smothered existing stream, swamp, estuaries, and shallow marine environments. Stripped from the Coast Range during the early part of the glaciation, Quadra Sand accumulated in front of and alongside the advancing apron of ice building deltas that pro-

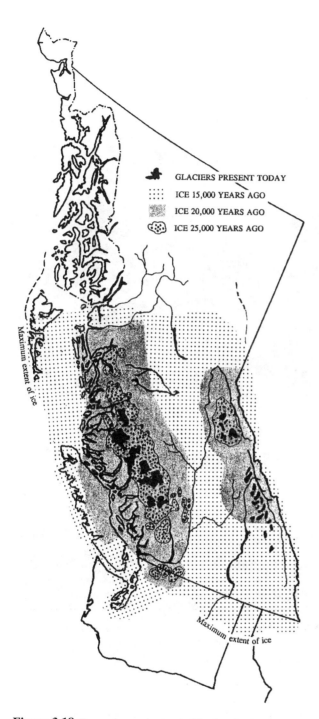

Figure 3.18 Expansion and retreat of the ice sheet over southern British Columbia (after Clague, 1981).

graded well out to sea. Ultimately these sands were collected and carried by moving glaciers to be redistributed in a patchwork pattern.

Following the Olympia interglacial period, the last significant interval of numerous ice advances, the Fraser Glaciation began about 26,000 years ago. As the ice sheet advanced into northern Washington at the height of the glaciation about 17,000 to 18,000 years ago, several thousand feet of Vashon sediments covered whole lowland forests of fir, pine, spruce, and yew. Sometime prior to 14,000 years ago, when temperatures rose, meltwater left layers of debris that marked the trail of retreating ice. Brief surges of ice took place, and a coalescing valley glacier, advancing from the mountains around 11,000 years ago, left Sumas till on the eastern lowland. As vegetation began to reestablish itself, a succession of herbs and shrubs followed by Douglas fir, western hemlock, and alder reflect progressively warmer and drier climates.

Because of the complex events of the Ice Ages, alternating cycles of deep erosion or momentous deposition affected the landscape of this province. Valley glaciers from the western slopes of the Coast Range coalesced into ice sheets that reached far out onto the continental shelf. The weight and bulk of glaciers can deeply entrench the mouth of a valley well below sea level, and tidewater glaciers gouged out deep straight fiords that today line the coast as long inlets to the sea. Howe Sound, a fiord northwest of Vancouver, is 25 miles long and over 1000 feet deep. Extending further inland and recut with each new glacial phase, Howe Sound was even deeper at the end of the Pleistocene before extensive deposits of silt, mud, gravel, and sand from the Squamish and Fraser rivers settled to its floor. By contrast, the Fraser Lowlands, that were beneath shallow marine water during much of the Pleistocene, accumulated almost 1000 feet of glacial and marine sediments. Once the ice disappeared, the landscape was elevated, and streams, augmented by glacial melt water and laden with gravel and sand, cut spectacular narrow canyons where the city of Vancouver is today.

THE IDEA OF AN ISOLATED ICE-FREE ZONE around the northeast Pacific margin during the Pleistocene has been debated at length. Like the biblical ark, this sheltered area, within a region otherwise buried by Pleistocene glaciers, would have served as a refuge for plants and animals. On the British Columbia coast, it has been postulated that Vancouver, Queen Charlotte, and the islands of southeast Alaska may have been such refuges or even provided continuous migration routes around the North Pacific Rim. Of these, Queen Charlotte Islands have been studied most intensively. Among the unusual plants living here are a number of distinctive species and subspecies of mammals and birds previously thought to be extinct or known only from Asian localities.

Recent analyses of fossil leaves and pollen at Cape Ball show a diverse flora existed on Graham Island between 16,000 and 11,000 years ago at the height of a glacial advance along the coast and mainland. Much of the

Figure 3.19 Tiedemann glacier, projecting westward toward Mt. Waddington, is streaked with gravel, mud, and other debris (B.C. Ministry of Environment, Lands, and Parks).

fossil-rich peat layer here extends to Hecate Strait where it is covered by ocean waters today. During glacial intervals, as ice is frozen in the polar regions and at high altitudes, global sea level is lowered, and a recorded drop of 300 feet during the last glacial advance would have left much of the Strait east of Graham Island above water, thereby providing a corridor for migrating animals.

MASSIVE LANDSLIDES AND RAPID ERO-SION, the inevitable result of oversteepened slopes and

poorly consolidated debris also pose special problems throughout British Columbia today. Once continual rainfall saturates the loose rock and soil, slides and slumping occur. This is especially true for cities like Vancouver, located on glacial fill. Within this century, several destructive landslides have taken place along Howe Sound from the entrance at Horseshoe Bay upriver to Britannia Beach where sudden earth movements have destroyed industrial and transportation facilities and killed 112 people. In 1915 a rock avalanche killed over 50 people near

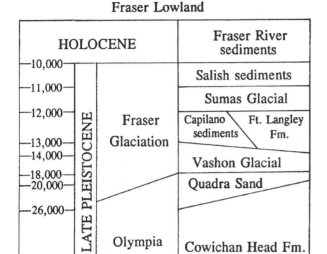

Fraser Lowland

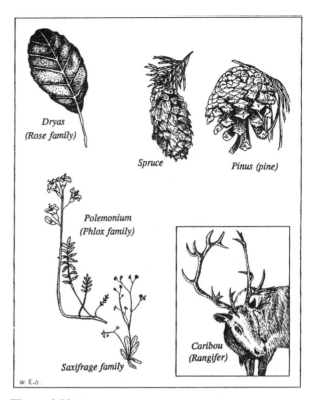

Figure 3.20 Pleistocene stratigraphy of the Fraser Lowland (after Fulton, 1989).

Figure 3.22 Plants and reindeer inhabiting Vancouver and Queen Charlotte Islands during the Pleistocene.

Figure 3.21 Waters of the meandering Alsek River southeast of Yakutat Bay, Alaska, are heavily laden with glacial debris (B.C. Ministry of Environment, Lands, and Parks).

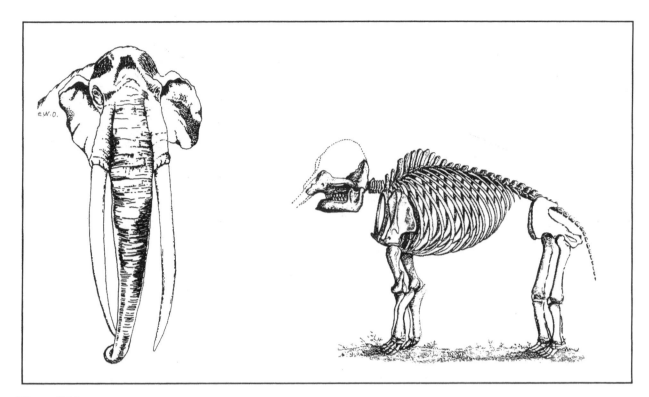

Figure 3.23 A charming rendering of a mastodon skeleton drawn in 1812 by M. Cuvier and a sketch of a mastodon head.

Figure 3.24 Enormously thick layers of till and glacial sediments of the Pleistocene Cowichan Head Formation and overlying Quadra Sand form cliffs in the Fraser Lowland (Geological Survey of Canada).

Figure 3.25 Glacial grooves and scratches in the bedrock are made by rocks carried in the moving ice mass (courtesy J.C. Reed, U.S. Geological Survey).

Figure 3.26 Steep valley walls of the Tracy Arm fiord in southeast Alaska show a glacial valley cut well below sea level (courtesy D.A. Drew, U.S. Geological Survey).

Figure 3.27 The June, 1948, Fraser River flood, which resulted when the river broke its dikes, damaged 2000 homes and caused losses of over $20 million (Geological Survey of Canada).

Britannia Beach, and in 1921 heavy rains triggered immense landslides that killed 37 and destroyed most of the town.

ALTHOUGH BRITISH COLUMBIA IS REGARDED AS SEISMICALLY active, only minimal damage has resulted from earthquakes because epicenters, for the most part, are offshore or in remote uninhabited areas. The zone of highest quake activity corresponds to the junctions between tectonic plates on the western side of Queen Charlotte and Vancouver islands. These are the Fairweather-Queen Charlotte fault zone, where the Pacific Plate meets the North American Plate, and the Cascadia fracture zone along Vancouver Island, where the eastbound Explorer and Juan de Fuca Plates are being subducted beneath the westbound American Plate.

Of the hundreds of earthquakes that have occurred on Vancouver Island in historic time, most were some distance from population centers, and few caused extensive damage. The June 23, 1946, event, northwest of Comox in the center of the island, had a magnitude of 7.3 and was the most destructive earthquake in western Canada in modern times.

There is increasing evidence that in the past earthquakes which caused subsidence of coastal regions also triggered offshore tsunamis or gigantic sea waves that suddenly innundated the Pacific coast. Each tsunami left a thin sheet of sand over nearshore soil and marshes. After the tsunami withdrew, the marshes recovered, but a layer of sand within the peat as well as undisturbed fossil plants, which have their roots in the peat but extend up into and through the sand, record each event. Muds with microscopic marine diatoms and foraminifera, which also covered the freshwater sediments, are additional indications of earthquakes and tsunamis.

Coastal sediments demonstrate that British Columbia was struck by tsunamis 3600, 1900, 800, and 500 years ago. The most recent sand layer from the 1964 Alaska quake is the thinnest of these suggesting that waves from the earlier tsunamis were 4 to 5 times higher. Large waves generated from faults close to the Pacific Northwest would be especially destructive since they could arrive before a warning or evacuation could take place.

COASTAL CITIES of British Columbia as well as those of southern and southeast Alaska are scattered across the terranes and superterranes that make up this province. A trip to Alaska on the Inside Passage threads through the picturesque Chatham and Clarence Straits. Multiple stops along the route provide a unique perspective on the complexity and formation of the numerous exotic slabs of this accretionary margin.

From Alaska to southwest British Columbia the Insular superterrane is a composite of the Alexander and

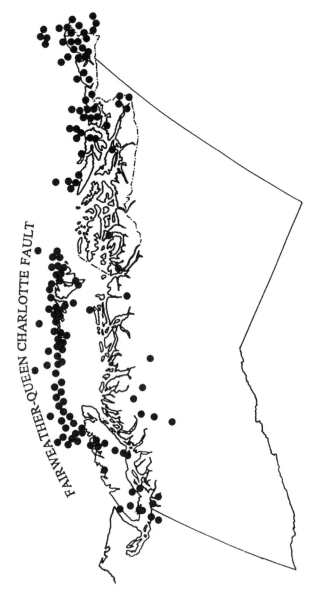

Figure 3.28 Patterns of recent earthquake occurrences along the Pacific coast of British Columbia trace the Queen Charlotte fault (after Wahlstrom and Rogers, 1992; Milne, et al., 1978).

Wrangellia terranes, while a half dozen smaller, loosely annealed terranes underlie the peripheral western border of islands and coastline. On southern Vancouver Island, the Upper Mesozoic Pacific Rim and Early Tertiary Crescent terranes have been thrust beneath rocks of Wrangellia. In southeast Alaska the Cretaceous Chugach and Yakutat terranes comprise the continental margin on the outboard side, while the Taku of the Coast belt forms the eastern boundary of the Insular belt.

Due to opposing plate motions along the Pacific Rim, terranes are typically sheared and offset by faults on a grand scale. These elongate fault systems displace terrane fragments hundreds of miles from their original

position. Between the Alexander and Taku terranes, the Clarence Strait follows a 200-mile long fault, and the 250-mile long Chatham Strait fault separates the Chugach terrane from the Alexander. The Yakutat terrane, beneath the coastline and shelf of the northeast Gulf of Alaska, represents a melange torn from its position west of Queen Charlotte Islands and shifted almost 500 miles northward by the Fairweather-Queen Charlotte transcurrent fault.

Travelling north by ship or ferry involves moving from one terrane to another. The Wrangellia terrane beneath Vancouver and Queen Charlotte Islands comprises the outboard strip of the Insular superterrane. In southeast Alaska, Ketchikan and Wrangell sit atop a thin silver of Taku terrane rocks, while Petersburg just to the northwest is on the Alexander terrane. North of Petersburg, Sitka is situated at the southernmost end of the Chugach, and Juneau straddles the border between the Taku and Gravina-Nutzotin belt. North of Juneau, Skagway lies on plutonic rocks between the older Stikinia and Taku terranes. Finally, Valdez, Seward, and the northern portion of Prince William Sound are part of the Chugach terrane.

MINERAL PRODUCTION

The mineral production of the Insular-Coast remains modest when compared to other regions of Canada. Even though only limited amounts of gold have been discovered in this province, the first account of gold retrieved in Canada was a nugget casually picked up on the western side of Queen Charlotte Islands by an Indian woman in 1851. The nugget eventually reached the Hudson Bay Company office at Victoria where interest was strong enough to trace the placer gravels back to the source vein. Gold in the vein was very limited and soon exhausted.

Copper, zinc, lead, gold, silver, and minor amounts of other metallic ores are concentrated on the east side of Howe Sound at Britannia Beach as well as on Vancouver and adjacent offshore islands. From 1905 to 1974 the long-lived Britannia Mine was one of the main sources of copper in Canada. Minerals, occurring as sulfide ores in Jurassic to Cretaceous volcanic rocks of the Gambier Group, were exploited until alternate copper sources in granitic intrusives were developed in the 1960s. Sulfides, originally deposited on the sea floor by superheated fluids laden with minerals, were later remobilized by intrusive activity. In all, the Britannia mines produced 48 million tons of ores.

At the northern tip of Vancouver Island, the Island Copper deposits have become an important source of this metal in recent years. Found in Wrangell terrane rocks that have been intruded by Middle to Late Jurassic plutons, this locality has yielded 114 million tons of ore

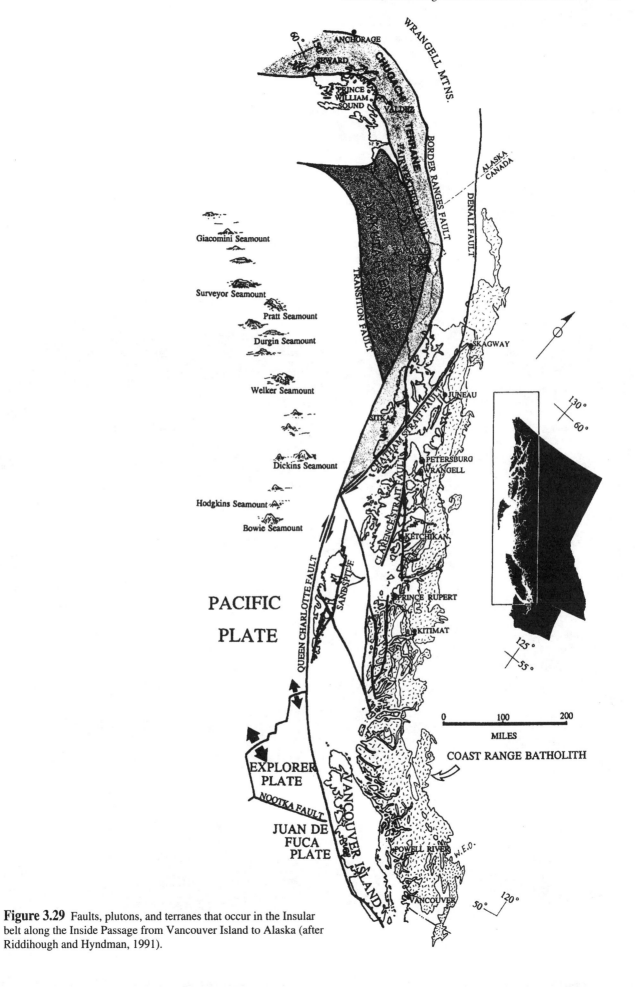

Figure 3.29 Faults, plutons, and terranes that occur in the Insular belt along the Inside Passage from Vancouver Island to Alaska (after Riddihough and Hyndman, 1991).

Figure 3.30 A close-up view of the spectacular caldera of Mt. Edgecumbe near Sitka, Alaska (courtesy B. Newton).

Figure 3.31 Thoroughly scalloped and carved up, Coast Range mountains in southeast Alaska bear witness to the incredible erosive power of moving ice (courtesy B. Newton).

Figure 3.32 Bear River valley, at the head of Portland Canal, is a ruler-straight, steep-sided canyon that was initially controlled by faulting (B.C. Ministry of Environment, Lands, and Parks).

and is still active. Significant amounts of gold are a by-product of the copper mining enterprise.

Gold from mines in the Zeballos River valley on the west coast of Vancouver Island and on Princess Royal Island to the north was always marginal. The mines closed during World War II due to lack of manpower and supplies, although currently there is renewed interest in vein gold from this region.

NATURALLY OUTCROPPING ON THE BEACH, coal from Vancouver Island was mined long before gold when it was used aboard ships of the Hud-son Bay Company as early as 1836. Coal seams near Nanaimo proved to be the most extensive, and by 1852 the trading company had brought in miners and set up collieries that shipped 2000 tons annually to San Francisco. Several coal mines were operated along the coast between Nanaimo and Comox until production dwindled to only 375 tons in 1967. Interest in coal resources here picked up in the 1980s, and several new mines were opened. Bituminous coal in the Nanaimo Group interfingers with marine sediments deposited in the Cretaceous basin, where most of the coal was derived from driftwood settling in shallow lagoons.

Figure 3.33 One of the many Japanese-owned copper, gold, and silver mines on Moresby Island, Queen Charlotte Islands, at the turn of the century (B.C. Archives and Records Service).

Figure 3.34 Iron, gold, and copper on Texada Island, adjacent to Vancouver Island, were mined from 1896 to 1976, yielding 21 million tons of ore (B.C. Archives and Records Service).

ADDITIONAL READINGS

Armstrong, J.E., 1977

Armstrong, J.E., 1981

Bamber, E.W., 1972

Brandon, M.T., Cowan, D.S., and Vance, J.A., 1988

Brown, A.S., 1968

Clague, J.J., and Bobrowsky, P.T., 1987

Clague, J.J., and Bobrowsky, P.T., 1990

Dawson, G.M., 1889

Eisbacher, G.H., 1977

England, T.D.J., 1991

Ettlinger, A.D., and Ray, G. E., 1988

Fulton, R.J., ed., 1989

Gabrielse, H., and Yorath, C.J., eds., 1991

Haggart, J.W., 1989

Haggart, J.W., 1993

Harrington, C.R., 1975

Hicock, S.R., Hobson, K., and Armstrong, J.E., 1982

Jackson, L.E., et al., 1985

Jones, D.L., Silberling, N.J., and Hillhouse, J.W., 1978

Keller, G., et al., 1984

LaMotte, R.S., 1936

Mathewes, R.W., 1989

Mathewes, R.W., and Clague, J.C., 1994

McLearn, F.H., 1972

Milne, W.G., et al., 1978

Monger, J.W.H., 1991

Monger, J.W.H., Price, R.A., and Tempelman-Kluit, D.J., 1982

Monger, J.W.H., et al., 1994

Muhs, D.R., 1987

Mustard, P.S., 1991

Pacht, J.A., 1987

Parrish, R.R., 1983

Riddihough, R.P., and Hyndman, R.D., 1989

Ross, C.A., and Ross, J.R.P., 1983

Von Huene, R., 1989

Wahlstrom, R., and Rogers, G.C., 1992

Wood, C.A., and Kienle, J., eds., 1990

Yorath, C.J., and Chase, R.L., 1981

Yorath, C.J., and Hyndman, R.D., 1983

4

CASCADE MOUNTAINS

A southern view along the axis of the Washington Cascades shows Mt. Rainier in the foreground, Mt. Adams to the left, and Mount St. Helens to the right (Washington Dept. of Natural Resources).

Beginning in California and extending into British Columbia, the Cascade Mountains, which form the backbone of the Pacific Northwest, are one of the most outstanding continental volcanic chains in the world. The southern end of the irregular Cascade province begins with Mounts Lassen and Shasta in northern California and traverses Oregon from Mt. McLoughlin to Mt. Hood. In Washington the range continues from Mt. Adams, Mount St. Helens, and Mt. Rainier in the south to Glacier Peak and Mt. Baker near the Canadian border. The British Columbia Cascades of Mt. Garibaldi and Mount Meager form the northern limit of the 600-mile long chain. In Oregon and California the range averages 30 miles in width, expanding to 150 miles in northern Washington, before constricting to narrow point in British Columbia. The overall trend is north-south except at the extremities, which curve westerly in Canada and eastward in California.

Highest peaks are Mt. Rainier in Washington at 14,410 feet in elevation, followed by Mt. Shasta in California at 14,162, Washington's Mt. Adams at 12,186, and Oregon's Mt. Hood reaching 11,245 feet. Mt. Lassen, California, the Three Sisters in central Oregon, and Mt. Baker of northern Washington are all around 10,000 feet in height. The highest peak in British Columbia is Mt. Garibaldi at 8787 feet. Overall the ranges of British Columbia tend to be lower in elevation than those to the south.

Such an extensive province touches on and overlaps with several other geologic areas, and it is often difficult to define the margins sharply. Elevated mountain blocks and alternating valleys of the Basin and Range project well into the Cascades of California. The Columbia River Basalts cover areas of the Cascade range in Washington, and in British Columbia the Coast mountains and Cascades overlap.

Broad valleys border the Cascade Mountains on their western flanks, and high arid plateaus lie to the east. The Sacramento and Shasta valleys in California along with the Willamette Valley, Puget Lowland, and Fraser Valley in Oregon, Washington, and British Columbia fall along the western margin. The Modoc Plateau, High Lava Plains, Columbia Plateau, Okanagan Plateau, and Fraser Plateau are on the east.

The Cascade Mountains create a climatic barrier intercepting atmospheric moisture from the Pacific Ocean before it reaches the eastern plateaus. Rainy maritime conditions exist west of the mountains, while to the east a dry desert-like environment is found. Annual rainfall in the Western Cascades, exceeding 100 inches annually, accumulates as heavy snow in the winter. Where the melting snowpack contributes to the volume of streams and rivers, erosion is actively dissecting the range.

Major streams incise the western slopes, while only small systems flow down the eastern face toward the drier plateaus. Of the four significant rivers traversing the range, the Fraser in British Columbia and the Columbia between Oregon and Washington are among the longest in North America. Originating in the Rocky Mountains of eastern British Columbia, the Fraser and Columbia follow similar drainage patterns before cutting through the Cascades. The smaller Klamath and Pit rivers traverse the Cascades in northern California.

INTRODUCTION

Imposing snow-capped volcanic summits of the Cascade Mountains that stretch from British Columbia, across Washington and Oregon, to northern California are only the concluding stage to a complex geologic sequence that built this province.

These young peaks rest on a foundation of accreted terrane rocks in the north and a platform of older volcanic ash and lava in the south. The oldest rocks within the province are found in the North Cascade range of British Columbia and Washington. Dating back over a billion years, terrane pieces, which make up the northern Cascades, were fastened to the North American landmass during the Mesozoic era. By contrast the Cascade mountains of southern Washington, Oregon, and California began as a continental volcanic archipelago situated along the North American coastline about 40 million years ago. Although cones of the chain erupted large volumes of lava and ash, they have been eroded to the subdued rounded rolling hills of the Western Cascades today.

A surge of explosive volcanism beginning 10 million years ago constructed the High Cascades. Today this narrow belt of prominent individual stratocones, mantled by a permanent cover of ice and snow, displays sharp serrated ridges and pinnacles cut by glaciers. Around these high peaks, a variety of low rounded domes, spatter cones, young rugged lava fields, and dark cinder-covered buttes are all part of the Cascade volcanic arc. Most of these volcanoes are dormant, but intermittent earthquakes and occasional eruptions are constant reminders of the origin of these mountains and their destructive power.

GEOLOGY OF THE CASCADE RANGE

Even though the Cascade geologic province is unified throughout its length by a line of Pliocene and Pleistocene volcanic peaks, contrasting older rocks divide the range into northern and southern portions at Snoqualmie Pass in Washington. The northern portion extends from Mount Meager in British Columbia and the southern portion from Snoqualmie Pass into northern California.

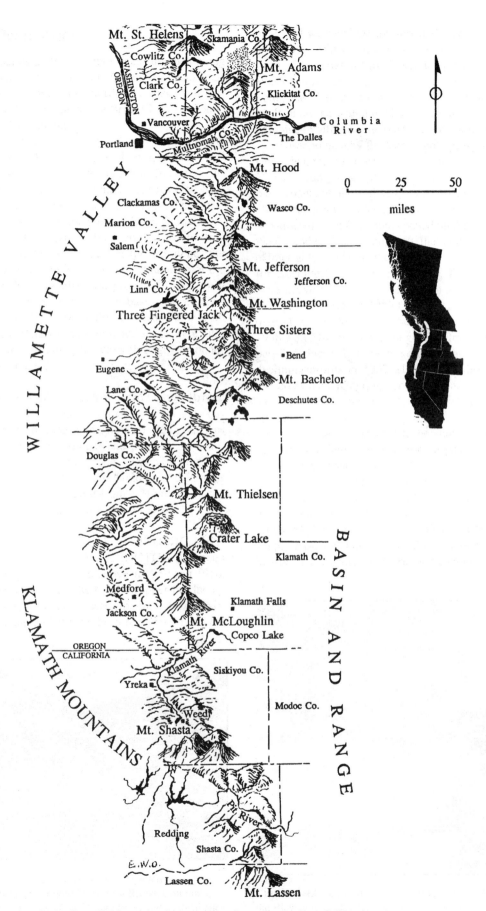

Figure 4.1 Locations in the Cascade Mountains from northern Oregon to northern California.

Figure 4.2 Locations in the Cascade Mountains from British Columbia to the Columbia River on the northern Oregon border.

Figure 4.3 Looking south in the High Cascades of central Oregon, Mt. Washington with Big Lake is in the foreground. Three Sisters, Broken Top, and Mt. Bachelor appear in the distance (Oregon Dept. of Highways).

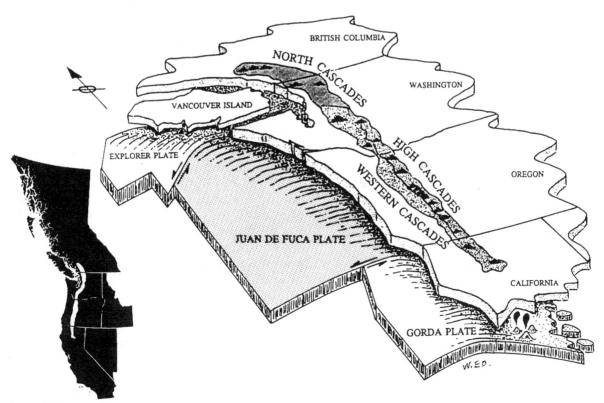

Figure 4.4 Sliding beneath North America, the trailing edge of the ancient Farallon tectonic plate has been subdivided into three smaller plates producing the modern day Cascade volcanic arc.

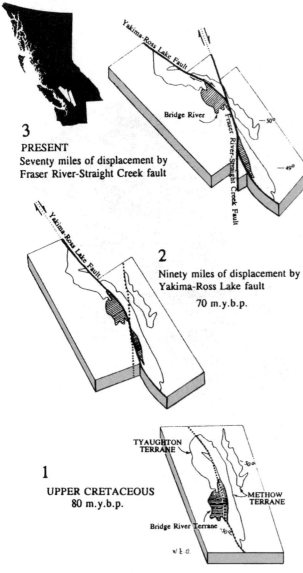

Figure 4.5 The Yakima-Ross Lake fault, which separated the Tyaughton-Methow basin, was, in turn, offset by the Fraser River-Straight Creek fault (after Kleinspehn, 1985; Monger, 1985).

Figure 4.6 Terrane rocks near Darrington in the North Cascades are adjacent to the Straight Creek fault (Washington Dept. of Transportation).

In British Columbia, the North Cascades merge with the coastal mountains that can be traced all the way to Alaska. Older rocks beneath the northern Cascade peaks are predominantly Paleozoic and Mesozoic exotic terranes, but the southern portion, called the Western Cascades, is a broad belt of andesitic and basaltic volcanic vents that span the Eocene through Miocene epochs.

THE NORTH CASCADES are a complex collage of folded, faulted, and metamorphically altered Precambrian through Lower Cretaceous accreted terranes intruded by plutons and perforated by Quaternary volcanoes. These multiple terrane fragments were assembled or amalgamated offshore into composite slabs or superterranes before being transported eastward and accreted to the North American continental plate during the Cre-

taceous. This newly fabricated patchwork crust formed the foundation for the northern Cascades.

Terrane rocks of the North Cascades are separated and dissected by a complex web of faults. During the Early Tertiary, the northward movement of the Kula Plate, where it merged with North America, wrenched the continental margin into a transcurrent fault system from the Puget Lowland through the North Cascades and into British Columbia. These transcurrent faults are "dextral" when the west side has shifted north and the east side moved south. Many of the faults display offsets of 50 to 100 miles or more where slivers and larger blocks of rock, torn from their source in the Cascades, have been moved far up into British Columbia. A similar situation exists today for faults off the coast of British Columbia and along the San Andreas system in central and southern California.

The north-south Fraser River-Straight Creek fault bisects the North Cascades. Terranes east of the fault formed along the margin of the North American continent, but those to the west developed in oceanic settings and volcanic islands out in the ancient Pacific. In Washington this Eocene transcurrent fault curves southeastward past Marblemount in Skagit County and Darrington in Snohomish County toward Yakima, while in British Columbia the feature projects north from Hope for several hundred miles along the Fraser River valley.

Because of its location on a major fault zone, the Fraser River canyon is subject to extensive rock slides. In recent years more than 100 landslides have blocked the river and disrupted railroads and highways that run through the valley. Slides are especially prevalent at meanders where water is undercutting steep cliffs. About 20 miles above Hope, the walls of Hells Gate are particularly susceptible to rock movement, a problem which

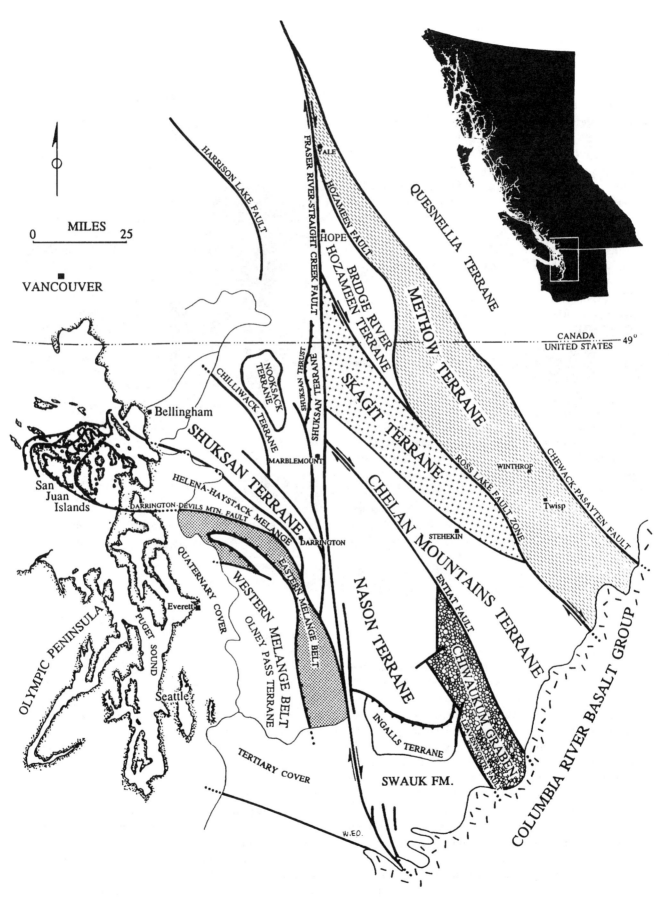

Figure 4.7 Terranes and faults of the North Cascades (after Miller, et al., 1994; Tabor, et. al., 1989).

was aggravated by construction of the railroad. Immense rockfalls that periodically restricted the river have even prevented migrating salmon from spawning.

EAST OF THE FRASER RIVER-STRAIGHT CREEK FAULT, terranes ranging from 250 million-year old Permian through 100 million-year old Cretaceous intervals. Rocks here have a North American complexion since they were derived in part from sources along the ancient continent. Rocks of the Tyaughton-Methow, Hozameen-Bridge River, Skagit gneiss, Chelan Mountains, and Nason terranes are laid out in broad northwest to southeast strips separated by faults. To the south, the terranes are overlapped by Miocene flood basalts of the Columbia River Group; however, the northwest edges terminate abruptly against the Fraser River-Straight Creek Fault.

The eastern-most Cascade structure is the Tyaughton-Methow basin across northcentral Washington and British Columbia. The basin was an elongated shallow inlet to the ancient Pacific, separated from the open ocean waters by a chain of volcanic islands. Oriented northwest by southeast, the sediment-filled basin is bounded by the Chewack-Pasayten and Ross Lake faults.

During the Late Mesozoic this basin was adjacent to the Omineca-Intermontane belt that formed the western margin of North America. At that time the Tyaughton-Methow was actually two subsiding basins, the Tyaughton in the Camelsfoot and Chilcotin ranges of southcentral British Columbia and the Methow in southern British Columbia and northern Washington. As the shallow floor of the seaway was thoroughly sliced up by faults, a chain of volcanic islands that make up the Bridge River terrane emerged separating the older basin into the larger eastern Methow and the smaller western Tyaughton troughs.

Late Jurassic to Early Cretaceous seas filled the trough with thick layers of sediments derived from three sources. On the east, uplifted mountains in the Omineca area shed arkosic sand into the surf and beaches of the Methow. Islands of the Bridge River terrane, across the axis of the basins, contributed cherty sediments into the Methow as well as into the Tyaughton. In addition, the Tyaughton depression received ash and lava from the island archipelago on its oceanic side.

In the Late Cretaceous, the Insular superterrane merged with the margin of North America to uplift, compress, and fold the Tyaughton-Methow feature. A network of Early Tertiary north-south faults, including the dominant Fraser River-Straight Creek fault, separated and shifted the basin, so that today the western Tyaughton portion is more than 60 miles north of the Methow.

Major terranes lying between the Methow basin and the Fraser River-Straight Creek fault include the Hoza-

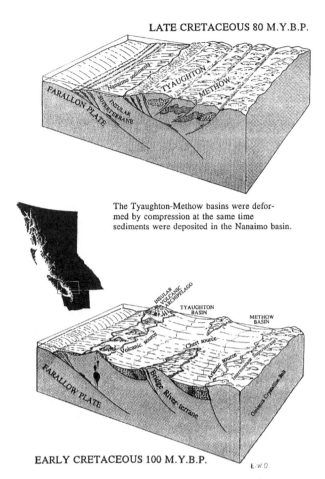

LATE CRETACEOUS 80 M.Y.B.P.

The Tyaughton-Methow basins were deformed by compression at the same time sediments were deposited in the Nanaimo basin.

EARLY CRETACEOUS 100 M.Y.B.P.

Figure 4.8 During the Late Jurassic to Early Cretaceous interval, the Tyaughton and Methow basins across northern Washington and southern British Columbia received sediment from numerous sources before the basin itself was folded and uplifted (after Garver, 1992).

meen, Skagit, Chelan Mountains, Nason, and Ingalls. Each can be identified by their unique characteristics. Of these, Skagit gneisses, severely altered by the heat and pressure of metamorphism, form the crystalline core of the North Cascades, whereas the Ingalls terrane, south of Mt. Stuart, Chelan County, represents the largest and most complete sequence of ophiolitic rocks in Washington. Formerly part of an ocean crust, this ophiolite covers 175 square miles.

ROCKS TO THE WEST OF THE FRASER RIVER-STRAIGHT CREEK FAULT contrast sharply with those to the east. Ranging in age from Precambrian, more than 1 billion years old, through Lower Cretaceous at 100 million years, the western rocks have been metamorphically altered and distorted by folding and faulting. Shale, chert, and pillow basalt of the Chilliwack, Nooksack, and Shuksan terranes, as well as the Eastern and Western melange belts are largely oceanic in com-

Figure 4.9 An eastward view across the Camelsfoot Range in southern British Columbia shows the Methow and Cache Creek terranes separated by the Fraser River. The river follows the trend of the Fraser River-Straight Creek fault (British Columbia Ministry of Environment, Lands, and Parks).

plexion, originating in volcanic island environments. This area of the Cascades has been interpreted as an ancient subduction zone along the collision border between two converging tectonic plates where the rocks were throughly fragmented before being jammed together in a chaotic mixture or melange.

Exposed at Twin Sisters Mountain in Whatcom and Skagit counties, Washington, the Shuksan terrane is a solid mass of weathered dunite. This dark rock is composed of iron-rich olivine that alters to a striking rust brown and makes the mountain unmistakable even from a distance. Dunite is normally associated with ocean crustal rocks at depths of 3 miles or more below the sea floor. Today the olivine at Twin Sisters is being quarried for use in the production of refractory slabs for waste incinerators.

In King County, from Snoqualmie Pass north along the Snoqualmie River, the Jurassic and Cretaceous Western melange belt, also called the Olney Pass terrane, is an incredibly coarse mixture of enormous sandstone blocks, some measuring thousands of feet across, set in a shaley matrix.

THE SAN JUAN ISLANDS, a panorama of narrow rocky beaches, promonitories, and highlands covered with Douglas fir, rise cleanly from the channel between Washington and Vancouver Island. Numerous islands of varying sizes are scattered through the Strait of Juan de Fuca, which separates the United States from Canada, but 80% of the total area is made up of the three largest, San Juan, Lopez, and Orcas. Although situated among the offshore islands of the Insular belt, the San Juans are

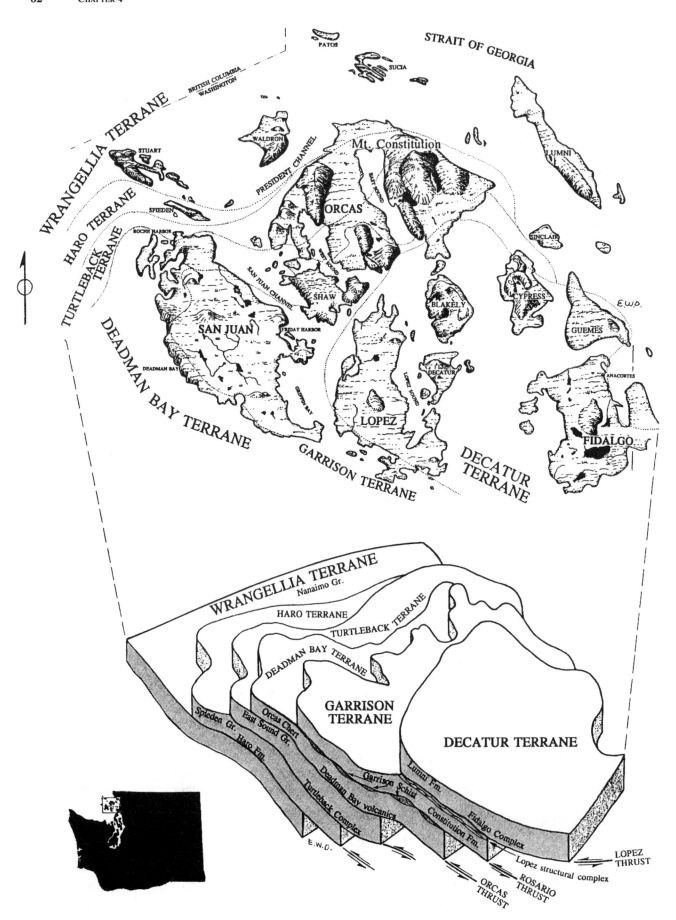

Figure 4.10 Stacked like dominos, six sheet-like terranes make up the subsurface structure of the San Juan Islands (after Brandon and Cowan, 1987; Brandon, Cowan, and Vance, 1988).

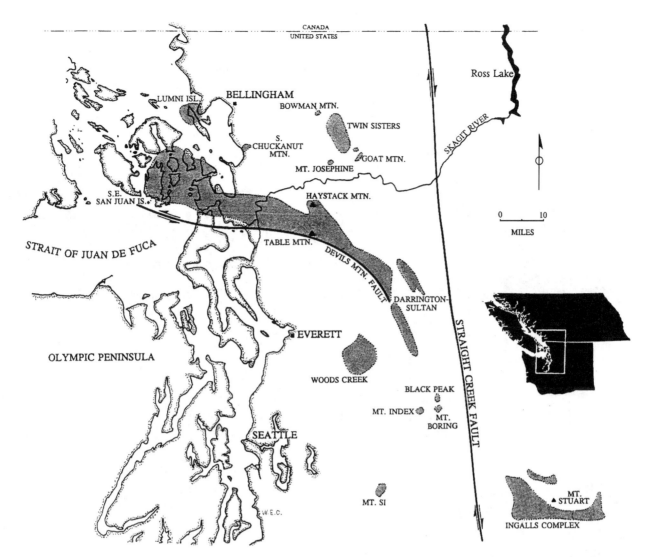

Figure 4.11 Part of the ocean crust during the Jurassic, 150 million years ago, rocks called ophiolites (dark color) link the San Juan Islands to the North Cascades (after Whetten, et al., 1980; Miller, 1985).

Figure 4.12 Limestones, frequently found in terrane rocks of the San Juan Islands, were processed to make cement in kilns like this one at East Sound on Orcas Island (Washington Dept. of Natural Resources).

a tongue of terrane fragments that extend for 50 miles from similar rocks in the northern Cascades.

Five separate terranes make up the San Juans giving them a remarkable geologic diversity in an area that covers only a few hundred square miles. Spanning the Early Paleozoic to Cretaceous, terrane rocks of the San Juans are a stack of thrust sheets shoved westward over the Wrangellia terrane as it, in turn, was being driven beneath the margin of North America. This short-lived event took place during the late Cretaceous, about 80 million years ago.

The oldest of the terranes are the Turtleback, Dead-man Bay, and Garrison representing intervals from Cambrian to Jurassic. Limestone pods within these terranes are dated by the abundant fossil remains of the sea lily crinoid, by the microscopic teeth of conodonts, a primitive fish, and by wheat grain-sized protozoan fusulinids. The fusulinids, belonging to a tropical Asiatic regime, are evidence that the island archipelago, in which the terranes formed, was foreign to North America.

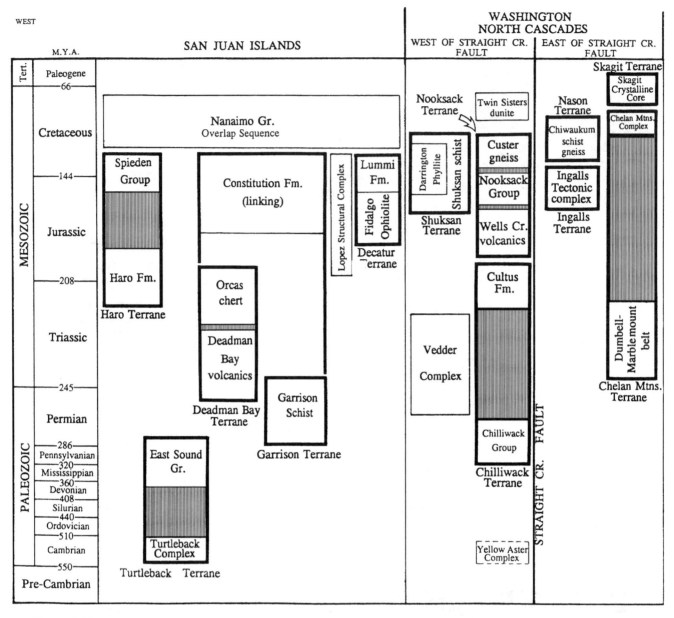

Figure 4.13 Stratigraphy of the North Cascades and San Juan Islands. Dark borders outline terranes (Brandon, Cowan, and Vance, 1988; Gabrielse and Brookfield, 1988; Wheeler and McFeely, 1991). (Continued on page 85.)

The Fidalgo ophiolite, on Fidalgo Island near Anacortes, Washington, is part of a sequence of Jurassic ocean crust rocks at the base of the Decatur terrane. Fossil belemnites, squid-like cephalopods, and abundant plant remains in the mudstone, sandstone, and conglomerate of the Decatur strata place it in the Later Jurassic to Early Cretaceous. Of Late Triassic to Early Cretaceous age, the uppermost Haro terrane is composed of sand and conglomerate derived from an offshore volcanic archipelago. Abundant and varied fossil molluscs within this terrane include clams, coiled ammonites, and bullet-shaped belemnites.

With the advancing Cretaceous seas that covered the San Juan area, fossiliferous Nanaimo sand accumulated atop the older terrane rocks. At Fossil Bay on picturesque Sucia Island, there are well-preserved molluscs in the Nanaimo layers.

Subjected to intense glacial erosion during the Pleistocene as great ice lobes ground south to deepen the straits, much of the bedrock of the islands is obscured by layers of glacial material. Carved largely by moving ice, the two sections of U-shaped Orcas Island are separated by a deep glacial sound. Mt. Constitution on Orcas, the highest point in the islands at 2409 feet above sea level, was beneath more than 1/2 mile of ice during the last Ice Age. Huge tracts of glacial debris on San Juan and Lopez islands are characterized by poor drainage, swampy depressions, and large boulders moved by glaciers called erratics.

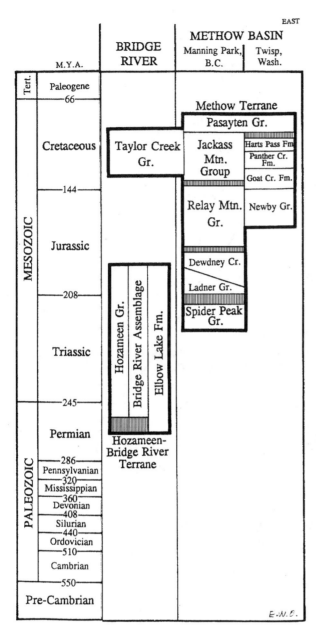

Figure 4.14 In the center of the photograph, Twin Sisters Mountain is an olivine-rich rock called dunite that formed in deep ocean crust during the Late Cretaceous. The Middle Fork of the Nooksack River is visible to the right of the mountain (Washington Dept. of Natural Resources).

(Continued from page 82.)

PLUTONS OR INTRUSIVE BODIES OF ROCK have played an important role in the geologic history of the North Cascades. The collision and accretion of the Insular superterrane in British Columbia generated extensive batholith intrusives as deep as 15 miles in older terrane rocks of this region. These crystalline granitic masses underlie much of this portion of the range and project as high sharp peaks where surrounding strata has been removed by erosion.

During the collision and subduction of crustal plates, the heavier slab, which is being subducted beneath the lighter continental landmass, thickens and produces heat that melts the surrounding rocks. This liquid magma is then injected into the adjacent strata as plutons or batholiths.

East of the Straight Creek fault, over a dozen large Cretaceous "stitching" plutons, intruded along fault zones, weld together the various terrane pieces. These batholiths represent the southern extent of the British Columbia Coast Range, and magnetic studies of the plutons indicate the granites originally crystallized as much as 1500 miles south of their present position.

In Washington the Mt. Stuart batholith is one of the oldest of the group at 88 million years. Close to the same age, the Entiat pluton, forming the Entiat Mountains, and the Chelan pluton, underlying the Chelan Mountains, are up to 50 miles long and parallel the regional northwest-southeast geologic trend.

To further complicate the geologic picture of the North Cascades, younger granitic plutons were intruded during several phases in the Tertiary period, reflecting the continued subduction of the Farallon plate beneath North America. East of the Straight Creek fault near Chelan, Washington, Eocene intrusions include the Duncan Hill and Cooper Mountain plutons. One of the few batholiths west of the fault is the small Eocene intrusion beneath Pilchuck Mountain east of Everett.

The 25 to 17 million year old Snoqualmie batholith and the 26 to 14 million year old Tatoosh pluton

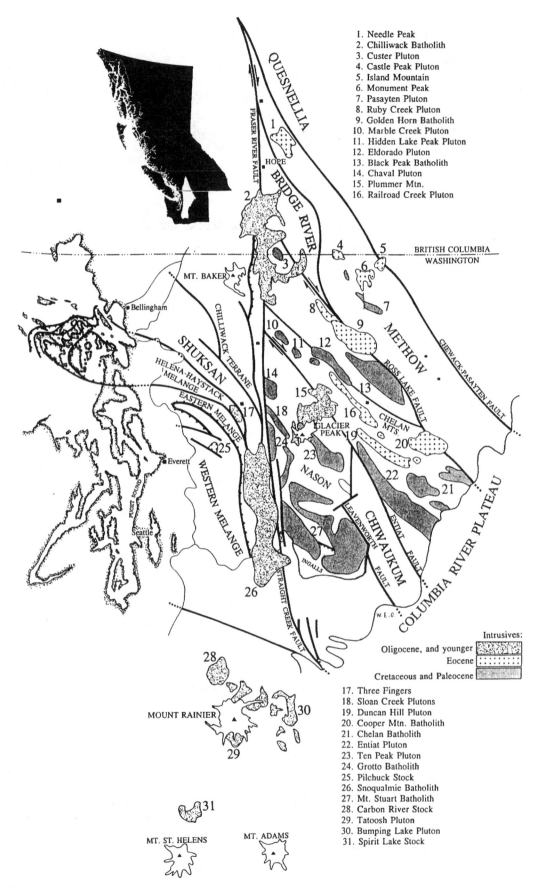

1. Needle Peak
2. Chilliwack Batholith
3. Custer Pluton
4. Castle Peak Pluton
5. Island Mountain
6. Monument Peak
7. Pasayten Pluton
8. Ruby Creek Pluton
9. Golden Horn Batholith
10. Marble Creek Pluton
11. Hidden Lake Peak Pluton
12. Eldorado Pluton
13. Black Peak Batholith
14. Chaval Pluton
15. Plummer Mtn.
16. Railroad Creek Pluton

17. Three Fingers
18. Sloan Creek Plutons
19. Duncan Hill Pluton
20. Cooper Mtn. Batholith
21. Chelan Batholith
22. Entiat Pluton
23. Ten Peak Pluton
24. Grotto Batholith
25. Pilchuck Stock
26. Snoqualmie Batholith
27. Mt. Stuart Batholith
28. Carbon River Stock
29. Tatoosh Pluton
30. Bumping Lake Pluton
31. Spirit Lake Stock

Intrusives:
Oligocene, and younger
Eocene
Cretaceous and Paleocene

Figure 4.15 The collage of accreted terranes, which form the foundation of the North Cascades, was extensively perforated by Cretaceous and Tertiary plutons.

Figure 4.16 East of the Straight Creek fault, rugged terrane rocks near Icicle Creek in Chelan County show long valleys controlled by faults (Washington Dept. of Transportation).

represent large shallow magma chambers, intruded in several phases only 1½ miles beneath the surface. The Snoqualmie underlies much of the central portion of the northern Cascades, and the Tatoosh lies beneath Mt. Rainier.

The core of the Cascades in British Columbia is defined by a linear system of Miocene plutons along the Fraser River-Straight Creek fault from the Chilliwack batholith at the International Border to the Eagle pluton at the core of Coquihalla Mountain and the Salal Creek pluton near Mount Meager. Part of the Pemberton volcanic belt, the Chilliwack batholith, dated at 29 to 26 million years ago, was active in at least four different intervals and predates the 21 to 16 million-year old Mt. Barr and even younger 8 million-year old Salal Creek plutons.

THE WESTERN CASCADE RANGE, which runs from Snoqualmie Pass, Washington, south to California, dominated the Cenozoic era. A coalescing line of broad volcanic cones, low domes, and fissures of the incipient Western Cascades developed parallel to the ancient Pacific coast about 40 million years ago. This Early Tertiary volcanic phase marked a collision boundary where the eastward-moving Farallon oceanic plate plunged beneath the westbound North American continent. As the slab descended, it began to melt at a depth of about 60 miles, providing heat and liquifying rocks, which rose through the overlying plate. Western Cascade volcanoes emerged directly above the melt zone that lay east of the suture between the two merging crustal sheets.

Along collision boundaries, the angle of the subducting slab beneath the overriding sheet and the style of volcanism are largely a function of the rate of convergence of the two slabs. During the Eocene epoch, the origin of the Cascade volcanic archipelago was

Figure 4.17 Intrusive rocks of the Tatoosh batholith lie beneath Mt. Rainier. The outline of the circular crater is kept free of snow by the steady emission of steam (Washington Dept. of Transportation).

controlled by the rapid rate of convergence. At the same time, the angle of the warm buoyant Farallon Plate beneath North America was shallow, producing the broad band of erupting vents of the older Western Cascades.

DURING THE CENOZOIC, the North American coastline was positioned along what is now the east margin of the Fraser, Puget, and Willamette lowlands of British Columbia, Washington, and Oregon, respectively. A broad swampy plain adjacent to the ocean extended eastward to meet the rising Western Cascade volcanoes. Ash and lava was spread over the coastal margin concealing the landscape and congesting streams. Draining the western flanks of the mountains, the heavily-laden rivers constructed wide deltas out into the shallow ocean.

THE EOCENE TO OLIGOCENE in the northwest was a time when dense tropical and subtropical forests grew along the shore. Fossil leaves, wood, and pollen of ferns, conifer, and deciduous plants from this setting are preserved today within the ash-rich deltaic rocks. Plant material that was buried and converted to thin lignitic coal beds trace a marshy floodplain from central Washington to California. The Montgomery Creek Formation along the Pit River in Shasta County and a 4000-foot thickness of Colestin volcanic debris, silt, and sand along the front range of the Western Cascades from Yreka, California, to Lane County in Oregon contain beautifully preserved plant remains as well as discontinuous layers of coal.

In central Washington, tensional stress and vertical movement along faults opened large-scale depressions or

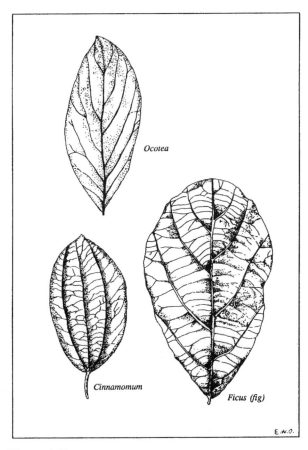

Figure 4.18 Large smooth-edged leaves of plants 50 million years ago, which grew in basins along the Western Cascades, were tropical in nature.

Figure 4.19 On the western flank of the growing Cascade range, swamps and lakes of the Eocene were the habitat of palmetto, swamp cypress, and heron-like birds.

grabens. The 12-mile wide Chiwankum graben in Chelan County was a long northwest by southeast trough bounded by the Entiat and Leavenworth faults. This graben was an active sinking basin for a period of 6 million years during the Middle Eocene when it accumulated an 18,000 foot sequence of sandstone and conglomerate of the Chumstick Formation, representing a variety of stream and lake environments. The Chumstick was subsequently covered by stream deposits of the Oligocene Wenatchee Formation, which, in turn, were succeeded by Miocene basalts of the Columbia River Group.

Delineated by the Leavenworth and Straight Creek faults, the Swauk basin, west of Wenatchee, built up 23,000 feet of sand, mud, and conglomerate. In addition to a tropical broadleaf flora, the Swauk basin contains abundant palmetto leaves identical to those in the Chuckanut Formation to the northwest near Bellingham. Thousands of dark-colored dikes of the Eocene Teanaway basalts, which cut through the light-colored Swauk sediments, average tens of feet thick.

Fossil plants along with coal-bearing shales of the Eocene to Oligocene Roslyn, Chumstick, and Wenatchee formations south and east of Wenatchee record

lake, stream, and swamp environments that also developed in basins related to faulting. Lake and marshy conditions in Kittitas County produced coal in the Roslyn Formation, which has been mined since 1885. Most of the coal in western Washington is of such poor quality and so difficult to extract that production has been limited to small operations.

Ash, volcanic sedimentary rocks, and andesitic and rhyolitic lavas of the Oligocene Ohanapecosh Formation in southcentral Washington exceed 10,000 feet in thickness near Mt. Rainier. This eruption marked the beginning of the Western Cascade volcanic arc in Washington. Volcanic cones of the Ohanapecosh were leveled by erosion, and the lavas gently folded before a second volcanic episode produced the lighter ash and rocks of the Stevens Ridge Formation.

In the Western Cascades of Oregon, the Eocene and Oligocene volcanic interval was spanned by the Little Butte lavas from numerous low andesitic vents east of Roseburg in Douglas County and along the front range to Salem in Marion County. Where they extended west to the shoreline, lavas and tuffs of the Little Butte interfingered with shallow marine sediments. The Scotts Mills and Eugene formations were the last to be deposited in the rapidly shrinking marine basin where molluscs and barnacles along with the remains of sharks and whales are preserved in the temperate nearshore sediments. Deposits of broken shells mixed with leaf and woody fragments trace the rocky headlands and old shorelines across Lane and Marion counties.

THE MIOCENE SAW RENEWED VOLCANIC ACTIVITY about 23 million-years ago as Fifes Peak stratovolcanoes east of Mt. Rainier, Washington, ejected ash and debris that flowed down the flanks and accumulated at

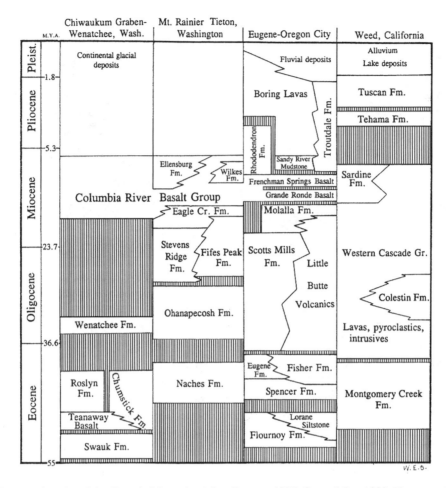

Figure 4.20 Tertiary stratigraphy of the Cascade Mountains (after Gresens, 1987; Orr and Orr, 1999; Vance, et al., 1987).

the base of the cone. Mudflows and lavas spread even further, some reaching as far south as the Columbia River. The large amount of volcanic material during this episode constructed the Fifes Peak edifices, which exceeded those built during the earlier Ohanapecosh interval.

Along the central part of the Western Cascade arc in Oregon, violent eruptions from composite cones between 13 and 9 million years ago produced basaltic and andesitic lava, tuff, and mudflows of the Sardine Formation. Following this activity, eruptions were chiefly limited to the region north and east of Salem.

Most of the volcanic cones of the older Western Cascades in northern California, dated at 31 to 20 million years ago, have been buried by younger lavas, although the range may extend beneath the Sierra Nevada Mountains. As much as 15,000 feet of Miocene lava, mudflows, ash, and tuff of the Western Cascade Group in California are nearly continuous for 45 miles south of the Oregon border. These andesitic flows average 20 feet in thickness, and some exceed 100 feet. Eroded domes and plugs of the older volcanoes form the landscape near Copco Lake in Siskiyou County and in Shasta Valley.

Even with the prodigious output of volcanic debris that accompanied the growth of the Western Cascades, the overall height of the range remained modest as it sank almost as fast as lavas accumulated atop the evacuating magma chamber. Near the end of the Miocene, uplift of the range accelerated erosion, which eventually reduced the volcanic cones of the older Cascades to low hills.

VOLCANISM SLOWED AND CEASED in the Western Cascades around 9 million years ago as the line of eruptive centers moved progressively eastward and narrowed to the present width of the High Cascades. The shift of volcanic eruptions from west to east was the result of ongoing tectonic plate movement. As the Farallon slab continued to dive beneath the margin of North America, the rate of collision changed along with the angle of the subducting plate, causing the centers of volcanism to sweep eastward.

This shift of volcanism coincided with a drastic reduction in the volume of material erupted and a profound constriction of the eruptive belt. Volcanic output is difficult to gauge in the older Cascades where exten-

Figure 4.21 Mt. Lassen in northern California erupted in the summer of 1917. Lava flows and explosive volcanic activity began several years prior to the final event (California Division of Mines and Geology).

sive erosion has taken place and much of the sequence is covered, but it is estimated that there was close to six times more volcanic debris in the Western mountains than in the younger range.

At intervals along the High Cascades, pronounced structural depressions or grabens occur on the older Cascade platform. These are well-developed east of the Sisters, Mt. Washington, and Mt. Jefferson, Oregon, where the 20-mile long Green Ridge fault is aligned along one margin of a north-south trough. A similar structural depression extends from Klamath Lake north to Crater Lake, and Mt. Hood sits in a graben formed, in part, by the Hood River fault. These separate depressions may relate to faulting along the eastern margin of the range that was caused by crustal spreading in the back-arc basin.

Atop the lava platform of the Western Cascades and the accreted terranes of the northern Cascades, conspicuous snowy peaks make up the High Cascades, which evolved during the latest Tertiary and Quaternary. Basaltic lavas, which covered hundreds of square miles, created dramatic cones and dark fresh cindery fields from California to British Columbia.

HIGH CASCADE VOLCANISM WITHIN NORTHERN CALIFORNIA was initiated with ash and

lava from four separate eruptive centers during the Pliocene and Pleistocene. Mt. Dittmar, Snow Mountain, Mt. Yana, and Mt. Maidu, along with numerous small vents exuded flows of debris that built a wedge-shaped 60-mile long field from Mt. Lassen to the Sacramento Valley. Combined flows are over 1500-feet thick in the north but thinner to the south. In all, eruptions from these vents covered an area of 2000 square miles with lava, ash, and mud of the Tuscan Formation. At the southern margin, lava is interbedded with fine-grained alluvial sediments of the Pliocene Tehama Formation. Following a period of extensive erosion, overlapping basalt flows in several episodes culminated with the building of Mounts Lassen and Shasta, the largest Cascade composite cones.

Mt. Lassen was erected on the wreckage of a large andesitic stratovolcano, Mt. Brokeoff, that was active as late as 400,000 years ago. After a period of preliminary explosions, the original cone was destroyed when enormous ash clouds and lava were suddenly expelled. As the magma chamber emptied and collapsed, a central caldera was created in much the same sequence of events responsible for the origin of Crater Lake, Oregon. Today the remnants of the decapitated volcano are seen as Brokeoff Mountain, Mt. Diller, Mt. Conard, and Diamond Peak, which encircle the caldera in Shasta County.

About 11,000 years ago when volcanic activity shifted to the north side of Brokeoff, stiff lava slowly oozed from vents on the flanks of the mountain producing numerous domes and emplacing a thick dacite pluton. Additional lava flows and dome fields around the caldera today make up the Mt. Lassen peak complex. While some of the domes have been extensively eroded and glaciated, the Chaos Crags domes, with extremely rough steep sides and piles of rubble around the base, are relatively fresh.

Historical eruptions of Mt. Lassen included a period between 1850 and 1851 from Cinder Cone, between 1854 and 1857 from Chaos Crags, and a series of violent eruptive and dome building events from 1914 to 1917. The May 22, 1915, explosive occurrence that sent a cloud of steam, ash, and fragments high into the atmosphere was visible in Sacramento over 100 miles away. Activity after this period was limited to steam clouds from fissures on the eastern slopes in 1921. Prior to the eruption of Mount St. Helens in 1980, the coal black lavas near the summit of Mt. Lassen represented the only 20th century eruption of a volcano in the coterminous United States. Mt. Lassen peak and the surrounding complex have been designated a National Volcanic Park.

At 14,162 feet in elevation, the beautiful snow-crowned Mt. Shasta in Siskiyou County is the second largest andesitic stratocone in the Cascades. Dominating the skyline for miles in northern California, the enormous bulk of Mt. Shasta, covering 80 cubic miles, rivals

YEARS AGO || YEARS A.D.

CALIFORNIA
Lassen Peak.. ....1780........................1850-51.......
Glass Mtn.. ....995...
Chaos Crags... ....995.......................................1854-57
Mt.Shasta....9,500...9,230.................................... ....1786.............................1855

OREGON
Mt. Mazama...........................7,000
South Sister..2,300.... 1853
Mt. Bachelor...2,000.
Sand Mtn.-
Belknap flow...............................4,000...3,000......... ....360....................1600
Mt.Hood.. ....995................1760..........1810............................

WASHINGTON
Indian Heaven................8,150
Mt.St.Helens.....................................4,000............ ......295.....795..........1480................1800-04....1831..1835.......1842-54..
Mt.Rainier..............8,750———to———2,200.| ....850.........................1820......................1854....
Mt. Baker..10,000.. ....1792.....1810.............1843-46..1852-58..

BRITISH
COLUMBIA
Meager Mtn.
Complex...2,490

Figure 4.22 Historic volcanic activity within the Cascades includes steam and ash as well as lava eruptions (after Harris, 1988; Wood and Kienle, 1990).

that of Mt. Fujiyama in Japan. Reaching 12,330 feet, the secondary cone, Shastina, on the west flank of the crest ranks as the fourth highest peak of the Cascades. Because of limited precipitation and the immense volume of lava that was extruded to construct this complex, Mt. Shasta and Shastina appear to be relatively untouched by erosion in comparison to cones at the northern end of the range.

Mt. Shasta was assembled in four brief stages lasting only a few hundred years. An upwelling of andesite lava, overlapping mudflows, and pyroclastic material around a central vent constructed smaller domes on the flanks of a central compound stratovolcano. Shastina and Hotlum cones took shape around 9500 years ago. Erupting on a 600 to 800 year cycle, Mt. Shasta was active as recently as 1786.

Today the 10 glaciers on Mt. Shasta are the largest in California, even though they are relatively small by Cascade standards. Whitney glacier, the largest and longest, has been increasing since the turn of the century and at present is approximately 120 feet thick. By contrast, Mt. Lassen has no permanent ice pack and is barren of snow by the middle of July.

DOMINANT PEAKS OF THE CASCADE PROVINCE IN SOUTHERN OREGON are Mt. McLoughlin and Crater Lake. At 9493 feet, Mt. McLoughlin has a basin carved into the northeast slope by glaciation. Crater Lake in Klamath County occupies a deep caldera formed when Mt. Mazama exploded and

Figure 4.23 About 9700 years old, the Military Pass lava flow on the northeast side of Mt. Shasta in Siskiyou County, California, is about 5 miles long and 500 feet thick near its nose. The smaller cone, Shastina, appears directly behind Mt. Shasta (courtesy C.D. Miller, U.S. Geological Survey).

......................................1914to 1917
....................................1910

......1859-1865.....................1907

.1857...1980
..........1873...1879..1882
.1860-1870.......1880.....................................mid 1900s....1975-1976

Figure 4.22 (continued here)

Figure 4.24 The 9178-foot high spire and broad base of Mt. Thielsen in southern Oregon resulted when glaciers carried loose volcanic debris down the slope to expose the sharp central plug (Washington Dept. of Transportation).

Figure 4.25 Crater Lake in southern Oregon formed about 7000 years ago when Mt. Mazama volcano collapsed inward. Wizard Island, seen in the center, is a more recent cinder cone (Oregon State Highway Dept).

collapsed. The history of Crater Lake began 400,000 years in the past with the construction of the Mt. Mazama volcanic complex of low domes and strato-cones from numerous overlapping lava flows. In an immense final eruption 6900 years ago, a tremendous magma chamber beneath the mountain was rapidly evacuated, propelling clouds of hot ash and rock fragments northward and piling debris over a foot deep as far as 75 miles away. Some of the ash can still be seen hundreds of miles distant in Montana and Seattle. As the central chamber emptied, it collapsed inward reducing the 12,000-foot high Mt. Mazama by 2500 feet. Wizard Island, a small cone built after the destruction of the stratovolcano, reflects continued but subdued activity.

Azure waters in the Crater Lake caldera owe their extreme clarity to the lack of algal blooms. With a water depth of 1996 feet, Crater Lake is the deepest freshwater body in the United States. The lake and surrounding 160,290 acres have been designated a National Park.

Ash and lava from Cascade volcanoes provide convenient markers to date more recent climatic events. The distinctive widespread ash from the eruption of Mt. Mazama has been found from northern California and Nevada to British Columbia and Saskatchewan, Canada, providing one of the most extensive marker beds in recent times. Specific optical and chemical properties distinguish ash of the Mazama eruption so that it can be used to define a brief period of glacial expansion prior to the Mazama event as well as a second glacial surge 2000 to 3000 years afterward. Moraines at the base of the Three Sisters in Deschutes County, along with lava and ash from Mt. Baker, and tephra sequences from Glacier Peak in Washington are all mantled by Mazama ash.

North of Crater Lake, the Three Sisters complex, Mt. Washington, Three Fingered Jack, and Mt. Jefferson, form a ridge of andesitic peaks overshadowing central Oregon. Rising from a broad base, these mountains stand out as sharp snow-covered spires extensively

Figure 4.26 Snow covered Mt. Bachelor in the central Oregon Cascades hosts a series of winter sports in addition to downhill ski runs (Oregon Dept. Geology and Mineral Industry).

sculpted by glaciers, which are active even today. Of the five significant glaciers on the Three Sisters, Collier Glacier, reaching across from the Middle to North Sister, is the largest with 30 million cubic feet of snow and ice. Shrinking at a fairly rapid pace, this glacier is currently only 300 feet thick.

Volcanic activity, as recently as 400 years ago, has spread dark cindery basaltic lava fields, conical buttes, and fresh flows at Sand Mountain and Belknap Crater between the North Sister and Three Fingered Jack. This region of the central Cascades is readily accessible and provides a scenic and recreational backdrop rarely matched in the Pacific Northwest.

The ice cream cone-shaped Mt. Hood, rising to an elevation of 11,235 feet on Oregon's northern border, looks across the Columbia River toward Mt. Adams and St. Helens in Washington and towers above the skyline of Portland to the west. In spite of its serene appearance, Mt. Hood had explosive periods that produced immense amounts of andesite and dacite lava followed by mudflows. Major eruptions 15,000 years ago built multiple layers of lava up to 500 feet thick on the south and southwest flanks of the cone. Explosions of hot ash and debris mixed with thin layers of lava, mud, and melted snow spread nearly 10 miles down stream valleys. Between 1760 and 1818, a late period of activity emplaced

Figure 4.27 In northern Oregon Mt. Hood towers above Timothy Lake, impounded by a dam on Oak Creek. The "bald" patches are where timber has been clearcut (Oregon State Highway Dept.).

Crater Rock, a dome that bulges from the west side of the mountain. At the same time a surge of thick mud and volcanic debris filled the Sandy River channel enveloping forests and burying trees standing upright.

Nine active glaciers mantle the slopes of Mt. Hood. The volume of the combined ice and snow here, totalling 12.2 billion cubic feet, would produce immense lahars or mudflows during an eruption.

NORTH OF THE COLUMBIA RIVER IN WASH-INGTON, Mt. Adams, at 12,286 feet, is the third highest andesitic composite cone in the Cascade range. Mt. Adams was active between 20,000 and 10,000 years ago when a great volume of lava, second only to that produced by Mt. Shasta, built the current volcano atop the ruins of two previous edifices.

Mount St. Helens, the most recent cone to erupt in the Cascade arc, lies about 50 miles east of Mt. Adams in Skamania County. Prior to 1980, the rounded cone of St. Helens reached 9677 feet, but since the eruption the crest is only 8357 feet. Originally the youthful St. Helens, which experienced very limited glacial erosion, had

a particularly symmetrical and pleasing appearance. The oldest lavas erupted only 40,000 years ago, and within the past 2000 years this volcano has been more active than any other in the United States. Of the 60 tephra layers and debris flows, at least 6 reached south to the Columbia River.

The eruption on the morning of May 18, 1980, although small when compared to previous events, was significant because it was so closely monitored from the air, by eyewitnesses, radar, satellite, and several geophysical techniques. Eyewitness accounts of the event provided valuable data and details not recorded on instruments. Careful post-eruption studies have even yielded figures on temperatures and supersonic velocities inside the ash cloud.

In addition to steam eruptions, a swarm of mini-earthquakes, which began in March, 1980, announced the movement of magma upward into a swelling dome beneath the crust of the mountain. At 8:32 AM on the morning of May 18 an explosion produced a lateral blast to the north destroying plants and animals within 5 miles of the peak. A massive landslide down the north face,

Figure 4.28 Smoke still poured from the top of Mount St. Helens during the summer of 1980, even after the May eruption (Washington Dept. of Transportation).

which was set off by a minor quake, triggered the actual eruption. Clouds of powdery grey ash that blanketed much of eastern Washington ultimately circled the globe. Syrupy mudflows or lahars poured down river valleys, choking the North Fork of the Toutle River. When activity ceased, 1320 feet had been removed from the mountain top, and in its place a new 1900-foot deep crater opened to the north was exposed in the cone. During that summer, later eruptions left thin ash layers across the northern Willamette Valley.

At 14,410 feet above sea level, Mt. Rainier in Pierce County is the highest and third most voluminous volcano in the Cascades behind Mt. Shasta and Mt. Adams. Because of its proximity to the mega-metropolis stretching from Olympia to Seattle, Mt. Rainier has the potential for causing unprecedented damage if an eruption were to take place today. In a volcanic event, landslides and mudflows down the slopes of Mt. Rainier, which support the largest glacier field in the continental United States, would, be fueled by melting snow. Gigantic lahars,

Figure 4.29 The once-rounded cone of Mount St. Helens, which was decapitated by the recent 1980 eruption, displays a massive dome in the center of the crater (Washington Dept. of Transportation).

a sticky, thick muddy mixture of ash, water, and volcanic debris, would engulf the area surrounding the mountain and spread far out into the Puget lowlands.

The history of this andesitic and basaltic stratovolcano is relatively obscure. Several early flows about 75,000 years ago constructed a thick lava sequence that served as the base for a small shield cone, which grew to over 15,000 feet in elevation. Repeated violent eruptions of lava and breccia, including over 10 pumice layers between 8000 to 2200 years ago, reduced the

cone by as much as 2000 feet. The eastern slope collapsed following explosions of steam. This loosened enormous blocks of rock and sent them cascading northwest down into the White River canyon toward the present site of Enumclaw in southern King County. An eruption 5000 years ago was followed by an incredible wall of mud, rock, and ash 100 feet high, moving over 40 miles per hour, that tore away even more of the mountain. The Osceola mudflow, one of the largest worldwide, enveloped an Indian encampment at Enum-

Figure 4.30 This photo, looking toward the northeast, was taken at 40,000 feet from a Boeing B-29 on a high altitude test flight in the 1940s. Mount St. Helens is in the foreground. Mt. Adams is in the background, and British Columbia is in the distance (Boeing Airplane Company).

claw. The summit of Mt. Rainier collapsed, adding to the landslide that today would have overwhelmed Kent, Auburn, Summer, and Puyallyp before entering the waters of Puget Sound. With the collapse and formation of the new caldera, the volcano had completely altered its appearance.

Even today periodic eruptive episodes continue. In the early 1900s, mudflows and tephra covered large areas of glacial ice, and in 1963 yet another slide moved down the White River valley. With more than 30 earthquakes annually Mt. Rainier remains the most seismically active in the Cascade volcanic archipelago.

Glacier Peak and Mt. Baker are the furthest north of the High Cascades in Washington. Projecting 10,451 feet, Glacier Peak is one of the most solitary of the range. Inaccessible by road, the mountain is reached only by hiking trails. This volcano produced vast quantities of air-borne tephra 11,250 years ago that spread over 500 miles to the east. Glacier Peak lies within the North Cascades Primitive Area, which encompasses 830,000 acres. Mount Shuksan, Mt. Redoubt, and Mt. Challenger project as high ridges of this dramatic mountainous area.

Known to the Indians as "White Steep Mountain" because it is almost completely blanketed by glacial ice, Mt. Baker looks down from its towering height of 10,775 feet in northern Washington. Over 20 square miles of ice, snow, and andesitic volcanic debris obscure early events in the history of this stratocone. Initial volcanism ended sometime prior to 10,000 years ago with an assortment of ash, rock, basaltic lava, and mudflows from Black Buttes, two eroded remnants of the original stratocone just west of the current peak. The 7-mile long flow and cinder cone at Schreibers Meadow along with Sherman Crater on the south side of the summit was erected since that time. The explosion that resulted in the 1500-foot wide Sherman Crater was so recent it is a feature in Indian legends. During the middle 1800s, eruptions melted ice packs, started large forest fires, and contributed to fish kills in the Baker River, while steam fumaroles appeared as late as 1976.

Figure 4.31 Gaping crevasses following a steep narrow channel down icy Mt. Baker signal the presence of active glaciers on its slopes (Washington Dept. of Transportation).

VOLCANIC ACTIVITY ACROSS BRITISH CO-LUMBIA is concentrated in five belts. Of these, the Garibaldi zone, from Mt. Garibaldi northeast of Vancouver to Mt. Meager along the upper Lillooet River, is an extension of the High Cascades from Washington. From south to north, the Garibaldi volcanic region can be broken into the Mt. Garibaldi complex, Mt. Cayley and Mt. Fee, and Mount Meager and the Bridge River cones. Eruptions within this region were a by-product of subduction of the Explorer Plate, a northern segment of the Juan de Fuca slab, beneath the North American crustal plate.

In the Mt. Garibaldi complex, at least twelve eruptive episodes took place between 1 million to within the last 100,000 years. This activity was punctuated by glacial periods, and Garibaldi Park is one of the few areas in the Pacific Northwest where volcanoes interacted continuously with glacial ice. Mt. Garibaldi itself is somewhat unusual because it partially rested on glacial ice where lavas erupted beneath, through, and on top of snow and ice.

Currently reaching 8787 feet in elevation, Garibaldi began to develop toward the end of the Pleistocene with clouds of hot ash, rock fragments, and flows of incandescent lava. During one stage of construction, a cone of glowing hot debris slumped and collapsed when the ice beneath it melted, washing away avalanches of mud and debris. The remaining partially destroyed cone, re-

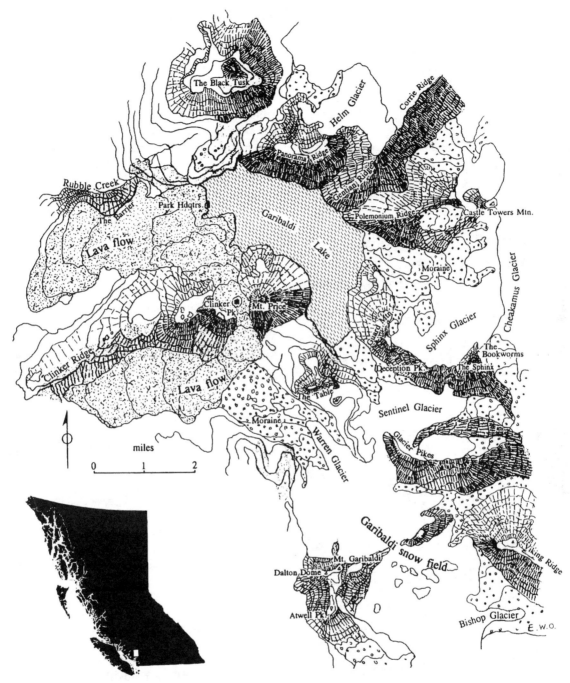

Figure 4.32 The Mt. Garibaldi volcanic complex of British Columbia includes lava flows, glaciers, cinder cones, and glacial lakes.

duced almost by half, has, in turn, been further eroded by glaciation. As much as 10 square miles of landslide rubble from the eruptions cover the floor of Squamish Valley to the west. A later thick stream of dacite lava formed the sharper pyramidal plug of the southern peak. Volcanism, which took place during a warming period, includes the eruption of Cinder Cone 100,00 years ago, the ring of lava around Clinker Peak, and the lava from Opal Cone.

High valleys and passes of the Garibaldi complex are covered year-round by ice caps. Warren, Sentinel,

Sphinx, Cheakamus, and Helm are among seventeen permanent glaciers remaining in the park.

Distinctive volcanic features in the Garibaldi complex include The Table and the Black Tusk. Originally interpreted as a volcanic plug, the Table is a flat topped 1000-foot high pillar composed of thick layers of andesitic lava that filled a pit within the ice sheet before cooling. Individual layers are delineated by vertical basalt columns. The striking Black Tusk is the remnant of a heavily dissected cone, and The Castle, an impressive

Figure 4.33 Lake Garibaldi was dammed by lava flows from Mt. Price a quarter of a million years ago. In this view to the south, Mt. Price is on the upper right of the lake and the dark-colored Table is just beyond (Geological Survey of Canada).

Figure 4.34 Aerial view of The Barrier, the source of the 1855 and 1856 Rubble Creek landslides just to the west of Lake Garibaldi, British Columbia (Geological Survey of Canada).

pinnacle near the town of Squamish, is a severely eroded volcanic spine.

A rough lava dam, The Barrier, is a series of flows broken into large jointed columns that contain the 1000-foot deep waters of Garibaldi Lake. Lavas of this feature have provided debris for spectacular rockslides that advanced almost 3 miles down Rubble Creek to the Cheakamus River. Following slides of historic proportion in 1855 and 1856, and more recent assessments of slide hazards, the British Columbia government prohibited housing below The Barrier in the 1970s.

A short distance from Vancouver, Garibaldi Provincial Park is popular for its scenery and recreational opportunities although not all of the area can be reached by automobile.

On the highlands north of Mt. Garibaldi, Mt. Cayley and Mt. Fee as well as a number of adjacent dacite domes began erupting during the Pliocene well before the advent of continental ice sheets. Flows of the Mt. Cayley eruption filled the Squamish River valley where remnants are still visible along the canyon walls. Much of the volcanic material from this long-lived center has been removed by erosion, and the original Mount Fee cone exists today only as a high sharp spine of dacitic lava.

At the northern tip of the Cascade chain, Mount Meager is composed of four overlapping andesite and dacite cones. Extruded atop crystalline granite intrusives of the Coast Mountains, the broad oval platform of lavas, mudflows, breccia, and ash, which make up the volcanic pile, has been heavily dissected by erosion. The final eruptive stage at this complex shifted north to the Bridge River volcanoes around 2490 years ago. During this event tephra, ash, and the lava engulfed conifer trees along Meager Creek and the Lillooet River where their charred remains are found today.

SEVERAL INTERVALS OF THE ICE AGES incised the magnificant sharp serrated ridges, deep valleys, and high clear mountain lakes of the Cascades. Today Pacific Northwest glaciers along the Cascade peaks are only fragments of ice fields that once extended the length of the range.

Increased precipitation and cooling temperatures of the final glacial period, about 25,000 years ago, formed ice masses that coalesced into continental glaciers over most of British Columbia. Conditions at that time in British Columbia were similar to those existing in Greenland today. As the ice sheet expanded south into Washington and Idaho, it brought extensive ice caps and alpine glaciers to the North Cascade mountains, but further south smaller valley glaciers predominated. At one point in the Late Pleistocene the summits from Mt. Baker in Washington to Mt. Mazama and Mt. McLoughlin in Oregon were covered by a continuous ice cap. Beyond the margin of the ice sheet from British Columbia,

Figure 4.35 Devastation Creek near Lillooet, British Columbia, shortly after the slide of July 22, 1975. Loose volcanic debris, mixed with water after heavy rains, swept down the valley killing four people (Geological Survey of Canada).

Mounts Lassen and Shasta in California experienced limited glaciation when ice descended along valleys from higher elevations to the base of the mountains.

IN WASHINGTON AN EXTENSIVE GLACIER cut its way through Chelan Valley in Chelan County, straightening the channel, polishing the granite walls, and scooping out the valley floor down to 426 feet below sea level. The glacier failed to reach the Columbia River, and a narrow neck still separates the Columbia and Chelan valleys. When the ice retreated it abandoned a pile of rubble as a natural dam to contain the

meltwater. Today Lake Chelan is a long sinuous body of water that extends northwestward for 50 miles from the edge of the Columbia plateau into the heart of the northern Cascades near Glacier Peak. A man-made dam, constructed here in 1927, raised the level of the lake, making it the third deepest in the United States.

Dramatic waterfalls are visible from the lake, and one of the most impressive, Rainbow Falls at the northern end, tumbles 300 feet. These cataracts drop from hanging river valleys left high above the canyon when it was deepened by the moving ice mass.

RECENT GLACIAL FLUCTUATIONS in the North Cascades have been carefully examined for comparison to worldwide trends. Three main glacial advances took place at intervals from 2500 to 3500 years ago, from 1500 years ago to 1800 A.D., and from 1944 to 1976. However, since the middle of the 1970s, glaciers have been withdrawing almost steadily with small temporary advances or stationary periods. Low winter precipitation and warmer global temperatures since this time have combined to accelerate the retreat of virtually all Cascade glaciers. The nine glaciers on Mt. Baker, for example, have receded an average of 165 feet annually, a trend that is apparent worldwide.

HAZARDS FROM FUTURE CASCADE ERUPTIONS must be assessed with respect to the volcanic history of the range as well as the growth of cities in the Pacific Northwest. Since over 20 centers within the Cascade volcanic chain have erupted for hundreds of thousands of years, the entire range is regarded as volcanically active and potentially hazardous. In the past 12,000 years, eruptions of Cascade volcanoes have occurred more than 200 times, or about one every 60 years. Mount St. Helens (1980), Mt. Lassen (1915), Mt. Hood (1865, 1859), and Mt. Shasta (1786) staged spectacular displays during historic times. Certainly Indians witnessed erupting Cascade peaks, and the peripatetic French explorer La Perouse, who happened to be sailing up the California coast in 1786, reported a large volcanic cloud, which may have been the last eruption of Mt. Shasta.

Damage and destruction from a major volcanic event today would depend on the proximity of towns, population density, direction of the blast, and the type of volcanism, but it is possible that eruptions similar to those of the past would seriously affect thousands of people miles from the main volcanic center.

Lava, mud, and ash have commonly accompanied the activity from Cascade volcanoes, and any of these eruptive modes could constitute a serious potential hazard. Lava tends to be slow moving but burns whatever is encountered, while mudflows or lahars move quickly down stream valleys with tremendous force. A mixture of mud, volcanic debris, and melting ice would carry

Figure 4.36 Originally blocked by glacial gravel and debris, the water level of Lake Chelan, Washington, was raised by construction of a dam in 1927 (Washington Dept. of Transportation).

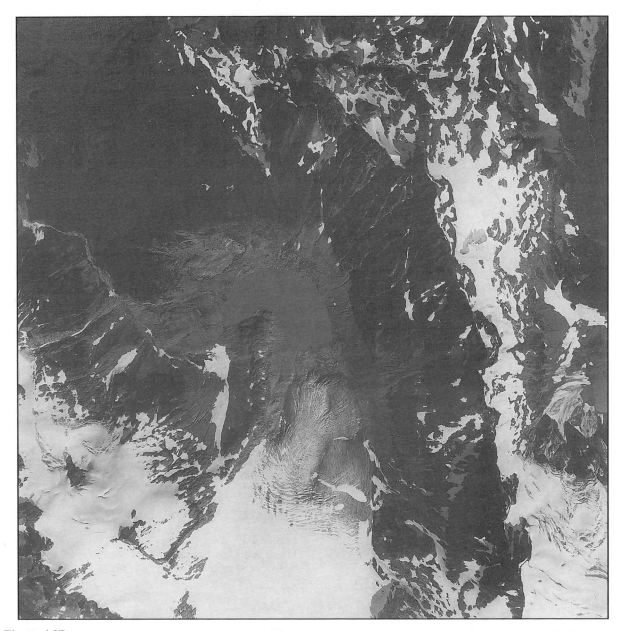

Figure 4.37 South Cascade glacier near Glacier Peak in Washington shows open crevasses that reflect moving ice (Washington Dept. of Transportation).

away bridges, buildings, dams, and roads. In the case of St. Helens, incandescent gas and rock fragments moved at incredible speed killing 57 people, destroying almost 300 square miles of forests, and causing over $950 million in losses to the state of Washington. The debris avalanche from Mt. Shasta 300,000 years ago reached more than 40 miles northward and covered over 250 square miles.

Fortunately prevailing winds in the Pacific Northwest are eastward, and nearby population centers over 100,000 are west of the Cascades; however, ash from previous eruptions has been carried hundreds of miles from the source blanketing large areas, damaging vegetation and machinery, and causing health problems or even death. Clouds of ash overwhelmed Spokane for over a month during the St. Helens episode.

It is worth noting that eruptions, as in the case of Mount St. Helens, are preceeded by clear warning signals such as small earthquakes and tremors, and it is difficult to imagine a catastrophic volcanic event coming as a complete surprise.

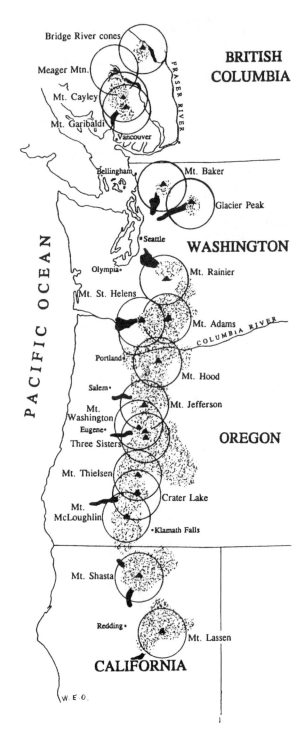

Figure 4.38 Estimates of volcanic hazard zones for the Cascade Mountains: the circles are 60 miles in diameter; shaded areas are potential lava flows; and black areas lie in the pathway of mudflows (after Miller; 1990).

Figure 4.39 Within the blast zone from the 1980 eruption of Mount St. Helens, bark was neatly peeled from the trees as they were blown over (courtesy C. McKillip).

Figure 4.40 A sluggish stream in the upper Toutle River watershed is choked with mud after the 1980 St. Helens volcanic eruption (courtesy C. McKillip).

MINERAL PRODUCTION

Economic minerals in this province occur in a belt along the Western Cascades from Lane County to Clackamas County in Oregon, and in the North Cascades from Kittitas County in Washington to the Bridge River volcanic region of British Columbia.

In both areas ores are found in veins and deposits around intrusive bodies. Gold and silver, lead, zinc, and copper in the Western Cascades occur in fractures cutting the older volcanic rocks, which were invaded by a number of plutons before faulting took place. These fissures provided an avenue for mineral-laden fluids to invade the older strata where they cooled at shallow depths. Since emplacement during the Miocene, the veins have been exposed by extensive weathering of the range. During the mineralization process in the North Cascades, hot springs, rich in both base and precious metals, precipitated deposits onto the ocean floor from undersea ridges. Metals in this enriched strata were later dissolved and moved along fractures through surrounding rock as they cooled.

IN BRITISH COLUMBIA the productive Bridge River district at the northern end of the Cascades has been a rich source of lode gold. Situated along margins of the Bridge River terrane, gold-laden fluids have invaded major fault and fracture zones. Exploiting the two most productive vein systems, the Bralorne and Pioneer mines yielded 130 tons of gold from 1899 to 1978, the largest producer of vein gold in western Canada.

THE PAST CENTURY OF ECONOMIC MINERAL PRODUCTION IN WASHINGTON has been dominated by copper, extracted between 1939 and 1956, when the state was among the top 10 copper-producing areas of the United States. Lesser amounts of gold, silver, zinc, lead, and nickel have been recovered from the

area, and weekend prospectors and private mine owners still work claims when the price of gold is sufficiently high. The outstanding Cascade mineral districts in Washington include those in Chelan, Snohomish, Whatcom, and Kittitas counties. Of these, Chelan County leads in output with a total of 1.5 million ounces of gold combined with lesser amounts of other minerals.

Disappointed by the lack of gold prospects in British Columbia, miners traveled south to the Blewett district in Chelan County where they were able to recover appreciable amounts of placer gold from Peshastin Creek in 1860. Once lode gold was located here in 1874, a wagon road constructed from Cle Elum opened up this remote area of the Cascades. A six-stamp mill, that pulverized the ore to extract gold, replaced the primitive stone arrastras used for crushing by earlier prospectors. The Blewett mines were active for about 50 years with an estimated output of 850,000 ounces. North of Blewett in the Chelan Lake district, the Holden Mine became the most prolific in 1938 with 514,500 ounces of gold. Extracting silver, copper, and zinc in addition to gold, operations ceased with declining profits in 1957 when the facilities were purchased by a church group for recreation.

Near Wenatchee the Cannon Mine was the state's top gold producer in the early 1990s, but with depleting resources it was scheduled to be shut down. However, this was not before production surpassed one million ounces in gold and silver in 1992.

Around 1874 rich gold placers were found on Silver Creek near Index in southern Snohomish County. With the building of a special branch of the railroad 5 miles up the Skykomish River valley, this district became the supply base for those surrounding mines that had been nearly inaccessible in the rugged mountains. Ore was

Figure 4.41 The Pioneer Mine in the Bridge River District was a prolific producer of gold in western British Columbia. Gold here occurs in rocks of the Bridge River and Cadwallader terranes (B.C. Archives and Records Service).

Figure 4.42 One of the oldest mining areas in Washington, the Blewett district in Chelan County was the scene of placer operations in 1860 (Washington Dept. of Natural Resources).

Figure 4.43 Buildings of the old Holden Mine at the base of Copper Peak in Chelan County, Washington, which produced large amounts of copper, gold, silver, and zinc between 1938 and 1957. Today an extensive tailings pile is being cleaned up and replanted (Washington Dept. Natural Resources).

Figure 4.44 In Snohomish County, the rugged Monte Cristo mining region of the North Cascades was the most active in Washington at the turn of the century. Indelible scars of glaciation are still visible on this landscape (Washington Dept. of Transportation).

brought to the railhead and then transported to a smelter in Tacoma. From 1897 to 1902 mining reached its highest point with the location of the Sunset Mining Company northeast of Index. This find proved to be the most consistant and lucrative for copper over the years. Mining fortunes rose and fell with the price of copper until the mill here closed. Prior to 1903, gold from this county, mainly from the Monte Cristo and Silverton districts, was estimated at $7 million, after which production dropped considerably.

Whatcom and Kittitas counties were the smallest producers in the Washington Cascades with 91,000 ounces and 10,000 ounces of placer and lode gold respectively. In the Whatcom district near the Canadian border, the Boundary Red Mountain and Lone Jack lode mines accounted for most of the tonnage since opening in 1916. In the Swauk district of Kittitas County, early gold placer operations at the turn of the century were worked in Pleistocene river gravels that sometimes reached up to 18 feet thick. Occasional rounded nuggets in these deposits were reportedly worth $1000 each.

OF THE MINING DISTRICTS IN THE OREGON CASCADES, the five prime producers of gold, silver, and

British Columbia
1. Bridge River
        (Bralorne; Pioneer)

Washington
2. Whatcom County
        (Mt. Baker; Slate Creek)
3. Chelan County
        (Blewett; Chelan Lake; Entiat; Wenatchee)
4. Snohomish County
        (Index; Monte Cristo; Silverton)
5. Kittitas County
        (Swauk)

Oregon
6. Marion County
        (North Santiam River)
7. Linn County
        (Quartzville)
8. Lane County
        (a. Blue River; b. Fall Creek; c. Bohemia)

Figure 4.45 Gold Mining Districts of the Cascades.

Figure 4.46 Glacially carved peaks in the North Cascades National Park of Washington average 7000 feet in elevation. The Cascade Mountain divide is to the upper left (courtesy M.F. Meier, U.S. Geological Survey).

associated lead, zinc, and copper are the Fall Creek and Bohemia in Lane County, the Quartzville in Linn and Lane counties, and the North Santiam in Clackamas and Marion counties. Located 25 miles southeast of Cottage Grove, the Bohemia mining area has been the largest and most profitable. Miners from California struck gold on Sharps Creek here in 1858 when they stopped to shoot and clean a deer. One of the prospectors, James Johnson, who was from Bohemia in eastern Europe, gave this name to the region. With the discovery of lode deposits in veins and the addition of a stamp mill, the Bohemia district

yielded about $1 million dollars in 1880s gold prices while in operation. In eastern Lane County, the Fall Creek district has seen only limited gold mining.

Located in the 1860s, the Quartzville and Blue River regions in the central range had the second highest output at $360,000 before operations ceased after 1913. Today the Quartzville Recreation Corridor is maintained for weekend mining and outdoor use.

Gold, silver, copper, zinc, and lead from the North Santiam River make this one of the most diverse mineral regions of the Oregon Cascades. Opened in the 1860s, activity has focused on the Ruth vein where the amount of zinc and lead far outstripped that of gold. Total dollar figure for the North Santiam is $25,000. Renewed explorations here in the 1980s and 1990s involved surveying and drilling for copper, gold, and silver.

Current proposals to mine copper in this watershed pose a serious threat to the water supply of Salem, parts of Eugene, and Portland as well. Copper mining is particularly toxic to vegetation, streams, and drinking water, and copper wastes are difficult to remove.

ADDITIONAL READINGS

Bailey, E.H., ed., 1966

Barnard, W.D., 1978

Brandon, M.T., and Cowan, D.S., 1987

Chesterman, C.W., and Saucedo, G.J., 1984

Crandell, D.R., 1965

Derkey, R.E., 1994

Doukas, M.P., and Swanson, D.A., 1987

Duncan, R.A., and Kulm, L.D., 1989

Easterbrook, D.J., 1986

Evans, J.E., and Johnson, S.Y., 1989

Frizzell, V.A., et al., 1987

Fulton, R.J., 1989

Gabrielse, H., and Yorath, C.J., eds., 1991

Garver, J.I., 1992

Gresens, R.L., 1987

Gresens, R.L., et al., 1977

Gresens, R.L., et al., 1990

Hammond, P.E., 1983

Hammond, P.E., 1989

Harris, S.L., 1988

Haugerud, R.A., 1989

Kleinspehn, K.L., 1985

Mathews, W.H., 1952

Mathews, W.H., 1987

McDonald, G.A., 1966

Miller, C.D., 1990

Miller, R.B., 1993

Miller, R.B., et al., 1994

Misch, P., 1977

Monger, J.W.H., Price, R.A., and Templeman-Kluit, D.J., 1982

Moore, G.W., 1994

Norris, R.M., and Webb, R.W., 1990

Peck, D.L., et al., 1964

Powers, H.A., and Wilcox, R.C., 1964

Priest, G.R., and Vogt, B.F., 1983

Read, P.B., 1990

Rhodes, P.T., 1987

Tabor, R.W., et al., 1987

Tabor, R.W., et al., 1989

Taylor, E.M., 1990

Trehu, A.M., et al., 1994

U.S. National Park Service, 1986

Vance, J.A., et al., 1987

Wells, R.E., and Heller, P.L., 1988

Whetten, J.T., et al., 1980

Williams, H., 1976

Wood, C.A., and Kienle, J., eds., 1990.

5

KLAMATH MOUNTAINS

The early settlement of the Klamath Mountains was launched by the discovery of gold. Placers, as this one near Ashland, Oregon, were responsible for three-fourths of the gold mined in this region since 1850 (Oregon Dept. Geology and Mineral Industries).

The Klamath Mountains province is definable from either a geologic or geographic perspective. Geologically, the Klamaths are successive sheets of older terrane rocks enclosed by younger strata of the adjoining regions. Encompassing about 11,800 square miles, the Klamaths extend beneath Tertiary rocks of the Oregon coast on the north and Cascade lavas to the east. South along the Sacramento Valley and California coast, overlying Cretaceous strata forms the periphery. Geographically, the Klamath Mountains form an accurate outline extending from coastal Oregon at Port Orford, northeastward to Roseberg, south to Medford, then across the border to Redding, California, where the boundary angles sharply northwest to the coast at Point St. George.

The region is a raised eroded plain, which has been tilted slightly toward the west and thoroughly dissected by streams. Mountains that dominate the province include the 9000-foot high Trinity and Salmon ranges of California, Preston Peak, at an elevation of 7310 feet in the Siskiyous, and Mt. Ashland at 7500 feet in Oregon. Straddling the border, the Siskiyou Mountains stretch from Oregon into California where they meet the Mar-

ble, Salmon, Trinity, and South Fork peaks in a continuous chain, the length of the province.

A heavy winter rainfall contributes to thick soils and an extensive stream network that follows precipitous gorges to empty into the Pacific Ocean. Only the Rogue and Klamath rivers reach entirely across the range following a winding course from the Oregon Cascades westward to the Pacific. Along its route the Klamath collects water from four major tributaries, the Shasta, Scott, Salmon, and Trinity in northern California, while in Oregon the Rogue River is augmented by the Applegate and Illinois streams.

INTRODUCTION

The layout of the Klamath Mountains of southwest Oregon and northern California mimics the ribbon-like bands of exotic terranes that characterize most of British Columbia and the Blue Mountains of Oregon. Transported to their present sites by moving crustal plates and faulting, the terrane rocks in all these areas are strikingly similar in composition and age. Microfossils of conodonts, foraminifera, and radiolaria, along with a variety

Figure 5.1 Coastal Port Orford, Oregon, is at the northwest apex of the Klamath Mountains province (courtesy E. Baldwin).

of invertebrates found throughout all three regions, help place the rocks in their proper time sequence as well as sketch out the ancient environments they occupied. Fossils in Klamath rocks suggest some terranes originated in locations as distant as China or Japan, but others formed in nearby marine settings off southern California or British Columbia.

Four major belts of accreted terranes, the essence of Klamath Mountains geology, are arranged in a curving north-south pattern across the province. Rocks within each represent Paleozoic to Mesozoic environments that encompass volcanic island archipelagos and adjacent oceanic basins. Late in their histories, terrane rocks were folded and faulted, intruded by plutons, and altered by heat and temperature of metamorphic processes as they were accreted to North America.

Near the base of the terranes, ophiolites—sequences of rocks that were once part of the ocean crust—have been enriched with minerals that played an important role in the economic history of this province.

GEOLOGY OF THE KLAMATH MOUNTAINS

The fabric of the Klamath Mountains is the multiple sheet-like terranes that originated as chains of volcanic islands and nearby ocean basins. Moving eastward as crustal tectonic plates, succeeding exotic slabs collided with North America where they were subducted below the continental plate. As the partially subducted terranes were accreted or affixed to the lower edge of the mainland, they assumed a stacked domino configuration, tilted eastward.

When tectonic plates merge, the heavier, thinner oceanic slab ordinarily slides beneath or is subducted below the continent. On this descending plate, intense compression and elevated temperatures caused changes to the rocks in a process called metamorphism. In this process rocks are partially melted, and often entirely new minerals crystallize. Rock texture and minerals in a metamorphic rock will indicate whether these changes were brought on by pressures or high temperature or a combination of the two. Many of the Klamath Mountains rocks have been metamorphically altered to what is known as the greenschist or blueschist level reflecting strong pressure but comparatively low temperatures. This type of metamorphic environment is typical of subduction zones where tectonic plates are in collision.

Terranes that make up the Klamath Mountains are of two varieties. The largest are slab-like sheets that moved eastward from sites out in the ancestral Pacific Ocean. Composed of pillow lavas, which erupted underwater, as well as limestones, fine-grained sediments, and cherts with deep-sea radiolarian microfossils, these exotic rock parcels may have traveled great distances after being deposited. A second, much smaller variety of terranes are usually thin linear slices, which were carried shorter distances north or south along the continental margin by faulting. Rocks within these smaller slivers are ordinarily greywacke, sandstone, siltstone, and mudstone, along with fragments of volcanic ocean crust. Both types of terranes are bounded by faults.

Clockwise rotations of terranes were typical along the Pacific margin where the movement was recorded by the orientation of mineral crystals in the rocks. As a molten rock cools, magnetic minerals within are frozen into place, recording not only latitude but magnetic north as well. Any later significant movement north or south or rotation will be reflected by the magnetic record that was fixed during crystallization. Rotations of large rock masses may take place when they are being affixed to the continent, but there is evidence that smaller terranes, caught between faulting slabs, simply turn like ball bearings.

In the Klamath Mountains, rotation of the terranes took place primarily during the Mesozoic, where up to 90 degrees of clockwise movement has been recorded. The best magnetic records are in the eastern Klamaths, but scattered data is increasingly available for other terranes. According to current magnetic readings, the exotic rocks do not show an appreciable shift from the north or south, although they do confirm that significant rotation of the terranes probably began during the Later Triassic or Early Jurassic as accretion was taking place. An interval of rapid rotation during the Jurassic ceased in the Early Cretaceous when accretion was completed. A smaller amount of rotation took place again in the Middle Tertiary triggered by oblique subduction of the Juan de Fuca plate and extensional stretching in the adjacent Basin and Range.

From east to west, the main divisions that make up the Klamaths are the Eastern Klamath belt, the Central Metamorphic belt, the Western Paleozoic and Triassic belt, and the Western Klamath belt.

THE EASTERN KLAMATH BELT, extending in an arcuate curve from Redding to Yreka, contains the oldest rocks within the province. These include Ordovician to Devonian Yreka terrane rocks, the Ordovician Trinity ophiolite complex, and the Early Devonian to Middle Jurassic Eastern Klamath terrane near Redding in Shasta County. In the Yreka terrane, volcanic rocks of the Silurian Duzel Formation, submarine fans of the Moffett Creek Formation, and continental slope fan environments of the Devonian Gazelle contain abundant microfossils, brachiopods, and corals. About 10 miles south of Yreka in Siskiyou County, the 6037-foot high, faulted monolith of Duzel Rock, part of the Yreka terrane, is composed of Duzel Formation at the base, covered by rocks of Moffett Creek, and a Duzel limestone cap at the top.

Figure 5.2 Principal locations within the Klamath Mountains province.

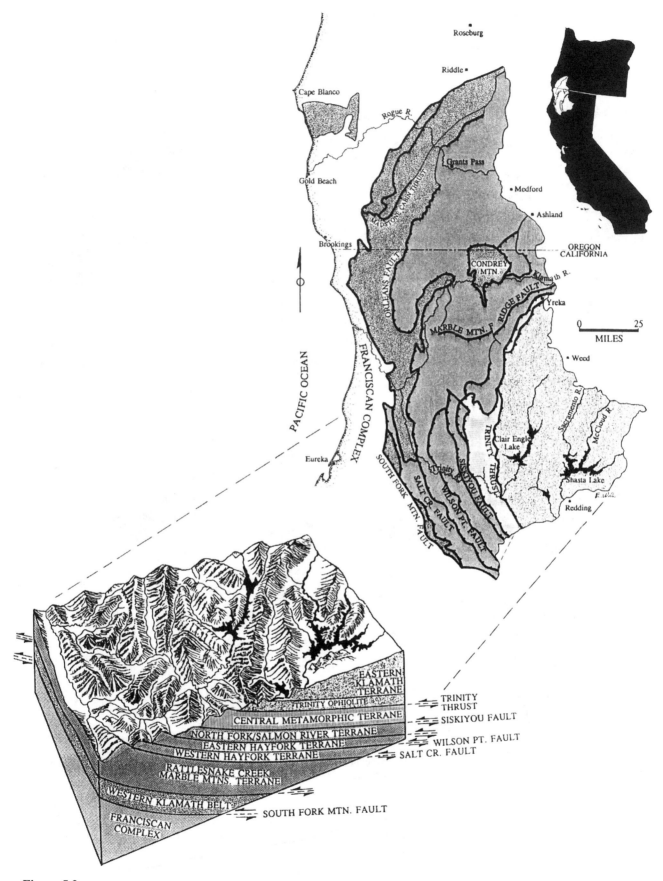

Figure 5.3 The Klamath Mountains are a series of multiple eastward-plunging slabs separated by thrust faults.

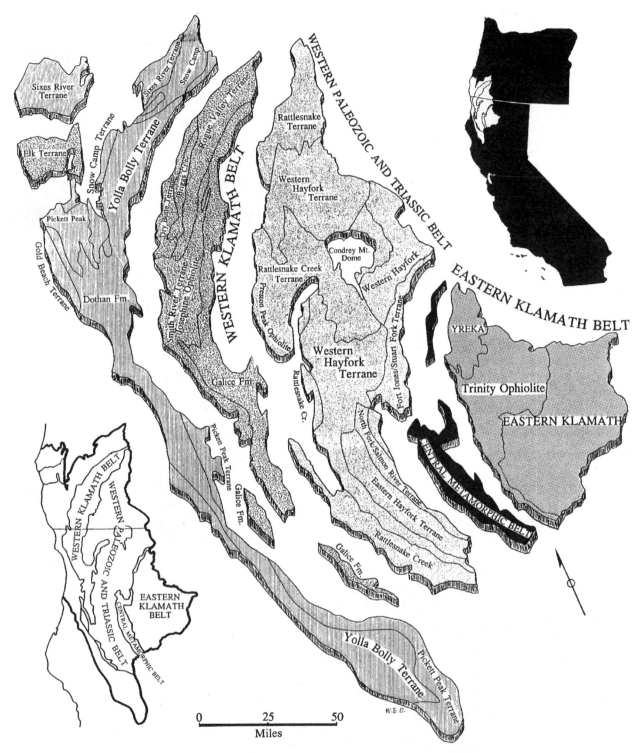

Figure 5.4 Major terranes that make up the Klamath Mountains province are the Eastern Klamath, the Central Metamorphic, the Western Paleozoic and Triassic, and the Western Klamath belts (after Burchfiel, Lipman, and Zoback, 1992; Silberling, et al., 1992).

Spread over an area 30 miles wide and 50 miles long in the east-central Klamaths, the Trinity ophiolite is one of the largest known exposures of ancient ocean crust. Originating in the vicinity of an island arc basin more than 475 million years ago, the Trinity ophiolite was emplaced along a spreading ocean ridge before being uplifted and eroded during the Early Devonian.

Strata of the Eastern Klamath terrane near Redding are represented by Devonian rocks of the Copley Greenstone, Balaklala rhyolite, and Kennett shale and lime-

Figure 5.5 Capped by a layer of limestone, Duzel Rock south of Yreka, California, is a monolith composed of mudstone and sandstone of the Eastern Klamath belt.

Figure 5.6 Pillow lava in the 400 million-year old Copley Greenstone is found near Castella, California. "Pillows" form when lava erupts under water (California Division of Mines and Geology).

stone that overlie the Trinity ophiolite. These rocks reflect environmental conditions from terrestrial to oceanic island arc lavas and shallow water limestones.

Permian McCloud Formation rocks form the main pre-Mesozoic basement in the eastern Klamaths near Redding, but characteristic McCloud fossils are recorded in scattered localities from the southern Yukon to central British Columbia, in the Blue Mountains of Oregon, in northwest Nevada, and in the northern Sierra Nevada Mountains of California. The McCloud, which evolved as long-lived island archipelagos, is defined by its distinctive fauna of wheat grain-shaped fusulinid microfossils, corals, and brachiopods that inhabited warm shallow limestone shelves fringing the islands.

Two environmental regions, separated by the paleo-Pacific Ocean, existed during the Permian. These were a North American province, which included the McCloud setting, and a tropical or Tethyan oceanic province further to the west. Although the McCloud volcanic arc originated closer to the North American continental shelf than that of the Tethyan fauna, the original geographic locations of each are not known, and the relationship of these two assemblages to the North American landmass is conjectural.

THE CENTRAL METAMORPHIC BELT consists of two major Devonian rock packages, the Salmon and Grouse Ridge schists derived from intense metamorphism of older ocean crust sediments. Alteration of rocks within this belt is closely related to the tectonic collision and emplacement of the Trinity ophiolite. The narrow discontinuous strip of Central Metamorphic rocks is separated from the Eastern Klamath by the Trinity thrust fault. Active during the Devonian, this fault plowed up a sizeable package of broken and mixed melange rocks when the Central Metamorphic belt was thrust beneath the Eastern Klamath plate. Unlike other major terranes of the Klamaths, the Central Metamorphic belt apparently lacks ophiolitic rocks.

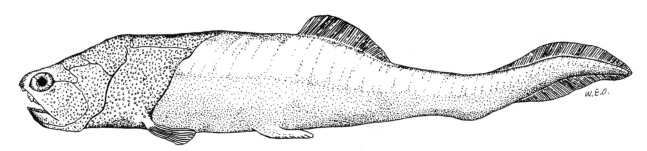

Figure 5.7 This Devonian fish, which featured an armor-plated head, dominated the seas over 350 million years ago. Armor plates belonging to a member of the Coccosteidae family were found in a tuff layer of the Balaklala rhyolitic lavas of northern California.

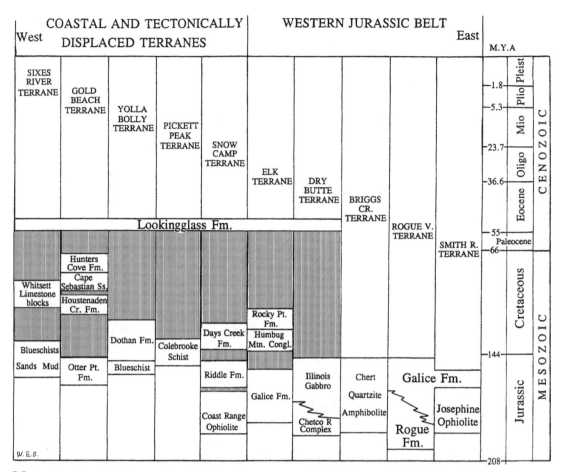

Figure 5.8 Stratigraphy of the Western Jurassic belt and coast terranes (after Blake, et al., 1985; Orr and Orr, 1999).

THE LARGEST AND MOST EXTENSIVE of all the rock suites found in the Klamath province is the Western Paleozoic and Triassic belt, which extends from near Canyonville, Oregon, to the South Fork of the Trinity River in California. At its widest on the Oregon-California border, the belt expands to 50 miles. Despite its name, this belt also contains thick sections of Jurassic rocks. The Jurassic Siskiyou fault separates this belt from the Central Metamorphic terrane to the west. With the exception of oceanic sediments, many of the clastic rocks in this belt were eroded from terranes of the Central Metamorphic and Eastern Klamath belts. Rocks within this terrane include intact island arc sequences as well as a jumbled mixture of fine-grained oceanic sediments, chert, ophiolite fragments, and volcanic debris. Invertebrates and microfossils from shelf limestone blocks as well as deeper marine rocks here reflect a variety of settings from the Late Paleozoic to Jurrasic interval.

At its northern extent in Oregon, the Western Paleozoic and Triassic belt was previously designated the Applegate Group, while the southern portion in California was divided into the Rattlesnake Creek, North Fork, and Hayfork terranes. Further subdivisions of the belt produced six additional terranes. In a stacked sequence from east to west these are the Rattlesnake Creek/Marble Mountains, Western Hayfork, Eastern Hayfork, North Fork ophiolite/Salmon River, North Fork, and the Fort Jones/Stuart Fork terranes. This diversity reflects conflicting views and many problems that are yet to be resolved within this complex province.

Rocks in the Rattlesnake Creek terrane are mainly fragments of Upper Triassic to Middle Jurassic ocean crust that have been thrust and telescoped together into a melange. Limestone coral reefs, which were locally altered to crystalline marble by metamorphism, are found in the Marble Mountains of northern California. In the

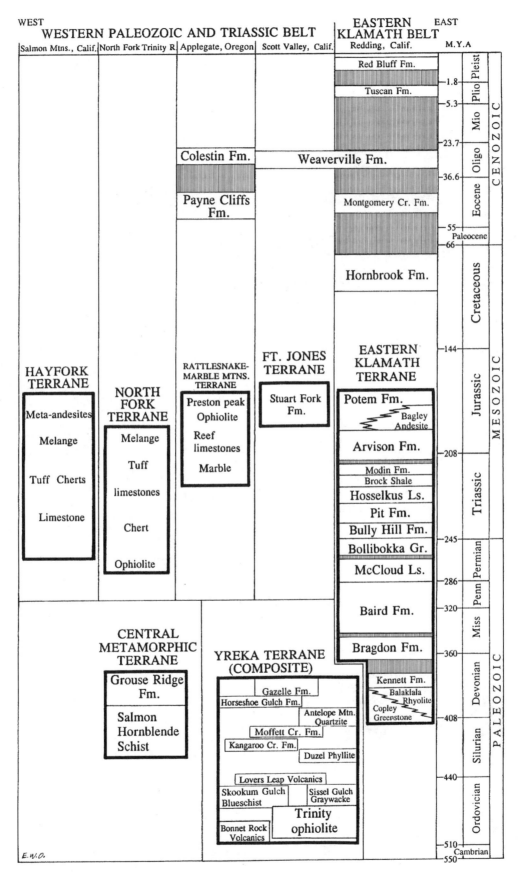

Figure 5.8 Continued. Stratigraphy of the eastern Klamath Mountains (after Miller, et al., 1992; Silberling, et al., 1992).

Figure 5.9 Marble Mountain, which rises to a height of 6880 feet in northern California, is composed of rocks of the Western Paleozoic and Triassic belt that have been altered from limestone to marble in the subduction process (Siskiyou County Historical Society).

same terrane cohesive ocean crust pieces, designated as the Preston Peak ophiolite, yield copper, zinc, gold, and silver. Developed in a volcanic archipelago or marginal basin in the Middle Jurassic about 165 million years ago, the Preston Peak complex was ultimately thrust into a position above the younger Josephine ophiolite.

Although isolated rocks as old as Silurian are known from the North Fork terrane, most are Late Paleozoic through Triassic. These rocks, which include ultramafics rich in iron and magnesium, form a nearly complete ophiolite ocean crust sequence from mantle and sheeted dikes to pillow lavas and deep-sea cherts. Dur-

Figure 5.11 A chaotic mixture or melange in the Hayfork terrane on the Salmon River in Trinity County, California, resulted from submarine landslides. The block of chert in the middle of the photo is 20 feet across and is imbedded in a shaley matrix (California Division of Mines and Geology).

Figure 5.10 The coral, *Waagenophyllum,* which lived in warm tropical waters, occurs in limestone near Redding in the eastern Klamath Mountains.

ing its accretion, much of the North Fork was peeled off and piled up into thick melange sequences. Early Permian microfossils in the North Fork are indicative of a tropical Tethyan setting from the equatorial Pacific. Similar rocks with identical tropical fossils appear in the Cache Creek terrane of British Columbia and the Baker terrane of northeast Oregon.

The melange mixture of chert, sandstone, tuff, and lava of the Hayfork terrane represents an ancient Jurassic volcanic island arc, even though loose blocks of older strata ranging in age from Silurian to Permian are scattered throughout. Imbedded within the younger Hayfork terrane, blocks of Silurian as well as lower Permian

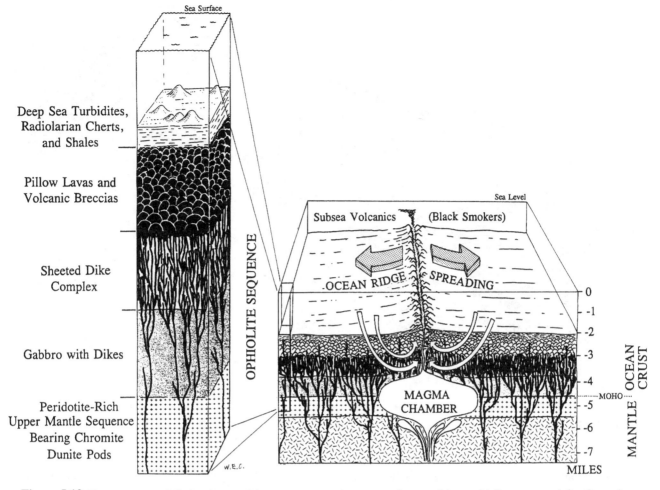

Deep Sea Turbidites, Radiolarian Cherts, and Shales

Pillow Lavas and Volcanic Breccias

Sheeted Dike Complex

Gabbro with Dikes

Peridotite-Rich Upper Mantle Sequence Bearing Chromite Dunite Pods

OPHIOLITE SEQUENCE

Sea Surface

Sea Level

Subsea Volcanics   (Black Smokers)

OCEAN RIDGE   SPREADING

MAGMA CHAMBER

MOHO

OCEAN CRUST   MANTLE

MILES

Figure 5.12 Five separate and distinct layers of the ocean crust and upper mantle comprise an ophiolite sequence (after Orr and Orr, 1999).

McCloud limestone with shallow-water fossils were deposited in environments quite distinct from the surrounding matrix. The presence of both McCloud and tropical Tethyan faunas within the Hayfork suggests mixed sources for the terrane.

ALSO KNOWN AS THE WESTERN JURASSIC BELT, the lengthy Western Klamath belt stretches for over 200 miles along the western edge of the Klamath Mountains in Oregon and California. Ophiolites, island arc volcanic rocks, and deep ocean basin sediments are all found in this area. The belt is separated from the overlying sheet of the Western Paleozoic and Triassic belt by the Orleans fault, while the base is marked by the younger South Fork Mountain thrust. Five small terranes presently recognized in the Western Klamath belt are the Smith River, Rogue Valley, Briggs Creek, Dry Butte, and Elk.

The easternmost Smith River terrane includes the celebrated Josephine ophiolite. Developed in a back-arc basin between the volcanic arc and the mainland, this ophiolite represents an almost complete slab of ocean crust that was thrust beneath the continental margin dur-

ing accretion. Ophiolites typically form the bottom of accreted terranes, acting as skid planes for faulting processes. This configuration is due to the nature of ophiolitic rocks, which readily alter to notoriously incompetent and easily ruptured serpentine.

Running along the western edge of the Klamaths for 120 miles, the 160 million-year old Josephine ophiolite extends from the South Fork River in Humbolt County, California, well into Josephine County, Oregon. Best known for its massive sulfide deposits, the Josephine ophiolite was impregnated with minerals from submarine hot springs. Such sulfides yield a variety of valuable ores.

Covering the ophiolite is a mile-thick layer of deepwater shales and turbidite sands of the Galice Formation. Within the Rouge Valley terrane, the shale and sand have been altered by metamorphic temperatures and pressure to slate that interfingers with the marine volcanic sediments of the Rogue Formation. Almost 20,000 feet of lavas and clastics of the Rogue accumulated adjacent to an island arc at the same time that Galice sediments formed offshore in a back-arc basin setting.

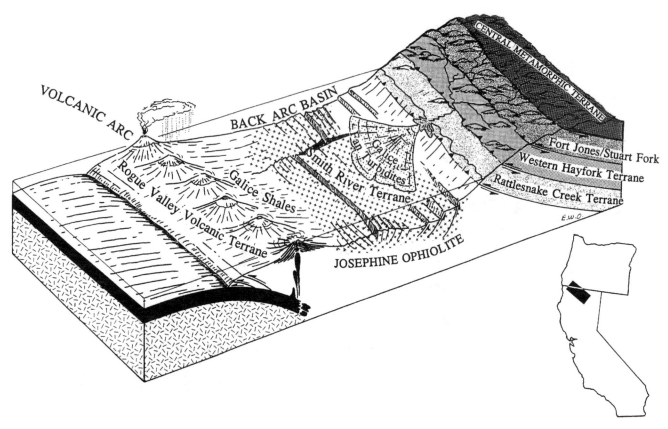

Figure 5.13 Environments of the Late Jurassic Rogue Valley and Smith River terranes within the Western Jurassic belt.

Similar to rocks of the Galice Formation, but 10 to 15 million years older, the Condrey Mountain dome extends from 30 miles west of Yreka, California, into Oregon. At the core of this mass, the Condrey Mountain schist forms a drop-shaped exposure bent upward. Tightly folded greenschist grading inward to blueschist, which make up Condrey Mountain, are exposed within an erosional window through the Rattlesnake Creek terrane. This window was produced when erosion stripped off overlying sediments.

Course-grained metamorphic rock of the Briggs Creek terrane forms a long structural belt beneath the Rogue Formation, while the Dry Butte terrane, is a Late Jurassic batholith known as the Chetco Complex.

Because it is well to the west of the Western Klamath belt, the Elk terrane has been called an erosional outlier. However, thick sequences of submarine slides of the Galice Formation found in the Elk, place it in the Western Klamath belt. The Elk has been interpreted as a tectonic slice displaced over 100 miles northward from California by faulting.

THE SNOW CAMP, PICKETT PEAK, YOLLA BOLLY, GOLD BEACH, AND SIXES RIVER TER-RANES are located on the southwest Oregon coast. This diverse group west of the Western Jurassic belt are primarily tectonic slices or pieces of larger terranes that

have been moved by faults. Some of these small fragments show evidence of being transported from sites in California to their present locations in southwest Oregon during Early Tertiary time.

Comparison with rock sequences in California indicates the Snow Camp terrane may have been shifted as much as 180 miles northward from the Sacramento Valley to southwest Oregon. Forming the base of the Snow Camp, the Coast Range ophiolite is covered by sands, silts, and conglomerates of the Riddle and Days Creek formations, which eroded from adjacent terranes. The Coast Range and Josephine ophiolites are the same age.

The highly folded Colebrooke schist within the Pickett Peak terrane in Oregon is similar in age and make-up to rocks of the South Fork Mountain schist in Trinity County, California, as well as to the Shuksan terrane of the North Cascades of Washington. Volcanic sediments along with shales and cherts in the Jurassic Colebrooke have been altered by high pressure but low temperatures to schists.

The Yolla Bolly terrane, like the Pickett Peak, contains blueschist, although much of the rock here is topped by a thick sequence of Upper Jurassic to Lower Cretaceous Dothan greywackes and turbidite sands mixed with deepwater shales and cherts. Possibly originating within a trench or fore-arc basin, Dothan sediments were derived from both a continental source on

Figure 5.14 The Rogue River, named after troublesome Indians, has its headwaters near Crater Lake in the Oregon Cascades, but most of the river bed has been eroded into 150 million-year old rocks of the Rogue Valley terrane (Oregon Dept. Geology and Mineral Industries).

the east and a volcanic island archipelago to the west. In California these sediments are equivalent to parts of the Franciscan complex.

Along the southern Oregon coast, melange rocks of the Upper Jurassic Otter Point Formation in the Gold Beach terrane are covered by sand and silt of the Cretaceous Houstenaden Creek, Cape Sebastian, and Hunters Cove formations. Island arc volcanic rocks and sediments of the Otter Point closely resemble parts of the Franciscan sequences in California, and at least 80 miles of northward displacement by faulting has been suggested for this terrane.

Northeast of Cape Blanco large blocks of blueschist and Whitsett limestone are imbedded in a heterogenous sequence of Jurassic and Cretaceous mud and sand of the Sixes River terrane. As with other terranes in the coastal region of the Klamaths, the Sixes has been identified as a probable tectonic sliver displaced to the north by faulting.

MESOZOIC BATHOLITH intrusions have thoroughly invaded tectonic slabs making up the Klamath

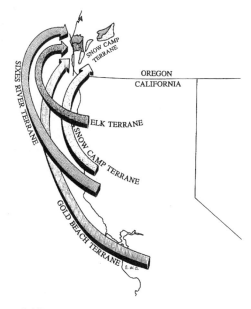

Figure 5.15 Several small terrane fragments were shifted northward by faulting along the coast from California to southwest Oregon (after Jayko and Blake, 1993).

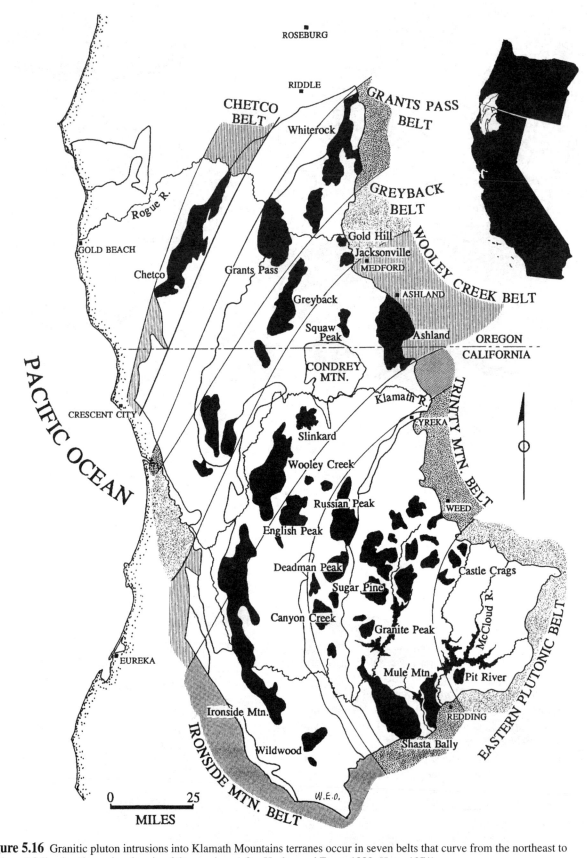

Figure 5.16 Granitic pluton intrusions into Klamath Mountains terranes occur in seven belts that curve from the northeast to southwest following the regional grain of the province (after Hacker and Ernst, 1993; Hotz, 1971).

Mountains. Emplaced from the Jurassic through Lower Cretaceous, between 170 and 130 million years ago, the intrusive bodies of plutons vary in size across the province. Arranged in seven arcuate zones, there is a significantly higher number of exposed batholiths in the older Eastern Klamath belt. This may simply be a function of age, and continued erosion will probably unearth additional intrusions on the west side of the province.

Intrusives are one of the end products of plate collision. As subduction takes place, the crust thickens while heat melts the rocks at depth. These liquids are either injected into the surrounding strata as batholiths or they make their way to the surface to be extruded as lava. When the subducting slab is frozen in place by accretion, a new subduction system immediately forms further oceanward, and the sites of melting and volcanism shift with it. The size of the crystallized batholith mass as well as the style and volume of volcanism depends on how fast or long subduction proceeded before accretion occurred. After terrane accretion, melting and volcanic activity may be briefly limited until magma from the new slab makes its way into the upper mantle.

Plutons as batholiths, at the core of many of the higher peaks of the California and Oregon Klamaths, range in size from less than one mile across to those exceeding 100 square miles. The largest of these bodies are the Chetco Complex in Josephine County, Oregon, and the Wooley Creek, Ironside Mountain, and Shasta Bally in California. One of the longest is Ironside Mountain, extending for 37 miles from Orleans Mountain in Humbolt County to Hayfork Creek in Trinity County.

At Castle Crags State Park in Siskiyou County, sharp pinnacles and crenellated towers have been carved into the deeply eroded Castle Crags granitic batholith, one of a score of intrusions into the Trinity ophiolite.

Contrasting with the surrounding landscape, this scenic feature has been fashioned by joints in the rock, and a hedge-like zone of quartz-rich dikes form resistant ridges within the more easily eroded granite. Sculpting by ice within the past 100,000 years polished and sharpened the Crags.

IN CRETACEOUS TIME the widespread seaways across the Klamaths were in concert with the global trend of rising sea levels. Shorelines advancing eastward over northern California and Oregon throughout this interval can be traced from the Ochoco depression of northcentral Oregon south into the Sacramento Valley. The similarity of fossils and rocks over this vast area indicates a continuous marine basin where much of the sediment was provided by erosion of the uplifted Klamath highlands. As the new seaway formed, a thick blanket of Cretaceous strata obscured rocks of the older terranes.

Within the Cretaceous basin of northern California and Oregon, up to 4000 feet of Hornbrook Formation deposits grade upward from early stream and estuarine environments to a deep open ocean setting. Beach sand with snails and clams is succeeded by deepwater sediments containing coiled ammonites. In the Ashland area resistant sandstone of the Hornbrook Formation forms distinctive hummocky ridges across the landscape.

THE RECORD OF CENOZOIC history in the Klamaths is poor because of intensive weathering and erosion. Once the area was uplifted and erosion accelerated, the Cretaceous seas retreated, and the present shoreline was established. The limited Tertiary deposits that survived these events are found around the margins of the province. East of Port Orford and along the Rogue River

Figure 5.17 The spires of Castle Crags, south of Weed, California, were a mass of molten rock, which crystallized to granite beneath the earth about 170 million years ago. Later erosion produced the sharp spines, which rise over 4000 feet above the valley floor (California Division of Mines and Geology).

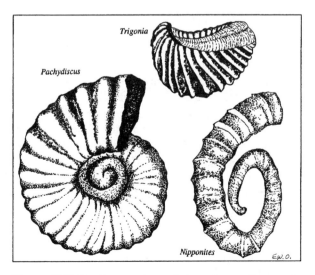

Figure 5.18 Fossil ammonites and clams from sediments of the Cretaceous Hornbook basin.

in Oregon, marine Eocene sediments of the Roseburg, Lookingglass, Flournoy, and Tyee overlap eroded Mesozoic terrane rocks of the Klamaths. To the southeast near Ashland, thick Late Eocene deposits of the Payne Cliffs Formation record an extensive braided river system. Choked with ash from the youthful Cascades, streams here drained northward out of the Klamaths. Silt of the Colestin Formation atop the Payne Cliffs is also derived from early Cascade volcanoes and includes tropical leaves and wood. Fossil plant remains and coal beds of the Weaverville Formation in Shasta County, California, occur in Oligocene lake, swamp, and stream deposits that spread out over a vast level floodplain. Swamp cypress and tupelo gum, found today along the Gulf Coast, were the most abundant trees. Coal was mined briefly from the Weaverville, while stream gravels yielded placer gold.

DURING THE COLDEST SPAN OF THE PLEISTOCENE most of the Klamath range escaped severe glaciation. In the Trinity Alps of northern California several glacial episodes took place from 15,000 to 10,000 years ago when at least 30 small glaciers occupied the valleys. Alpine glaciers above 5000 feet advanced along streams to incise a sharp landscape and leave irregular mounds of gravel, boulders, and sand in the lower reaches of the canyons. Along the Trinity River most of these glacial deposits were later examined for gold. Mining operations turned up fossil freshwater mussel shells mixed in with bones of deer, horses, mammoth, mastodon, and sloths. Today small glaciers are sustained only on the shaded north side of some of the higher ridgelines.

THE KLAMATH COASTLINE is notable for its myriad of sea stacks, islands, and rocky headlands. Most are erosionally resistant knobs of Jurassic and Cretaceous rocks left as remnants offshore. Many long chains of rocks or reefs appear at the surface of the water where a hard ridge of strata once stood. Today some areas of the shorelines are being eroded at the astonishing rate of up to 2 feet per year.

Along the coastal region from Port Orford, Oregon, southward, prominences at Sisters Rocks, Crook Point, and Cape Ferrelo are braced by metamorphic rocks of the Jurassic Otter Point Formation. The dome-like monolith of Humbug Mountain and the cliffs at Cape Sebastian that enclose Hunters Cove are Cretaceous sedimentary rock. Coarse boulders of the Cretaceous Humbug Mountain Conglomerate grade upward into sandstone, making Humbug Mountain the highest point on the southern coast. Landslides are common in this conglomerate, and large tracts of loose rock and soil are indications of a coastline in constant motion. The curving promontory at Cape Sebastian is one of the few exposures of Late Cretaceous rocks on this stretch of the

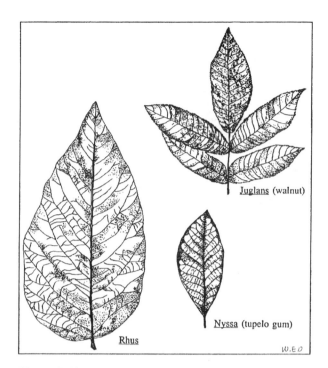

Figure 5.19 The Oligocene flora near Weaverville, California, reflects a more temperate climate than that of the earlier Eocene interval.

Figure 5.20 Sawtooth Ridge in the background and Thompson Peak (center) in the Salmon-Trinity Alps of California show unmistakable signs of glaciation as U-shaped valleys and small lakes (California Division of Mines and Geology).

Figure 5.21 Sea stacks at Cape Sebastian on the Oregon coast are remnants of the former land, which has been eroded by the ocean (Oregon Dept. Geology and Mineral Industries).

Figure 5.22 Offshore stacks and submerged rocks of the Otter Point Formation in a belt along the coastline of southern Oregon and northern California are part of an extensive reef system that includes Mack Reef near Crook Point (courtesy E. Baldwin).

Figure 5.23 At Port Orford, Oregon, Battle Rock is an historic landmark that provided a refuge for a party of enterprising businessmen who were besieged by Indians in June, 1851. In the ensuing battle seventeen Indians were killed, and the settlers escaped after several days. Projecting well above water, the rock is Mesozoic basalt (Oregon Dept. Geology and Mineral Industries).

coast, where massive 900-foot thick Cape Sebastian and Hunters Cove sandstone and siltstone predominate.

South of Brookings near the California border, notable stacks, as 100-foot high Hastings Rock, McVay Rock, and Goat Island, the largest on the Oregon coast, are all cut from dark sandstone, siltstone, and volcanic rocks of the Jurassic to Early Cretaceous Dothan Formation. As part of the same formation, the numerous stacks and submerged islands in this vicinity form a continuous system that includes Mack Reef, Rogue River, Port Orford, and Cape Blanco reefs. Most of these have been dedicated as wildlife refuges.

Over the border in California, sea stacks like Hunter Rock, Prince Island, and those clustered at Point St. George are Jurassic to Cretaceous Franciscan Formation. Lakes Earl and adjoining Talawa north of Crescent City, now trapped by high sand dunes, were open to the ocean but bordered by long spits when pioneer geologist Joseph Diller passed through on a trip across the coast range in 1889.

Figure 5.24 On the southern Oregon coast, Humbug Mountain is a familiar sight. This promontory, composed of Cretaceous sandstone and conglomerate, is subject to frequent slides (Oregon State Highway Dept.).

Figure 5.25 Working its way through limestone to carve out a cavern, groundwater often festoons the cave with dripstone creations that defy imagination. These cave formations can be found at Oregon Caves National Monument in the Siskiyou Mountains (Oregon State Highway Dept.).

AS WATER WORKS ITS WAY underground through limestone layers and lenses, extensive caverns and passageways with intricate shapes and patterns slowly evolve. When limestone or marble is bent and folded, it breaks leaving small cracks and fissures throughout the layers. The openings act as avenues for the passage of groundwater dissolving the calcium carbonate to widen the fractures into passageways. Over time, these may be enlarged into caves, and continued percolation of water through the underground caverns deposits a beautiful veneer of calcite on the ceiling, walls, and floor as dripstone. Elaborate icicle-like stalactites and stalagmites on the ceilings and floors are precipitated by calcite-saturated waters moving downward through the cave. Underground streams also create subterranean standing pools and a network of flowing water within cave systems.

The slow percolation of groundwater through 190 million-year old Triassic limestones within the Applegate Group of the Hayfork terrane widened cracks and fractures to create the elaborate passages of the Oregon Caves National Monument. Located about 20 miles southeast of Cave Junction in Josephine County, the Monument is actually one cave, 1600 feet long, containing graceful shapes from calcium carbonate deposits. First discovered in 1874 by a deer hunter, the Oregon Caves Monument is now publicly operated.

Between the McCloud and Pit rivers in California a number of caverns, widened within the lower Permian McCloud Limestone, contained Pleistocene fossils. Over half of the 50 species identified were of extinct mammals, including large bears, lion-like cats, and wolves. Other fragmentary bones represent mountain goats, deer, camels, ground sloths, and mastodons. Most of these caves were later flooded by water of Lake Shasta. One of these remaining, Lake Shasta Caverns, also in McCloud Limestone, is a privately operated facility that displays multicolored columns, white dripstones, and draped creations throughout. In 1878 the first recorded white explorer was James Richardson, a federal fisheries employee, who wrote his name on the cavern wall with carbon from his miners lantern.

MINERAL PRODUCTION

Economic mineral deposits were emplaced in rocks of the Klamath Mountains at several stages during their history. Gold and silver, copper, zinc, lead, chromium, iron, nickel, platinum, and manganese are among the metallic resources present here.

Major riverways in the province yielded placer gold in Tertiary and Quaternary gravels, whereas lode gold and other minerals are found in massive sulfides deposited during the formation of ocean crust as well as in association with plutonic intrusions. Slabs of oceanic crust or ophiolites were enriched with minerals long before being raised above sea level by tectonic plate movement. During the sea-floor spreading process, seawater circulated along faults and through pillow lavas near ridges where it was heated, dissolving minerals. Percolating onto the sea floor as hot springs, the cooling water, in turn, precipitated the minerals as sulfides. Iron, copper, cobalt, zinc, gold, and silver accumulated in this way near the tops of ophiolitic sequences, while chromium and nickel were associated with rocks much deeper in the crust.

Granitic bodies or plutons are also important for the role they play in the emplacement of economic minerals. Once magma has intruded into strata, it hardens forming mineral crystals in successive stages. In the early stage, high temperature minerals begin to crystallize, making the magma increasingly slushy. Toward the end of the

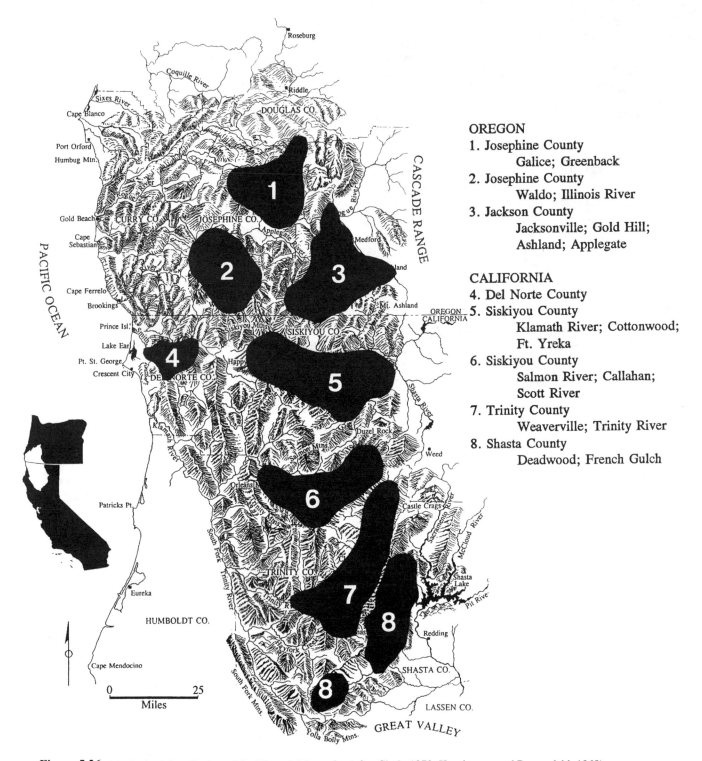

Figure 5.26 Principal mining districts of the Klamath Mountains (after Clark, 1970; Koschmann and Bergendahl, 1968).

OREGON
1. Josephine County
        Galice; Greenback
2. Josephine County
        Waldo; Illinois River
3. Jackson County
        Jacksonville; Gold Hill;
        Ashland; Applegate

CALIFORNIA
4. Del Norte County
5. Siskiyou County
        Klamath River; Cottonwood;
        Ft. Yreka
6. Siskiyou County
        Salmon River; Callahan;
        Scott River
7. Trinity County
        Weaverville; Trinity River
8. Shasta County
        Deadwood; French Gulch

cooling process, the pluton is somewhat like a sponge with a rigid framework around pockets of superheated water saturated with dissolved minerals. As the entire mass solidified, these fluids are driven to the outer margins of the pluton where they invade and crystallize in the enclosing rock as gold-bearing dikes and veins.

GOLD HAS ALWAYS HAD A REMARKABLE ABILITY to lure people into vast untamed areas, and the remote Klamaths were no exception. In 1848 placer gold was discovered by Major Pierson Reading in Pleistocene gravels on Clear Creek near Whiskeytown in Shasta County, California. That event drew miners to

HYDRAULIC MINING.

Figure 5.27 Using powerful jets of water, river gravels, impregnated with gold, were washed out of cliffs, much to the detriment of the environment (Idaho State Historical Society).

explore terraces on the nearby Trinity, Salmon, Scott, and Klamath rivers in northern California and on the Illinois and Rogue rivers of Oregon. By the late 1850s, prospectors were dispersed along waterways from the Trinity to the Rogue. Rich gold finds on many sandbars of the Salmon and Klamath rivers yielded as much as $1.00 to $3.00 a pan. Sometimes the profit amounted to $40.00 a day. The ever-present Chinese did much of the rough mining work, transporting goods and cutting lumber at a wage of $1.50 per day plus food. Shopkeepers did a booming business in supplying miners, and a shop might take in $1000 daily.

After placer gravels had been thoroughly exploited, miners traced the gold back to its lode sources. Vein gold soon gave out, and miners who came in the early 1850s were all but gone by 1856, even though many districts continued to yield smaller amounts from both placers and lode veins well into the 20th century.

Over 75% of Klamath gold was provided by placers. Of the three main gold-producing areas of California in Shasta, Trinity, and Siskiyou counties, twice as much has been extracted from placers as compared to lode, whereas the opposite was true of Jackson and Josephine counties in Oregon. In all, this province has produced close to 9,000,000 ounces of both placer and lode gold. Following the overall grain of the region, the principal gold regions lie in an arcuate strip extending from

Shasta and Trinity counties of California to Josephine and Jackson counties of Oregon.

IN NORTHERN CALIFORNIA, the French Gulch-Whiskeytown-Deadwood district, about 20 miles northwest of Redding in Shasta County, proved to be the most lucrative. The total gold output here was 2,400,000 ounces. Lode gold in this region was primarily from quartz veins of the Shasta Bally and Mule Mountain batholiths penetrating Paleozoic sedimentary and volcanic rocks of the Bragdon Formation. Miners quickly learned to recognize gold veins associated with the distinctive large "birdseye" crystals of the surrounding rocks.

To the west, the Weaverville district of Trinity County included both lode and placer mining, but the greatest yields were about 1,750,000 ounces from thick Pleistocene gravels of the Weaverville Formation on the Trinity River and its tributaries. The famous LaGrange hydraulic mine here, one of the largest in California, excavated 100 million cubic yards of gravel after beginning operation in 1851.

From gravel bars near Happy Camp and Yreka on the Salmon, Scott, and Klamath rivers in Siskiyou County, 1,775,000 ounces of placer gold far exceeded that of lode gold. Unusually extensive dredging occurred in Scott Valley near Callahan. Once rumors of rich sandbars circulated through the gold fields, thousands of

Figure 5.28 Quaternary placers between Sawyers Bar and Forks of Salmon in Siskiyou County, California, within the Salmon River drainage, were responsible for an estimated 15,000 ounces of gold before 1959 (Siskiyou County Historical Society).

Figure 5.29 Nearly two dozen gold mines dotted the Liberty area in Scott Valley, Siskiyou County, California. Although mining has proceeded here since the late 1800s, this region still has a high potential for future discoveries (California Division of Mines and Geology).

Figure 5.30 Most of the gold from Orofino in Siskiyou County, California, was extracted from river placers (Siskiyou County Historical Society).

Figure 5.31 As seen in this photo, mining in Jacksonville, Oregon, even involved city streets. Downtown on C Street, $40,000 in gold was recovered. Today historic Jacksonville has been preserved (Oregon Dept. Geology and Mineral Industries).

miners flocked to Happy Camp during the middle 1800s, a community named following a particularly boisterous celebration. Even today small gold washing and dredging operations dot the local streams, and renewed commercial exploration was begun in western Siskiyou County during the 1990s.

Placers on the Smith River in Del Norte County, which were active until 1959, produced small amounts of gold. Many of these claims on streams and rivers are worked today on a weekend basis.

IN THE OREGON MINES, lode gold in Josephine County surpassed that of placer gold in Jackson. Following the initial discovery on Josephine Creek in 1852, gold in stream beds was depleted shortly after the middle 1800s leading to the search for lode sources. Most of the 1,234,000 ounces of lode gold recovered here came from the Galice and Greenback districts. Placer gold also played an important role in the Greenback area. From 1935 to 1938 the largest dredge in the history of Josephine County worked 115 acres of gravel in the Greenback district, halting only when the enormous pile of loose spoils made fresh gravel inaccessible. Gold from these districts occurred in volcanic rocks of the Applegate and Galice formations where it is associated with the Josephine ophiolite.

In southern Josephine County, placer gold was the mainstay of the Waldo district. Beginning with modest sluices on rivers and creeks in the 1850s, expansion to large-scale hydraulic works at the beginning of the 1900s revived production. Powerful streams of water were directed at old gravel terraces washing out 218,000 ounces in gold northwest of Takilma until the resource was depleted.

The discovery of gold along Jackson Creek in 1851 brought miners into this southern Oregon county where placer gold from the Applegate and Gold Hill districts far outstripped that of lode gold from the Ashland mine. About one-half of the estimated 500,000 ounces of placer gold from Jackson County originated in the Applegate district. Placers, first discovered in 1853 on Forest Creek, led to the examination and exploitation of streams throughout the area. Of these, the Sterling mine was the most successful, yielding $3 million. Within the Gold Hill-Jacksonville district, the origin of gold in discrete near-the-surface pockets is not well understood but may relate to extensive leaching of sediments by acidic groundwater. Here several lucky gold seekers uncovered small rich hordes of gold

Figure 5.32 A bucket-line dredge, shown here on Grave Creek, Oregon, was in operation from 1935 to 1939 (Oregon Dept. of Geology and Mineral Industries).

worth up to $3 million. The Ashland district is well known for the Ashland vein, worked from 1886 to 1939 along a mile-long underground tunnel. The Jurassic Ashland pluton intruded rocks of the Triassic Applegate Group to precipitate gold here.

COPPER ALONG WITH ZINC, lead, and gold is one of the dominant mineral extracts from massive sulfide deposits near the Oregon border at Waldo and Takilma and northwest of Redding, California. Most of the copper produced in Oregon was from the Queen of Bronze mine opened in 1860. The second largest copper producer in Oregon behind the Iron Dyke mine in the Blue Mountains, over 25,000 tons of ore was mined and smelted at the Queen of Bronze between 1903 and 1915. The Turner-Albright sulfide deposit, part of the ocean crust sequence of the Josephine ophiolite, supplied the copper and gold. During the 1990s there has been increased interest in exploring this deposit.

In California, massive sulfide deposits in volcanic rocks from the West Shasta and East Shasta copper-zinc region were mined from the Devonian Balaklala and the Triassic Bully Hill rhyolites and marine sediments of the Pit Formation. Over 90% of the copper, zinc, and lead from this area was from the West Shasta district, totaling 340,000 tons of copper, 30,000 tons zinc, and smaller amounts of lead. Currently operations have ceased at both districts.

CHROMITE DEPOSITS are spread throughout the Klamath region, but, unlike copper that is precipitated in the upper ophiolite layers, chromite occurs in isolated pods near the base of ocean crust sequences. Production of chromite in the United States only took place during both World Wars when metals were difficult to obtain and in the 1950s when the government purchased it for a national strategic reserve. Numerous small mines opened in this province, but the largest were in Del

Figure 5.33 The Hanna Nickel Mine near Riddle, Oregon, when in operation, removes nickel ore from ophiolite rocks, which have been extensively weathered (Oregon Dept. Geology and Mineral Industries).

Norte County, California, where close to 50 pods yielded up to 20,000 tons of ore each. In Josephine County, Oregon, a total of 50,000 tons of ore was mined up to 1958. Sources for both deposits were rocks of the Josephine ophiolite.

OPERATING SPORADICALLY in recent years, the only extant nickel mine in the United States is located at Nickel Mountain just northwest of Riddle in Douglas County, Oregon. Nickel is found as an insoluble residue after intensive weathering of ophiolitic rocks, and mining companies have extracted the pale green ore by cutting parallel benches around the sides of the mountain. The quality of the ore is only moderate, and often less than half of the yield is acceptable for further processing. Extensive undeveloped deposits also occur on the eroded Klamath plateau where the Josephine ophiolite is found.

ADDITIONAL READINGS

Aune, Q.A., 1970

Bailey, E.H., ed., 1966

Brooks, H.C., and Ramp, L., 1968

Burchfiel, B.C., Lipman, P.W., and Zoback, M.L., eds., 1992

Burnett, J.L., 1990

Charvet, J., et al., 1990

Clark, W.B., 1970

Diller, J.S., 1914

Diller, J.S., et al., 1915

Hacker, B.R., and Ernst, W.G., 1993

Harper, G.D., 1984

Harper, G.D., Grady, K., and Wakabayashi J., 1990

Harwood, D.S., and Miller, M.M., eds., 1990

Hotz, P E., 1971

Howell. D.G., ed., 1985

Hutchinson, R.W., and Albers, J.P., 1992

Irwin, W.P., 1966

Jayko, A.S., 1990

Jayko, A.S., and Blake, M.C., 1993

Koschmann, A.H., and Bergendahl, M.H., 1968

Lindsley-Griffin, N., et al., 1993

Lund, E.H., 1973

Mankinen, E.A., and Irwin, W.P., 1990

McGinitie, H.D., 1937

Miller, E.L., et. al., 1992

Miller, M.M., and Wright, J.E., 1987

Murchey, B.L., and Blake, M.C., 1993

Nilsen, T.H., 1984

Norris, R.M., and Webb, R.W., 1990

Orr, E.L., and Orr, W.N., 1999

Poole, F.G., 1992

Saleeby, J.B., and Harper, G.D., 1993

Silberman, M.L., and Danielson, J., 1993

Stevens, C.H., Yancey, T.E., and Hanger, R.A., 1990

Vennum, W., 1994

Wagner, D.L., 1988

Woods, M.C., 1988

6

BLUE MOUNTAINS

A view looking west at the South Fork of the Imnaha River, which drains the Wallowa Mountains of Oregon, clearly shows evidence of glacial activity (U.S. Forest Service).

Except for small exposures in the southeast corner of Washington and a narrow discontinuous strip from Lewiston to Riggins, Idaho, the Blue Mountains lie almost entirely within northeast Oregon. Covering over 50,000 square miles in a roughly triangular shape, the province is bounded on the south by the High Lava Plains and to the west by the Deschutes River. Lavas of the Columbia Plateau overlap most of the province on the north and northeast, while the Idaho batholith forms the eastern margin.

The Blue Mountains platform has been gently tilted to the west, resulting in the subdued western hills and open landscape of the Ochoco and Aldrich mountains, which rise to the 10,004-foot high glaciated Matterhorn Mountain of the Wallowas opposite the 9800-foot high Sacajawea Peak in the Seven Devils range.

Of the stream drainages in the Blue Mountains, the John Day is the most extensive with a watershed that begins in the Strawberry and Greenhorn mountains meandering west across the province before turning abruptly to the north near Clarno to enter the Columbia River. Flowing northward as well, the Grande Ronde and Imnaha rivers merge with the Snake, while the smaller Willow and Butter creeks and Umatilla and Walla Walla rivers enter the Columbia from the south.

Within Hells Canyon on the eastern margin of the province, the Snake River cuts an 8000-foot deep gorge separating the Wallowa Mountains of Oregon and the Seven Devils Mountains of Idaho. Measured from the summit of He Devil Peak in Idaho to the riverbed below, Hells Canyon is deeper than the Grand Canyon of the Colorado River. A total of 215,233 acres on both sides

of the river have been designated as a National Recreation area, and sections of the river have Wild and Scenic restrictions.

INTRODUCTION

In spite of its limited size, the Blue Mountains province displays more geologic diversity than other provinces of the Pacific Northwest. Perhaps this is because the Paleozoic and Mesozoic rocks here are something of a bridge between the Klamath Mountains accreted terranes to the south and those of Washington and British Columbia to the north.

The foundation of the province is a collage of imported oceanic terranes that were part of the Blue Mountains volcanic island chain lying to the west of the North American landmass. Carried along on moving tectonic plates, successive volcanic archipelagos collided with the edge of North America where individual pieces were annealed together. Upon making contact with the mainland, terrane rocks underwent folding and changes by temperature and pressure of metamorphism as well as intrusion by plutonic bodies.

Many of the rocks and fossils of these exotic terranes reflect tropical marine settings of the ancestral Pacific Ocean. Deep ocean waters, shallow shelf, and fringing coral reefs around the volcanic island chains supported a mixed fauna of large marine reptiles as well as molluscs, sponges, and sea urchins.

After the Late Mesozoic accretion of Blue Mountains terranes, ocean shorelines retreated far to the west. During the Cenozoic era, thick sequences of terrestrial sediments and a long volcanic history of lava and ash

Figure 6.1 In the foreground, Heavens Gate overlook provides a view across Hells Canyon to Seven Devils Mountains in Idaho as well as into Oregon, Washington, and Montana (Idaho Dept. of Commerce and Development).

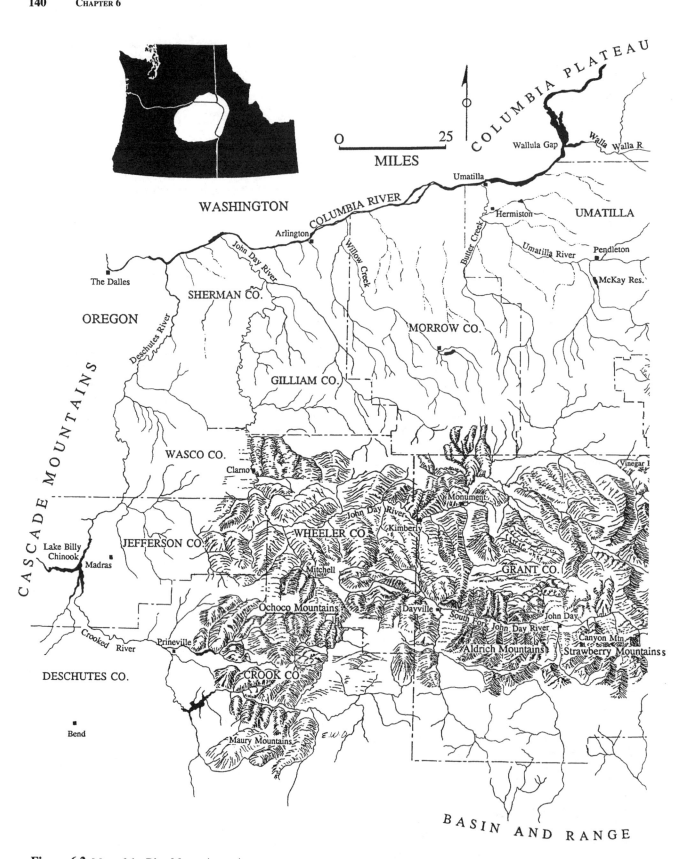

Figure 6.2 Map of the Blue Mountains region.

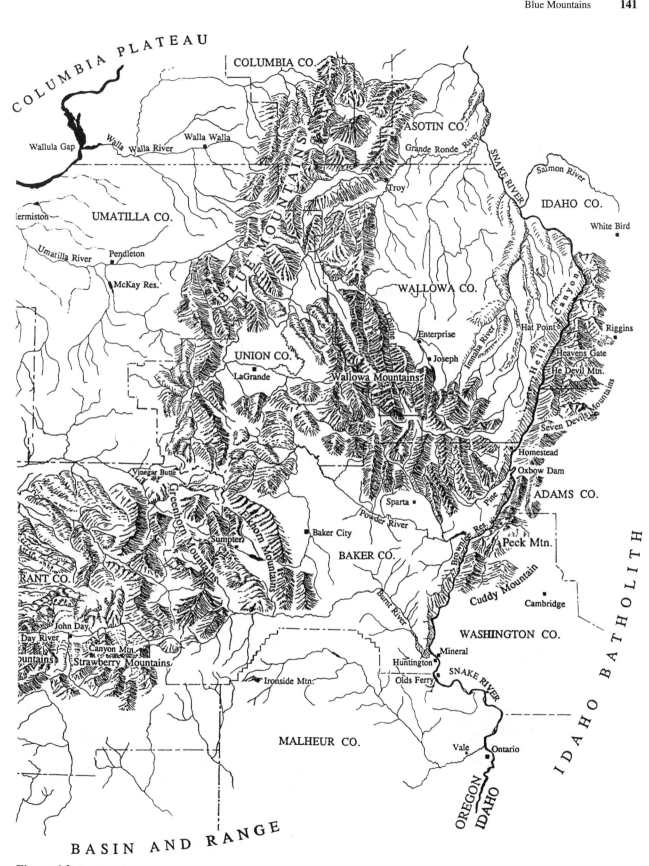

Figure 6.2 Continued.

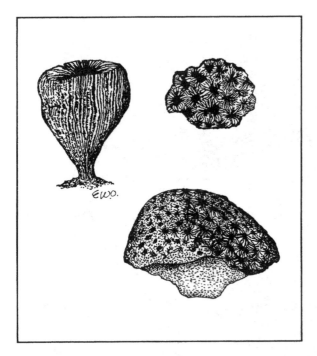

Figure 6.3 A variety of fossil corals from reef environments of the Blue Mountains volcanic island archipelagos.

flows, spanning 50 million years, added to the rich geologic history here. Streams that steadily eroded the highlands collected volcanic detritus, which was deposited as sediments in basins throughout the province. The great diversity of tropical and temperate plants and mammals entombed in these basins are indicative of the varied past of the Blue Mountains.

GEOLOGY OF THE BLUE MOUNTAINS PROVINCE

In the Blue Mountains, exotic accreted terranes of the Late Paleozoic and Early Mesozoic date back as much as 400 million years. At that time Pacific Northwest shorelines formed the margin of a vast seaway across British Columbia, eastern Washington, Oregon, and Idaho. Out in the proto-Pacific Ocean the volcanic chain of the Blue Mountains island archipelago rose from the deep sea floor like present-day Hawaii.

Volcanic activity from this chain continued intermittently for over 100 million years from the Permian period through the Triassic and Jurassic. During this in-

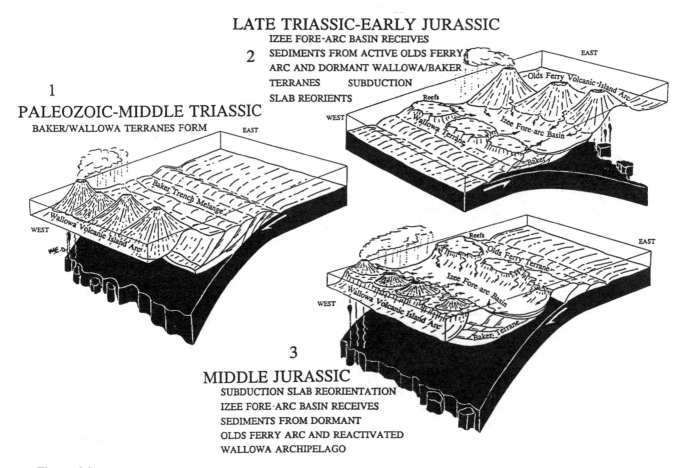

LATE TRIASSIC-EARLY JURASSIC

2 IZEE FORE-ARC BASIN RECEIVES SEDIMENTS FROM ACTIVE OLDS FERRY ARC AND DORMANT WALLOWA/BAKER TERRANES   SUBDUCTION SLAB REORIENTS

1
PALEOZOIC-MIDDLE TRIASSIC
BAKER/WALLOWA TERRANES FORM

3
MIDDLE JURASSIC
SUBDUCTION SLAB REORIENTATION
IZEE FORE-ARC BASIN RECEIVES
SEDIMENTS FROM DORMANT
OLDS FERRY ARC AND REACTIVATED
WALLOWA ARCHIPELAGO

Figure 6.4 Model for evolution of the Blue Mountains volcanic arc (after White, et al., 1992).

terval the volcanic arc was carried eastward atop a moving tectonic plate eventually colliding with the western edge of the North American landmass. Here rocks of the archipelago were swept up and accreted or welded to the mainland to become the foundation of the Blue Mountains province.

During accretion, terranes were compressed to half their original width generating folds and faults on a grand scale. Intense pressure and elevated temperatures from this process metamorphically altered and partly melted the rocks so that many are almost unrecognizable today. To gain an understanding of the original size and shapes of the terranes it is necessary to stretch and restore them graphically to their precollision shape.

In the long interval of time when the volcanic islands were forming, a multitude of environmental settings evolved in and around them. Each of these environments left a decipherable record of volcanic and fossiliferous sedimentary rocks. Within the Blue Mountains, belts of these rock formations and fossils have been traced across the province, helping in the reconstruction of ancient environments.

TERRANE ROCKS ARE DATED primarily by using fossils so small that a microscope is necessary to study them. Although microfossils are ordinarily difficult to see in a rock with the naked eye, they have the advantage of being abundant in comparison to larger invertebrate fossils such as corals and molluscs. A small chip of limestone or chert, for example, may contain hundreds or even thousands of microfossils.

Three major groups of microfossils—fusulinids, conodonts, and radiolaria—are traditionally employed in dating exotic accreted terranes. Fusulinids, single-celled animals with a calcareous shell about the size and shape of a grain of rice, have a fusiform coiled shell. As they grow, the shells of these bottom-dwelling marine animals coil or enroll like a tamale. To be studied adequately, fusulinids must be sliced along two axes and examined in very thin pieces. Known from Late Paleozoic, Mississippian, Pennsylvanian, and Permian rocks, fusulinids have been used to date parts of the Grindstone and Baker terranes in the Blue Mountains.

Conodonts are very small tooth-like microfossils that formed the jaw elements of fish-like unknown organisms. These fossils of calcium phosphate are particularly useful since their composition resists fluids that frequently dissolve and destroy fusulinids. Conodonts are important in dating Triassic and older rocks.

Like fusulinids, radiolaria are single-celled protozoa that construct glassy skeletons of opaline silica and live as plankton suspended in open ocean currents. Their abundance in rocks of the Devonian through Cretaceous periods makes them extremely significant in determining the age and environments of ocean settings from all of the terranes.

INDIVIDUAL TERRANES of distinctive rocks run in crescent-shaped curves from the southwest to northeast across the province. By grouping related rock formations together, five characteristic belts have been identified. These strips are separated by faults and overlap with

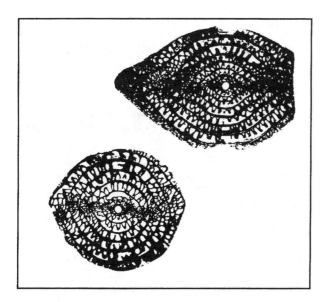

Figure 6.5 Distinctive Permian fusulinid foraminifera from the Coyote Butte Formation within the Grindstone terrane reflect a western Pacific origin (Oregon Dept. of Geology and Mineral Industries).

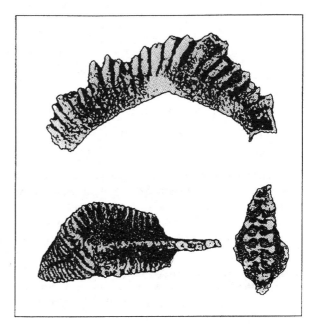

Figure 6.6 Conodonts, belonging to a primitive extinct fish-like organism, are a major aid in age-dating older terrane rocks.

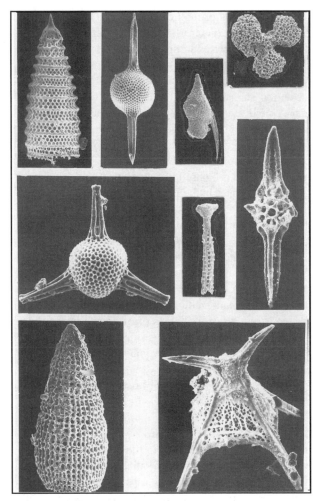

Figure 6.7 Glassy skeletons of Permian and Mesozoic radiolaria are common in most terranes of the Blue Mountains (courtesy E. Pessagno and C. D. Blome).

During two separate episodes of the Permian and Triassic periods, volcanic islands erupted lava and ash into surrounding ocean basins. Of these, the most extensive are the Seven Devils Group, representing flows and slides of volcanic debris into adjacent lowlands. Occasional limestone reefs around the volcanic platforms supported crinoids (sea lilies), corals, and molluscs.

As subsidence occurred, limestones and shales of the Martin Bridge and Hurwal formations accumulated atop the volcanic material during the Late Triassic and Early Jurassic. Fossils from these rocks include ammonites, corals, molluscs, crinoids, sea urchins, and sponges that lived in a variety of tropical ocean conditions between 200 and 150 million years ago. Dense masses of corals and molluscs found in the Martin Bridge Formation of Hells Canyon and the Wallowa Mountains are evidence of reefs that grew as fringing atolls around shallow volcanic platforms. Remains of ichthyosaurs, streamlined fish-like reptiles that inhabited deeper ocean waters, have been recovered from these rocks.

After uplifting and shallowing of the volcanic islands, conglomerate and sandstone of the Jurassic Coon Hollow Formation were deposited in nearshore deltas by braided streams and as alluvial fans of a deeper basin. Seed ferns, cycads, primitive conifers, horsetails, and ginkgos of the Coon Hollow along the Snake River are similar to a Mesozoic flora that spread across the Pacific Northwest to Alaska, Asia, Europe, Africa, and Australia.

each other in time. Representing specific environments and episodes of time within the history of the Blue Mountains archipelago, these belts of exotic terranes include the Wallowa, Baker, Olds Ferry, Izee, and Grindstone.

THE WALLOWA TERRANE is the largest remnant of the ancestral Blue Mountains volcanic arc. This package of rocks in Oregon takes in all of the Wallowa Mountains, the northern half of the Elkhorn Range, and a small piece of the Greenhorn Mountains, as well as isolated exposures south of Pendleton. A thick continuous sequence of Wallowa terrane volcanics is displayed along the steep walls of Hells Canyon of the Snake River, in the Seven Devils Mountains of western Idaho, and in limited areas within southeast Washington. Lava, ash, and sedimentary rocks of the Wallowa terrane are strikingly similar to those of a widely scattered terrane called Wrangellia found in Alaska and on Queen Charlotte and Vancouver islands in Canada.

Figure 6.8 Ferns, quillworts, and tall ginkgo trees of the Coon Hollow Formation, exposed along the Snake River, lived during the warmer Jurassic interval.

Figure 6.9 Five separate terranes make up the Blue Mountains province (after Orr and Orr, 1999).

Figure 6.10 Treacherous switchbacks (at the top of the photo) on White Bird Hill, Idaho, traverse rocks of the Wallowa terrane. Here Chief Joseph and his handful of Nez Perce Indians defeated U.S. Army troops in battle on June 17, 1877 (Idaho Dept. of Commerce and Development).

LYING IN A TAPERING BELT BETWEEN THE OLDS FERRY and Wallowa terranes, rocks of the Baker terrane occur in the Snake River Canyon near Brownlee Dam as well as between Riggins and White Bird, Idaho. West of this point, the Baker terrane spreads out like a fan to include large tracts of the southern Greenhorn Mountains and virtually all of the Elkhorn Mountains of Oregon. Limited exposures of Baker terrane rocks are displayed in erosional windows cut through younger Tertiary sediments and lava flows in the divide between watersheds of the Deschutes and John Day rivers. Rocks and fossils of the Baker terrane are remarkably similar to those of the Cache Creek terrane, stretching in a broken belt for over 1000 miles across British Columbia.

Within the Blue Mountains island arc setting, Baker terrane rocks formed offshore along a subduction zone. In a strip near the suture between colliding tectonic plates, sediments piled up against the leading edge of the continent in what is called a melange or mixture. Here deep-sea deposits from the ocean floor, pillow basalt, lower crust, and upper mantle rocks were converted by heat and pressure of metamorphism into greenstone and serpentine.

Major groups that make up this terrane are the Burnt River Schist, the Elkhorn Ridge Argillites (shales), and the Canyon Mountain complex. Each of these is itself a melange of a great variety of rocks imbedded within the layers of the terrane. The high pressure and temperatures produced in the subduction process converted parts of the terrane to metamorphic rocks, and the Burnt River Schist may be a thoroughly cooked portion of the Elkhorn Ridge Argillite.

Broken, shattered rocks of the Canyon Mountain complex, located south of John Day in Grant County, and the Sparta complex, a broad area 20 miles east of Baker, have been identified as ophiolites. From the Greek word meaning "serpent," an ophiolite displays smooth slick surfaces of the rock serpentine that have been polished by faulting. An ophiolite is a predictable series of layers that make up ocean crust beginning with deep-sea clays, cherts, and pillow lavas extending as much as 3 or 4 miles below the sea floor into rocks of the upper mantle. Normally ophiolitic rocks are carried down and recycled within the mantle in the subduction process, but occasionally, as in the Canyon Mountain and Sparta areas, they are peeled up to be retained and preserved in the accretionary wedge. In the northeast corner of Grant County, the Greenhorn Mountains are so named because of a 300-foot high monolith of greenish serpentine on Vinegar Butte.

Figure 6.11 Brownlee dam, here under construction in the late 1950s, was built on an active fault through Baker terrane rocks. It is estimated that if the dam fails, the resulting flood would take out all structures downstream and cover Lewiston, Idaho, with 100 feet of water (Idaho State Historical Society).

Figure 6.12 In the deep Snake River canyon near Eagle Bar, Oregon, rocks of the Seven Devils volcanics are displayed (Oregon State Highway Dept.).

Ophiolites such as those in the Klamath Mountains and on the eastern Mediterranean island of Cyprus contain abundant economic mineral deposits, ranging from chrome near the base to iron, copper, and zinc near the top. Chromite deposits have been recognized within Canyon Mountain complex.

A mixture of tropical and temperate fossils as well as those of both North American and Asiatic affinities found in the shale and limestone fragments within the Baker terrane reflect the varied origins of this terrane. Corals, molluscs, crinoids (sea lilies), fish fragments, fusulinids, and conodonts range from Devonian, almost 400 million years old, through Jurassic, 200 million years, making the Baker terrane one of the oldest in the Blue Mountains arc.

ORIGINALLY KNOWN AS THE HUNTINGTON ARC, rocks of the Olds Ferry terrane lie in a curved east-west belt from Huntington, Oregon, to Mineral, Idaho. Although the Olds Ferry is composed of quartz-rich rhyolitic lavas in contrast to the basalts of the Wallowa terrane, the Olds Ferry may simply be a younger phase of the Wallowa volcanic arc. Ammonites and the flat clam, *Halobia,* scattered widely throughout the terrane, reflect a shift from tropical to cooler temperate conditions within the Huntington and Weatherby formations. Peck and Cuddy mountains in Washington County, Idaho, are remnants of the Olds Ferry terrane, which project through the later covering of Columbia River basalts.

The island arc rocks of the extensive Quesnellia terrane, extending over 1200 miles across British Columbia and into northcentral Washington, are similar in composition and age to those of the Olds Ferry.

ENVIRONMENTS OF THE IZEE TERRANE contrast with those of the Wallowa, Baker, and Olds Ferry. While the Wallowa and Olds Ferry terranes are the by-products of different episodes of the volcanic arc setting, and the Baker terrane reflects the trench and accretionary wedge, the Izee terrane represents fragments of a linear fore-arc basin that separated the volcanic chain from the offshore subduction zone. As the youngest of the five terranes, the Izee blankets the older rock assemblages.

Sediments making up the Izee were eroded then redeposited from the older volcanic rocks of the Blue Mountains complex. Formations within the Izee are particularly rich in fossils, tracing the gradual shallowing of the seaway in response to Mesozoic compression and folding. Belemnites, cephalopods, corals, clams, and snails from open ocean conditions in the Begg Formation are overlain by deposits of the Brisbois containing oyster-like clams from shallow shelf waters and corals that built up in nearshore reefs. The Rail Cabin shales in central Oregon contain pod-like concentrations of limestone representing offshore coral knolls, which supported dense populations of cephalopods and bottom-dwelling clams and snails.

Figure 6.13 Blocks of the Permian Coyote Butte limestone, within the Grindstone terrane, may be the products of a submarine slide (Oregon Dept. of Geology and Mineral Industries).

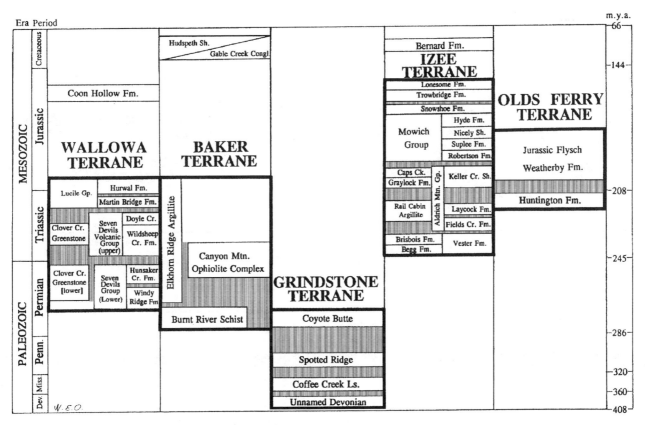

Figure 6.14 Pre-Cenozoic stratigraphy of the Blue Mountains (after Armentrout, et al., 1988; Vallier and Brooks, eds., 1986).

THE GRINDSTONE TERRANE, barely adhering to the western edge of the Izee terrane, is the smallest of the five. Since rocks and fossils of this peculiar terrane are similar to fragments of limestone and sand imbedded in the Baker terrane, the Grindstone is regarded as a coherent block of the Baker that was spared the destructive effects of the subduction zone.

Spanning the Devonian through Permian periods, the Grindstone is composed of separate massive blocks set in sand and mud. A recent interpretation suggests that these huge monoliths were shaken loose from an upper shelf environment to slide en masse into deeper water. Such submarine landslides of immense slabs are termed "olistostromes" or "olistoliths," meaning a mixture of rocks. Certainly, the jumbled appearance of these separate blocks imbedded in a shale matrix lend themselves to interpretation as a deep-sea slide.

The 200-foot thick unnamed Devonian limestones of the Grindstone terrane in Crook County, dated by fishtooth-like conodonts at 380 million years old, are the oldest in Oregon. Fossiliferous sediments of the succeeding Mississippian Coffee Creek limestone, the Pennsylvanian Spotted Ridge sandstone, and the Permian Coyote Butte limestones make up the remainder of this terrane. Fossil corals, molluscs, and crinoids from these formations suggest shallow ocean environments. Lush tropical fern and fern-like foliage from the Spotted Ridge was typical of warm and humid late Paleozoic climates that generated coal in the eastern United States.

MESOZOIC BATHOLITHS INTRUDED then slowly hardened and crystallized deep in terrane rocks with the collision and accretion of the exotic blocks to North America. As these chambers of liquid magma cooled, fluids from the melted rocks invaded surrounding layers to concentrate economic minerals in veins. During the Jurassic and Early Cretaceous periods, 160 to 120 million years ago, granitic batholith intrusives were emplaced into the northern part of the Wallowa Mountains, in the Elkhorn and Greenhorn ranges, and in several small areas south and east of Baker. Although batholiths typically crystallize several miles deep in the earth's crust, uplift, faulting, and glacial erosion have laid bare many of these granites today.

Batholiths vary in size, and in the Elkhorn Mountains the Bald Mountain batholith covers over 144 square miles. However, the Wallowa batholith at the heart of the Wallowa Mountains is the largest in Oregon at 324 square miles.

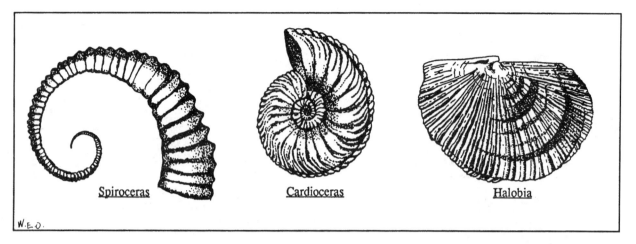

Figure 6.15 Ammonites and the flat clam, *Halobia,* from exotic terrane rocks.

THE CEONZOIC WAS USHERED in over 50 million years ago by rotation of the Blue Mountains block and eruptions of the Eocene Clarno volcanics. The shoreline had retreated westward beyond the present day Cascades, and the Cascade Mountains, Coast Range, and Willamette-Puget lowland had yet to form.

During the Tertiary, severe distortion and faulting of the exotic terranes in the Blue Mountains began as they were thrust up against North America. Continued movement of large tectonic plates rotated the Blue Mountains block clockwise almost at right angles and gently folded the strata within the province. The immense diagonal upfold of the Blue Mountains anticline runs northeast to southwest across the northern margin of the province exposing older rocks along its axis. Crossing the anticline from northwest to southeast, the Olympic-Wallowa lineament, a fault-related feature so large it can be seen on photos from space, projects from Seattle, Washington, to Payette, Idaho. Similar parallel fault systems all across the Pacific Northwest resulted from a shearing action by opposing motions of major tectonic plates.

AFTER A PERIOD OF INTENSIVE UP-LIFT AND EROSION, dozens of Early Tertiary volcanic vents across northcentral Oregon issued thick layers of andesitic and rhyolitic lava and ash of the Clarno Formation. When the ashfalls combined with water, a thick fluid mixture poured across the land as mudflows or lahars. Plugging stream valleys and smoothing topography, these volcanic muds provided perfect conditions to preserve plants and animals of the period.

Over 1000 feet thick in Wheeler County, the Clarno Formation contains a unique collection of wood, fruits,

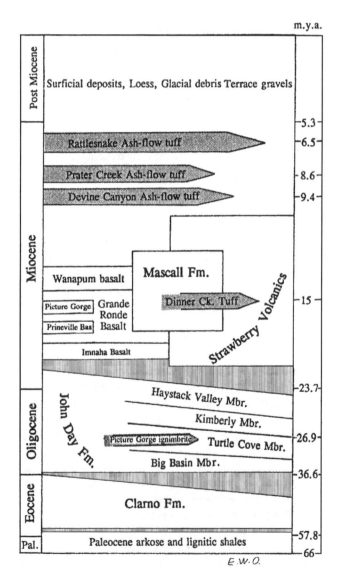

Figure 6.16 Tertiary stratigraphy of the Blue Mountains (after Armentrout, et al., 1983; Walker, 1990).

Figure 6.17 Lahars or volcanic mudflows of the Eocene Clarno Formation weather into characteristic forms in the arid climate of eastern Oregon (courtesy E. Bushby, Oregon Dept. of Geology and Mineral Industries).

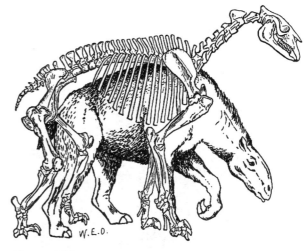

Figure 6.19 First cousin to the horse, the chalicothere (*Moropus*) reached a height of about 9 feet before it became extinct in the Late Miocene (after Markman, 1952).

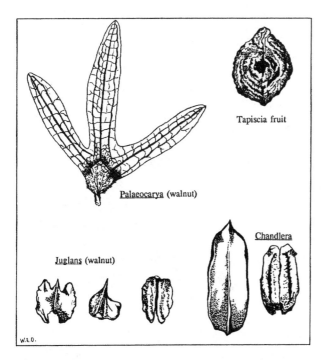

Figure 6.18 A rich and unusual collection of nuts, leaves, and animals of the Clarno Formation have been invaluable in reconstructing the environment of eastern Oregon 45 million years ago.

seeds, nuts, and leaves for which the formation is famous. Tropical Eocene palms, avocado, walnut, camphorwood, aralia, magnolia, sequoia, bald cypress, and even bananas grew in the warm wet Clarno climate of lakes and grassy plains. Large mammals such as titanotheres and brontotheres existed alongside miniature horses, camels, and *Moropus,* a muscular clawed horse-like creature. Graceful, plains-dwelling, running rhinoceros as well as a heavier aquatic species similar to a hippopotamus shared the environment with crocodiles.

Volcanic activity of the Clarno was followed by that of the John Day interval 36 million years ago when enormous clouds of rhyolitic ash and dust were blown over eastern Oregon from volcanoes near the present-day Cascade Mountain range. Ash-flows—glowing incandescent clouds or ignimbrites, which flowed down the sides of volcanic cones and across the landscape for hundreds of miles—cooled and annealed into glassy masses. The mantle of fine volcanic ash, collected by streams, was redeposited on alluvial plains and in lakes, providing an excellent medium to entomb and preserve plants and animals of this time.

The colorful John Day Formation, a succession of thick soils atop ash beds, is made up of 4 intervals, the lower reddish, Big Basin member with few fossils, a middle pale-green Turtle Cove member, rich in fossil plant and animal remains, and the upper buff and cream-colored Kimberly and Haystack Rock members, which also contain abundant fossils. Famous for its wide variety of mammal remains that include dogs, cats, rodents, the three-toed horse, and the sheep-like oreodon, John Day animals would have looked vaguely familiar today. The *Metasequoia,* or dawn redwood, is the most prominent plant from what was a temperate climate. Once

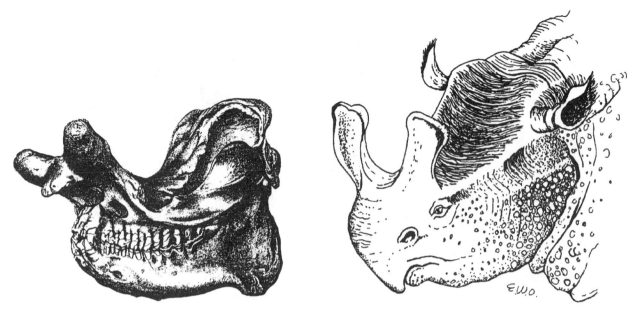

Figure 6.20 Ponderous titanothere was one of the enormous Eocene mammals to inhabit eastern Oregon (after Markman, 1952; skull from Osborn, 1929).

Figure 6.21 Sheep Rock, a monolith of John Day sediments capped by Miocene basalts, is one of the most familiar Tertiary scenes of eastern Oregon (Oregon Dept. of Geology and Mineral Industries).

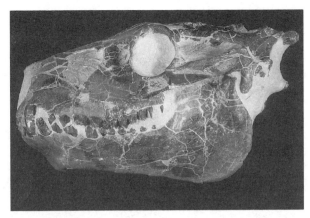

Figure 6.22 Small, pig-like oreodons were numerous in the Oligocene of the John Day region (photo courtesy of Condon Museum, University of Oregon).

Figure 6.23 *Metasequoia,* with a ropey bark and feathery leaves, was common in the Pacific Northwest forests of the Oligocene and Miocene (photo by G.E. Beacraft, U.S. Geological Survey).

thought to be extinct, a *Metasequoia* was found living in China during the 1930s by a Chinese Forester, Tsang Wang. Wang sent specimens to Elmer Merrill, thereby reintroducing the tree into North America.

Severe oxidation, weathering, and erosion of the soft iron-rich clay and ash layers of the Clarno and John Day formations have carved them into a colorful badlands topography of spires, columns, and mesas displayed in the 14,012 acres of the John Day Fossil Beds National Monument in Wheeler and Grant counties.

FLOWS OF MIOCENE COLUMBIA RIVER
BASALTS built a vast plateau of lava that covered the Blue Mountains. Exuded from fissures and broad low vents in northeast Oregon, nearby Washington, and Idaho, layer upon layer of these fluid basalts innundated and subdued previous erosional topography until only the tallest peaks projected above level black sheets of lava. Most of these basalts originated from dikes near Monument and Kimberly in Wheeler County and only travelled about 20 miles from this source. Eruptions lasted just a few weeks, even though a single flow might cover tens of thousands of square miles. Of the six separate lava formations, the Grande Ronde, Picture Gorge, and Imnaha basalts brought about the most drastic changes to physiography in the Blue Mountains.

Figure 6.24 View downstream in the Imnaha Canyon on the northeast flank of the Wallowa Mountains displays multiple flows of Miocene Imnaha basalts (courtesy U.S. Forest Service).

In Hells Canyon, the Snake River exposes layers of older Paleozoic terrane rocks, Mesozoic granites, and Tertiary Columbia River basalts. After the ancestral river and its tributaries had cut channels into the folded and sheared rocks of the exotic terranes and intruded granites, outpourings of Miocene Columbia River lavas overwhelmed the landscape. As the area was uplifted, the river, in turn, repeatedly incised a narrow channel through the basalts, exposing the dark-colored successive layers. Overlooks at Heavens Gate and Kinney Point in Idaho or Hat Point in Oregon provide spectacular views of the contrast between the younger basalts and older terranes.

In addition to basalts of the Columbia River Group, explosive episodes of andesitic volcanism in the Middle and Late Miocene spilled lava from vents aligned northeastward across Baker and Grant counties. Stiff flows of the Strawberry volcanics covered around 1500 square miles of eastern Oregon, ponding to a thickness of almost one mile at Ironside Mountain in Malheur County. Toward the end of the Miocene epoch, clouds of billowing ash from young Cascade volcanoes again blanketed the landscape. This fine residue was picked up and carried by streams to fill local depressions with deposits of the Mascall and Rattlesnake formations. The last major ash flow in the Blue Mountains province was a distinctive glassy ignimbrite in the Rattlesnake that caps isolated buttes and mesas as far away as Harney County.

Figure 6.25 Hells Canyon of the Snake River looking south from near Hat Point. Steep canyon walls are cut into Seven Devils volcanic rocks of the Wallowa terrane (U.S. Forest Service).

Both the Mascall and Rattlesnake formations contain a varied fauna of Miocene fossil mammals and broadleaf plants reflective of a cool semiarid climate that followed the moist John Day environment. This climate change is correlated with the Miocene uplift and tilting of the Western Cascades that cast a rainshadow over eastern Oregon, resulting in the present high desert.

THE PLEISTOCENE EPOCH BROUGHT ICE to the crests of the Wallowa Mountains in northeast Oregon, the most heavily glaciated in the province. Called

the Oregon Alps, the Wallowas were sculpted by valley glaciers beginning less than 2 million years ago. Straight trough-like valleys, clear glacial lakes, and winding moraines of sand and gravel are the residue of a number of glaciers, some of which stretched over 20 miles down the canyons. The largest of these in the Lostine River basin was an ice sheet up to 1/2 mile thick. Although today there are no active glaciers in the Wallowas, moving ice masses were recorded as late as 1929.

A most striking effect of glaciation can be seen at Wallowa Lake near Joseph in Wallowa County. Curving moraine deposits, some projecting 2000 feet above the

Figure 6.26 Ash flow tuff of the Rattlesnake Formation caps a thick layer of Miocene fossil-laden sediments in Grant County (Oregon Dept. of Geology and Mineral Industries).

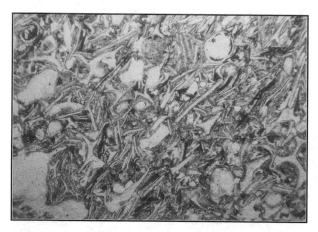

Figure 6.27 A thin section through a welded tuff shows gas bubbles set in a mixture of sharp pieces of volcanic glass (Oregon Dept. of Geology and Mineral Industries).

valley floor, trap the lake waters within the steep-sided valley.

MINERAL PRODUCTION

Gold was extracted from the Blue Mountains province in a strip 50 miles wide and 100 miles long from Umatilla and Grant counties to the Snake River. Gold deposits in this belt are associated with granitic batholiths intruded during the Late Jurassic and Early Cretaceous. Gold and other ores, which had precipitated onto the ocean floor from deep sea ridges, were remelted and distributed in veins around the margins of cooling batholiths.

Treking across eastern Oregon in 1845, settlers, who inadvertently picked up nuggets while looking for straying cattle, may have been the first to discover gold in this province. The immigrants tossed the pieces of soft metal into a tool kit, where they were forgotten as the party reached the Willamette Valley. It was not until five years later that gold discoveries in Griffin Gulch in Baker County and near John Day in Grant County sparked the beginning of the gold era here and rekindled the settlers' interest in the spot where they had found the nuggets. Now known as the Blue Bucket mine, the exact locality has never been determined.

In spite of the dry hot climate and less-than-friendly Indians, 1000 men were soon living in hastily erected tents under rough conditions as they found gold nuggets plentiful in river gravels of eastern Oregon. Before long, however, prospectors concentrated their efforts on uncovering lode veins. Once the sources were pinpointed, tunnels were laborously dug into bedrock to extract the ore.

Lode mines at Cracker Creek, Greenhorn, and on Conner Creek on the Oregon-Idaho border dominated production until the early 1900s when the lucrative Cornucopia vein was discovered in northeast Baker County. Before closing in 1941, the Cornucopia mines were responsible for more than half of Oregon's annual lode gold output. The combined amount from both placer and

Figure 6.28 A southerly view of moraines that trap Wallowa Lake at the foot of the snow-covered Wallowa peaks near Joseph, Oregon (courtesy E.M. Baldwin).

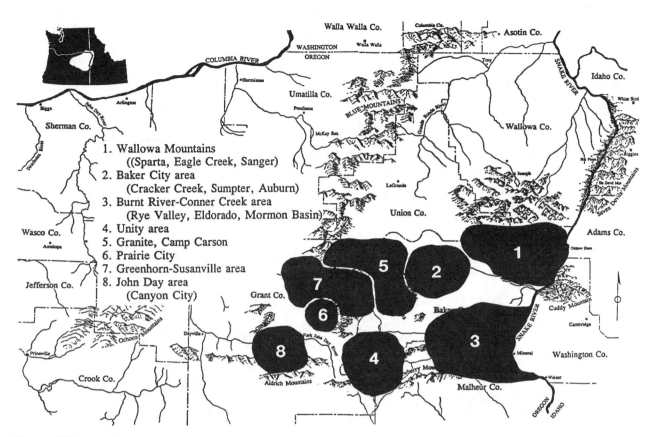

Figure 6.29 Significant mineral districts of the Blue Mountains (after Brooks and Ramp, 1968).

1. Wallowa Mountains
   ((Sparta, Eagle Creek, Sanger)
2. Baker City area
   (Cracker Creek, Sumpter, Auburn)
3. Burnt River-Conner Creek area
   (Rye Valley, Eldorado, Mormon Basin)
4. Unity area
5. Granite, Camp Carson
6. Prairie City
7. Greenhorn-Susanville area
8. John Day area
   (Canyon City)

Figure 6.30 Bucket dredges, such as this one near Mt. Vernon in Grant County, operated on the John Day River in the early 1900s (Oregon Dept. of Geology and Mineral Industries).

lode deposits in this region totalled 36,000 ounces until 1965 when production dropped off drastically.

Because of elevated values for gold in recent years, these districts are targets for new drilling programs and mining techniques using sodium cyanide to extract gold from the rocks. Even with stricter environmental laws, the use of cyanide solutions has left large open pits with hazardous toxic material.

SILVER AND COPPER frequently occur with gold, and silver ores worth over $2 million were extracted from districts of the Blue Mountains between 1900 and 1965. In the northeast corner of Baker County, the Iron Dyke Mine opened in 1897 and was responsible for most of the local copper ore, totalling close to $3 million. Copper deposits here are found in faulted, shattered rocks of the Permian Hunsaker Creek Formation in the Wallowa terrane.

Figure 6.31 In the gold camps, arrastras, turned by horse, donkey, or water, were used to crush gold ore (Oregon Dept. of Geology and Mineral Industries).

ADDITIONAL READINGS

Alpha, Tau Rho, and Vallier, T.L., 1994

Ash, S.R., 1991

Blome, C.D., and Nestell, M.K., 1991

Blome C.D., Nestell M.K., et al., 1986

Brooks, H.C., 1979

Brooks, H.C., and Ramp, L., 1968

Brooks, H.C., Ramp, L., and Vallier, T.L., 1978

Fisher, R.V., and Rensberger, J.M., 1972

Goldstrand, P.M., 1994

Jones, D.L., Silberling, N.J., and Hillhouse, J., 1977

Morris, E.M., and Wardlaw, B.R., 1986

Orr, E.L., and Orr, W.N., 1999

Orr, W.N., and Orr, E.L., 1999

Pessagno, E.A., and Blome, C.D., 1986

Robinson, P.T., Walker, G.W., and McKee, E.H., 1990

Stanley, G.D., 1986

Stanley, G.D., and Beauvais, L., 1990

Vallier, T.L., and Brooks, H.C., eds., 1986

Vallier, T.L., and Brooks, H.C., eds., 1994

Vallier, T.L., and Brooks, H.C., eds., 1995

Vallier, T.L., Brooks, H.C., and Thayer, T.P., 1977

Walker, G. W., 1990

Walker, G. W., ed., 1990

Walker, G. W., and Robinson, P.T., 1990

White, D.L., and Vallier, T.L., 1994

White, J.D.L., 1994

White, J.D.L., et al., 1992

CENTRAL IDAHO

At the headwaters of the Clearwater River, the cascade of Selway Falls on the Selway River in northcentral Idaho is more impressive during the early summer because of snow melt (Idaho Dept. of Commerce and Development).

Central Idaho, situated between the Snake River Plain, Rocky Mountains, and Blue Mountains, is characterized by deep valleys and high serrated divides that include the Clearwater and Bitterroot to the north and the Salmon River and Sawtooth mountains to the south. Within this rugged topography, some of the highest points in the state are Hyndman Peak at 12,009 feet, Twin Peaks at 10,328 feet, McGuire at 10,082 feet, and Trinity Mountain at 9451 feet. Between these high crests, a network of steep narrow canyons have been incised by the Clearwater, Salmon, Payette, and Boise rivers.

The lengthy Salmon and Clearwater drainages stretch across the state from east to west, separating the panhandle from the southern portion. Winding through 425 miles of wilderness, the Salmon is the largest tributary of the Snake River, draining 14,000 square miles. From its beginning in the Sawtooth and Salmon River mountains, the river drops from an elevation of 8000 feet to 903 feet at its mouth, flowing between precipitous canyon walls that are sometimes 5000 to 6000 feet high. These canyons are only exceeded in depth by Hells Canyon on the Snake.

With headwaters in the mountainous plateaus of the Bitterroot Range, the Clearwater River, joined by the junction of the Middle and South Forks near Kamiah, drains 10,000 square miles before meeting the Snake. The North Fork of the Clearwater defines the northern boundary of this province.

Over 4 million acres of central Idaho are part of the Salmon River Wilderness and Idaho Primitive Area. Additionally, the Middle Fork of the Salmon was preserved in its natural state when it was declared to be Wild and Scenic in 1968.

INTRODUCTION

At its center, Idaho is anchored by the monolith of the Idaho batholith, forming a crystalline granitic basement to the Clearwater, Salmon River, and Sawtooth ranges. This large plutonic body is the product of ongoing tectonic plate movements that have dominated Pacific Northwest geology since late Paleozoic time. Beginning 100 million years ago, multiple chambers of liquid rocks called magma, introduced into ancient marine sedimentary rocks, severely altered the surrounding layers in the process. The comparatively large size of mineral crystals in the batholith suggests that the rocks cooled deep within the earth's crust. Intensive heat from the crystallizing magma forced mineral-laden fluids through adjacent rocks to concentrate in veins that would later produce much of Idaho's mineral wealth. The numerous gold-seekers during the middle and late 1800s played a significant role in opening the interior of this province to settlement.

Tectonic activity did not cease after the emplacement of the Idaho batholith but continued with the Challis volcanic episode that began 52 million years ago. This second phase of plutonic intrusions was accompanied by immense quantities of lava and ash that erupted from vents across central Idaho.

The presence of gold and abundance of hot springs characterize this province and are reflections of its igneous origins.

GEOLOGY OF CENTRAL IDAHO

Precambrian metamorphic rocks, 1.5 billion years old, are exposed in northcentral Idaho. Called core complexes, these ancient rocks are thought to be native to the landmass of ancient North America, but their extreme antiquity as well as metamorphic alteration and folding obscure their origin. Highly deformed metamorphic or plutonic core complexes are typically separated from the enclosing layers by faults. Domed exposures of these rocks are scattered across a 100-mile wide strip through British Columbia, Washington, Idaho, and the eastern Great Basin into Arizona and Mexico.

Although the formation of core complexes is not well understood, it seems clear they were raised from deep to middle crustal depths, thus providing a look at buried older rocks without the expense of drilling or the tedium of waiting for slow erosional processes to unearth them. As the crust of the earth was pulled apart and stretched, it thinned and in some spots broke to create windows into core rocks below. These tensional episodes in the northern Rockies took place during the Eocene epoch, while in the Great Basin and southern Rockies crustal stretching took place in the Oligocene and Early Miocene. All of the domed core complex structures predate the pervasive extensional faulting that characterized the Middle Miocene of the Great Basin.

Small to moderate size core complexes have been identified in a rough north-south trend across Idaho and into Washington. In the mountainous Priest River and Sandpoint areas of the Idaho panhandle, older metamorphic rocks, altered by granite intrusions, are the eastern members of a number of core complexes that extend to the Okanogan dome in northeast Washington. The Bitterroot dome in northcentral Idaho is capped by a thoroughly broken and sheared "mylonite" zone that separates the core rocks from a younger sequence above.

Two north-trending elongate ridges of the Pioneer Mountains east of Ketchum display yet another metamorphic core complex. The Pioneer Mountains may be as old as Precambrian, but their exact age is not known. The metamorphic action that altered these rocks dates

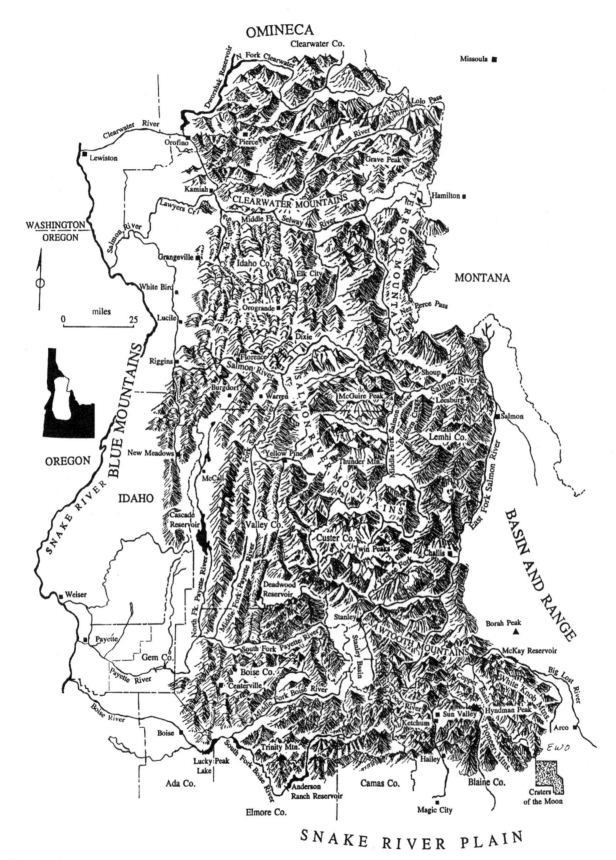

Figure 7.1 Map of localities within the region of central Idaho.

Figure 7.2 Variously composed of Cretaceous and Eocene granitic rocks, the Sawtooth Range, viewed here across Stanley basin, was heavily glaciated during the Ice Ages (Idaho Dept. of Transportation).

back to the Devonian time, 400 million years ago, although the complex was finally exposed during extensional movement that took place in association with the eruption of Eocene Challis volcanics.

Core complexes of Precambrian metamorphic and sedimentary rocks in southern Idaho in the Sublett Mountains were raised close to the surface by thrusting and extensional faulting. Similar deep crustal rocks in the Albion Range of Cassia County, dated at 2.5 billion years, are considered to be the oldest in Idaho.

DURING THE MIDDLE PALEOZOIC a broad seaway across western North America lay behind an offshore volcanic archipelago. As extensive limestone banks spread over the shelf of this back-arc ocean basin, fine silts and shales accumulated on the outer margins,

and submarine fans spread out at depth. A thick belt of siltstones and limestones, 30 miles in length in the Wood River region of Idaho, records the Devonian Milligen Formation associated with a deep continental slope environment. Mineral-rich black shales, within the Milligen, yielded silver, lead, and zinc ores, mined during the 1880s.

Above the Milligen, the Pennsylvanian-Permian Wood River Formation is distinguished by dark siltstone and sandy limestone that represent an early environment of fans and deltas followed by turbidity currents when the basin deepened. As these sediments were folded and uplifted, a deep marine trough was created to the east of the raised belt.

The Late Devonian to Early Mississippian Antler mountain-building event, which took place between 350

Figure 7.3 Ancient core complex rocks form the foundation of the scenic jagged Pioneer Mountains (Idaho State Historical Society).

Figure 7.4 The Paleozoic shelf across central Idaho was inhabited by crinoids or sea lilies, corals, algae, and swimming ammonites.

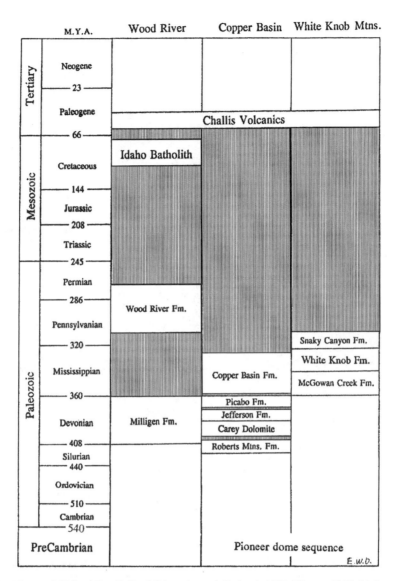

Figure 7.5 Stratigraphy of central Idaho (after Ballard, Bluemle, and Gerhard, 1983; Hintze, 1985; Link, et al., 1988).

to 380 million years ago, created a highlands across eastern Idaho and into central Nevada. The uplift simultaneously trapped an inland sea along the eastern edge of the uplands. Sandy turbidite limestone and mudstones carried into the basin from the raised Antler region formed the Mississippian Copper Basin, McGowan Creek, and White Knob formations from Montana westward across the White Knob and Pioneer mountains of Idaho. Over 10,000 feet of the Copper Basin Formation records a deep-sea fan environment adjacent to a highland. Rocks of the Copper Basin were later uplifted so that today they form the core of the mountains across the southern end of the province.

THE DEVELOPMENT AND EMPLACEMENT OF THE IDAHO BATHOLITH during the Late Mesozoic is the common denominator to the geologic history of central Idaho. Like a massive inkspill across the center of the state, the batholith is anything but the simple intrusive it appears to be.

Batholiths, literally meaning "deep rock," are thickened crustal areas where large deep magma chambers slowly cooled to produce crystalline rock. Scattered over western North America, batholiths are defined as intrusive bodies that cover over 40 square miles. Once erosion strips off the surface cover of rocks over the dome, the crystalline batholith is exposed in a progressively wider area. In the Pacific Northwest these crystalline granitic bodies underlie broad areas of the Coast Range of British Columbia, the Sierra Nevada Mountains of California, and all of central Idaho.

Originally it was thought that batholiths were virtually bottomless, but using modern geophysical methods deep crustal rocks can now be "viewed" without drilling,

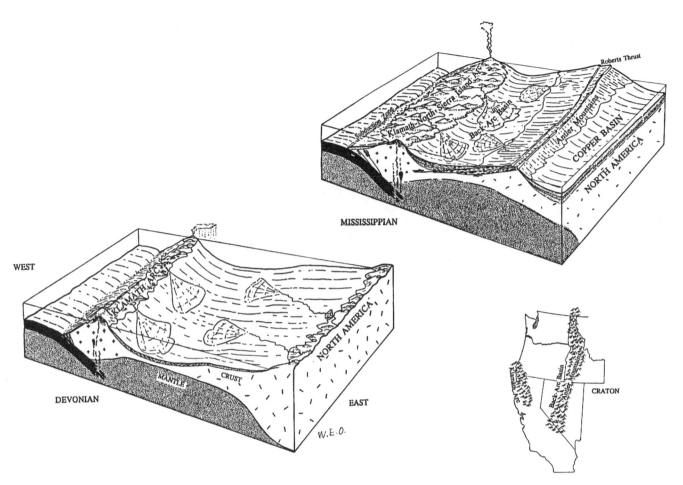

Figure 7.6 In the Middle Paleozoic, a volcanic island chain lay offshore, and uplift of the Antler highlands depressed the Copper basin inland (after Poole and Sandberg, 1977).

giving a three-dimensional picture of the intrusion. These observations suggest that while some batholiths are blimp-shaped, as in the Sierras, others are more like an inverted drop of water tapering downward from a round upper dome. Early in its emplacement history the Idaho batholith may have been sausage-shaped, but with time it began to mushroom out, developing a dome-like roof and flattened bottom.

Beginning in the Jurassic, over 180 million years ago, the North American continent began to separate from Europe along the proto-Atlantic Ocean, developing a subduction zone along its leading western edge. Subsequently, a succession of exotic terranes, carried eastward by the ocean spreading process, were sutured to the West Coast of the North American plate in the Late Jurassic and Cretaceous, 160 to 85 million years ago. Following collision, these heavier oceanic plates were engulfed beneath the lighter North American slab along the subduction zone.

In the subduction process, the crust of the heavier slab thickened at the same time that heat was generated, melting rocks in the suture zone. This liquid magma, which would invade the surrounding strata, later cooled to granite batholiths. Batholiths formed with each colli-

sion before the subduction zone between the two plates was abandoned and a new one developed on the oceanic side of the freshly accreted pieces. As the old slab ceased to move, the source of heat was cut off, and the magma would begin to cool. Shortly after this, a new subduction zone opened further oceanward, and intense heat began to generate a younger batholith. With this cycle, successive batholiths were placed to the west of the older chambers. An alternative to the subduction-related origin of the Idaho batholith suggests that this mass crystallized in place when a greatly thickened section of crust began to melt in its lower portion.

A problem that arises with the study of large plutons is understanding the method by which they are emplaced. What structure occupied that spot before the batholith? Did the batholith push aside the existing rocks or were they simply melted in place? Currently, with respect to the Idaho batholith, it is believed that heat and fluids slowly liquified existing rocks that were then incorporated into the ascending batholith. This suggests a succession of smaller pockets of magma so the entire mass was never simultaneously liquid. The Sierra Nevada batholith in California has been shown to be composed of a large number of smaller discrete bodies of varying composition that cooled at different times to make up the whole. By contrast, although the Idaho batholith displays different crystalline rocks and textures over its breadth, the boundaries between the various magma parcels are indistinct.

The Idaho batholith crystallized slowly at depths up to 10 miles during the Late Mesozoic. Even though this

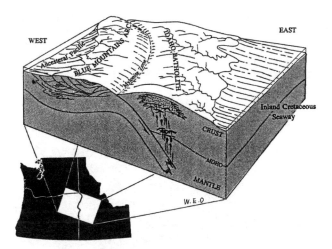

Figure 7.7 During the Late Cretaceous, the Blue Mountains volcanic archipelago accreted to North America. Subduction of this island chain triggered the Idaho batholith.

pluton predates the tensional stretching of the Basin and Range by a wide margin, it is almost synchronous in time with the late Mesozoic arrival of exotic terranes. Between the Idaho batholith and the exotic terranes of the Blue Mountains, an interval of shattered rocks over a mile wide form what is called a mylonite zone. This zone apparently represents the suture point where terranes were affixed to North America.

Intense eastward pressure on the earth's crust in Idaho and adjacent Montana and Wyoming folded the Paleozoic strata into tight wrinkles that eventually broke and thrust up over each other in a shingled configuration. The role of the Idaho batholith in the thrusting episode is controversial. One suggestion is that the eastward overthrusting resulted from pressure exerted by colliding exotic plates. This is presently taking place in India where the Himalaya Mountains are piling up. An earlier notion proposes that the Idaho batholith rose rapidly through the crust like a balloon as it was being emplaced to form a large surface mound. Once the sides of this behemoth steepened, rocks along the margins began to spall off and slide downward. A problem with this theory arises from the discovery that most batholiths postdate thrusting events.

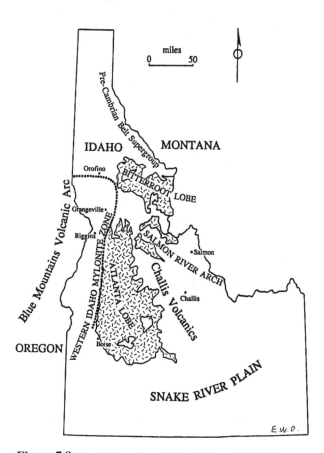

Figure 7.8 The Bitterroot and Atlanta lobes of the Idaho batholith in central Idaho (after Lewis, et al., 1987).

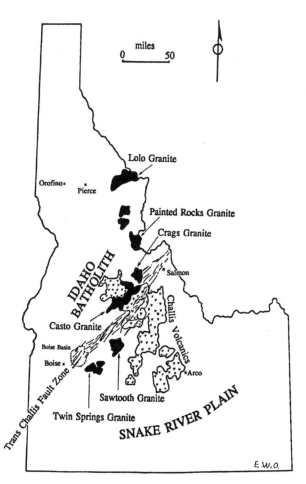

Figure 7.9 Challis volcanic area and plutons of central Idaho (after Kiilsgaard, et al., 1986; McIntyre, et al., 1982).

Figure 7.10 The imposing Bitterroot Mountains form the boundary between Idaho and Montana. In 1832 explorer Captain Bonneville camped at the base of the mountains during his travels through the Northwest (Idaho Dept. of Commerce and Development).

The bulbous mass of the Idaho batholith, 200 miles long, 75 miles wide, and 5 miles thick, can be divided by age and rock composition into a larger southern portion called the Atlanta lobe and a northern somewhat smaller Bitterroot lobe. The Atlanta lobe is older than the Bitterroot, but the two overlap slightly in their cooling histories. The Atlanta lobe is estimated to have been emplaced between 100 to 75 million years ago, while the Bitterroot enlargement cooled between 85 and 65 million years ago. These two lobes are separated by an upward bend in subsurface rocks called the Salmon River arch that dates back into the Precambrian, over 1.5 billion years in time.

THE CENOZOIC OPENED around 51 million years ago with the Challis volcanic activity that persisted for approximately 10 million years. Eocene Challis volcanic rocks extend in a loosely defined belt from northeast Washington across Idaho and into the Absaroka Mountains of Wyoming. Today, in Idaho this volcanic suite covers close to 2000 square miles, but originally Challis ash and tuff blanketed over half of the state. This episode coincided in time with the more andesitic Clarno eruptions of eastern Oregon, and if both volcanic episodes arose from the same magmatic source, it would mean the activity was spread over an area 300 to 500 miles wide.

Figure 7.11 Fishfin Ridge, part of the rough Bighorn Crags 30 miles west of Salmon, towers above Lake Wilson. The spires and ridges have been cut into Eocene granites (Idaho Dept. of Commerce and Development).

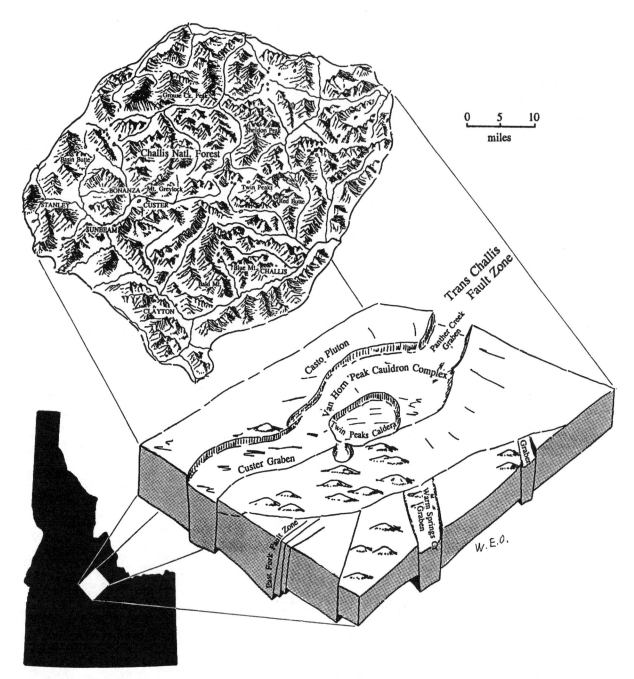

Figure 7.12 Sitting astride the Trans-Challis fault zone in central Idaho, massive calderas mark the site of Eocene rhyolitic volcanoes (after Kiilsgaard, et al., 1986; McIntyre, Ekren, and Hardyman, 1992).

The Eocene Challis and Clarno time interval was a period of major tectonic plate adjustment in the Pacific basin. Prior to 56 million years ago, a slowing in the spreading rates may have caused the Farallon subducting slab beneath North America to nose down into a steep angle decreasing the amount of volcanism and restricting it to a narrow belt. After this interval, a reorganization of the major plates in the Pacific greatly accelerated the spreading and collision rates. As warm young crust was subducted, increased buoyancy of the slab caused it

to rise to a flat angle. This extremely low angle of the subducting crust corresponds relatively well to the period of the widespread Challis and Clarno eruptions between 52 and 44 million years ago.

Challis igneous activity occurred in two phases beginning with intrusive plutons followed by the eruption of very siliceous lava and ash. Granites appear as small bodies or plutons just east of the Atlanta lobe where they show evidence of having been emplaced at much shallower depths in the crust than granites of the adjacent

Figure 7.13 Early Tertiary igneous rocks north of Salmon in Lemhi County weather into peculiar shapes (Idaho Dept. of Commerce and Development).

larger Idaho batholith. Whereas the crystalline Idaho batholith granites cooled at depths up to 10 miles, the Challis granites and andesites crystallized only 3 to 4 miles below the surface. Even though the granites of the Challis and Idaho batholiths are found in the same vicinity, the two are easily distinguished from each other. The pink color and large crystals characteristic of Challis rocks contrast readily to the gray-hued rocks and somewhat smaller crystals in the Idaho batholith granites.

During the eruptive phase, Challis lavas and ash were extruded along a broad belt of faults running southwest by northeast from Boise well into western Montana. Numerous vents and fractures in this Trans-Challis fault zone produced voluminous rhyolitic lavas that were up to 2 miles thick. With extensional stresses along the fault zone, volcanic activity resulted in at least 7 major caldera complexes from Stanley in Custer County to

Salmon in Lemhi County. These eruptive centers include grabens or faulted depressions, lava flows, and domes. Of these, the Van Horn Peak volcano began about 49 million years ago with rhyolitic ash flows that ended when the caldera collapsed inward. Within this complex, eruptions from the Twin Peaks caldera marked the final stage of activity. As the caldera subsided, the linear depressions of the Custer and Panther Creek grabens, extending northeast and southwest from the Van Horn Peak complex, were the sites of individual lava flows and domes.

In a similar sequence of events, Challis ash and lava built the nearby Thunder Mountain field in Valley County to a height of almost a mile before it subsided into a circular caldera 40 miles in diameter. A second Cougar Basin eruption and caldera was even larger, enveloping the Thunder Mountain field. The emplacement

Figure 7.14 With headwaters in central Idaho, the Salmon "River of No Return" is the only stream lying completely within the state (Idaho Dept. of Commerce and Development).

of numerous intrusives, including the Casto pluton along the northwest margin of the Van Horn Peak complex, tilted the floor of the adjacent Thunder Mountain caldera to the southwest.

HAVING ITS ORIGINS IN THE MOUNTAINS OF EASTERN IDAHO, the west-flowing Salmon River, joined by its Middle and South Forks from the south, stretches across Idaho, slicing directly through the granitic mass of the Idaho batholith and Challis volcanic rocks. Much of the river follows a zigzag course in response to a pattern of faults and fractures in the granite. Small meanders occur at the bottom of steep canyons, and there are little or no flat uplands that have not been dissected by erosion. Along the meanders of the rivers, bars of sand and gravel, which occur at every turn, move downstream a few feet each year to make their way to the Snake River. Somewhat younger geologic features are terrace deposits left "high and dry" in the canyon walls when the river began downcutting. Gold in these terraces is relatively inaccessible because

of the difficulty of getting water up to them to wash out the gravels.

The deep canyon of the upper Salmon remained largely unmapped and unexplored as late as 1949 in spite of a 1935 National Geographic expedition that traveled the length of the river beginning at Shoup and ending at Lewiston on the Snake. Expedition members included a forester, geologist, and State Representative, well as the Geographic crew. During this 253-mile long trip, the canyon was surveyed and photographed.

Tricky whitewater rapids gave the Salmon its name, The River of No Return. Because its rapid turbulent flow makes the river difficult to navigate, boats were abandoned at the end of a trip, and many people lost their lives while attempting the journey.

FROM THE MIOCENE WELL INTO THE HOLOCENE, volcanic eruptions continued to play an important role in shaping the central Idaho landscape. Over the past 2 million years, the Boise River and its South Fork have been dammed five times by lava flows.

Figure 7.15 A view westward of the Boise River curving through flat-topped basalt buttes (Idaho Dept. of Transportation).

The oldest eruption was from Lucky Peak on the border of Ada County, and the most recent was from a vent north of Anderson Ranch Dam in central Elmore County. Lava flows, filling the canyon for 40 miles, were repeatedly breached by the river as it cut through and retrenched, deepening and steepening the canyon walls. Today remnants of older basalts may be seen as ledges high above the river while the younger flows are lower in the canyon walls.

ALTHOUGH PLEISTOCENE CONTINENTAL ICE SHEETS only reached as far south as the Idaho panhandle, periods of low temperatures and heavy precipitation brought multiple cycles of ice accumulation and extensive glaciation to the central mountainous regions.

In the northern section of the province, ice caps and merging valley glaciers characterized the earliest period. An exact chronology of ice age events here has not been formulated because dating information is lacking. The most extensive ice mass, reaching thicknesses of 1200 feet, covered Grave Peak in the Selway-Bitterroot Range. Numerous alpine glaciers, present in stream drainages, coalesced to form ice sheets over hundreds of square miles, and glaciers eroding back-to-back along

the ridgeline at Lily Lake near Lolo Pass on the continental divide cut through the rock to form a continuous layer of ice. A second glacial phase was less extensive than the first, deepening bowl-shaped valleys, while a third diminished episode left glacial lakes behind dams of gravel, sand, and clay.

Across the southern part of the province, beyond the margin of the continental ice sheet, smaller valley glaciers created the beautifully sculpted serrated ridges, steep horns, and curved valleys of the Salmon River, Sawtooth, Pioneer, and White Knob mountains. Within these valleys, hummocky surfaces from poorly drained moraines record the advances and retreats of the ice.

GEOTHERMAL POTENTIAL IS HIGH in central Idaho, an area characterized by numerous hot springs. Of the 200 warm water springs known in Idaho, over half emerge from the granites and related rocks of the Idaho batholith. Groundwater, moving through fractures and along cracks, is heated at depth to appear at the surface as hot springs that mark the southern boundary between the Idaho batholith and Snake River Plain. Most thermal water contains abundant dissolved minerals such as calcium, magnesium, or silica collected when it percolates through the host rock.

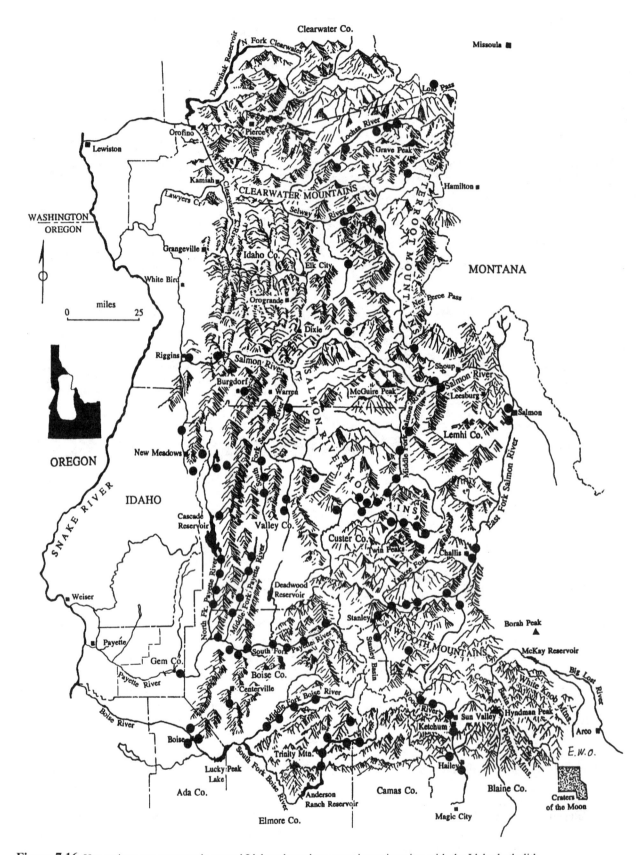

Figure 7.16 Hot springs are common in central Idaho where they occur in conjunction with the Idaho batholith.

Many of the Idaho springs are operated for recreational use, but aside from swimming and bathing, warm waters are used for residential heating. In Sun Valley and the Boise basin, a number of private and public accommodations have been heated with thermal well water since 1890. Near the bustling and famous Sun Valley-Ketchum-Hailey resort complex in Blaine County virtually all of the hot springs have commercial and recreational facilities. Water temperatures in the resorts generally reach 170 degrees Fahrenheit and are high in dissolved salts. Thermal waters near Boise have also been used in greenhouses and for irrigating crops.

Along the western margin of the province, springs occur near Riggins, Burgdorf, New Meadows, Cascade, and along high river valleys. Resorts at both Riggins and Burgdorf in western Idaho County record water temperatures up to 135 degrees Fahrenheit. A continuous discharge of warm waters for over a mile near the South Fork of the Salmon reaches over 180 degrees Fahrenheit at Vulcan Hot Springs in Valley County, and temperatures are similar at Boiling Springs on the Payette River in Boise County.

Although geothermal energy is already being utilized for commercial and domestic purposes, this is a region of particularly high future potential.

MINERAL PRODUCTION

Mineral resources accelerated settlement of the remote valleys of central Idaho, which was opened for mining following the discovery of gold and the decisive Indian wars in the late 1800s. The promise of vast wealth initially excited the imagination and encouraged swarms of eager prospectors even though ultimately gold here proved to be much less an economic asset than silver and lead. Some stories of great wealth were true, but many were exaggerated. In the most productive operations, sluices might yield $75 to $100 per day and even an exceptional $20,000 in one week of work. However, prospectors generally only broke even, or worse, lost money on their investments.

GOLD EXPLORATION IN CLEARWATER AND IDAHO COUNTIES began along the Clearwater River in 1860 despite vigorous efforts of both the Nez Perce Indians and the United States government to keep prospectors out. Heavy winter snows in the mountains did more to hold back the rush than U.S. agents or Indian treaties.

Determined to seek his fortune, E.D. Pierce, with a small band of men, furtively entered Idaho by an obscure northern route. Exploring stream gravels, the men found gold on Orofino Creek, a branch of the Clearwater. News of the strike spread throughout the Pacific Northwest and into California, and in 1861 over 71 claims were filed. From claims near the newly established city of Pierce, prospectors systematically probed south to Idaho County, following creeks to Elk City, Florence, and Warren where rich new finds were recorded in late 1861 and 1862. In spite of an extremely cold winter in 1862, when sections of the Columbia River froze and snow in the high valleys around Florence was still 4 feet deep in

Figure 7.17 Emerging from intrusive rocks in the Wood River valley west of Ketchum, Guyer Hot Springs heats a number of homes locally (Idaho State Historical Society).

Figure 7.18 Placer mining near Elk City in Idaho County employed a powerful hydraulic stream of water to wash out gold-laden river gravels (Idaho Dept. of Transportation).

May, the gold season reached an all time high at $50,000 a day. Even though placer mining continued at Florence for 30 years, these fabulous early strikes were never matched, and nearby districts proved to be much less productive.

In all, around 1 million ounces of placer gold were extracted from the Florence district followed by that from the Warren area with 900,000 ounces of stream gold. It is interesting to note that while the lode sources in these counties were fissure and vein deposits associated with mineralization during emplacement of the Idaho batholith, placers were located in old channels on terraces as much as 500 feet above the present stream surface. The flow of these ancient waterways was blocked by the Columbia River basalts that diverted them from the old beds where the gold is found.

Today a number of placers and mining operations are active in the state. In the 1990s, the largest placer was near Lucile in Idaho County, and just south of War-

ren in Valley County a heap-leaching gold mine at Yellow Pine had the highest production figures. Gold here was recovered from a large sulfide deposit where the minerals had been emplaced by hot springs on the ocean floor. Reclamation is ongoing at this site.

FURTHER DISCOVERIES WERE MADE IN LEMHI COUNTY in 1866 and 1867 where rich gold-bearing gravels were located on Napias Creek in the Leesburg basin. Founded by ex-confederate Army soldiers, Leesburg, at its height, was a settlement of 26 houses, 6 stores, 2 butcher shops, and 7000 miners before the population dwindled to 25 hardy persons by 1900, who were able to make only a modest living. Worked for about 14 years, placers in Pleistocene gravels accounted for most of the gold in this district. Smaller lode veins along shear zones were exploited for a few years in the 1870s, and both sources, operated intermittently, yielded approximately 271,000 ounces.

1. Clearwater and Idaho Counties
     (Orofino, Pierce, Elk City, Florence, Warren, Lucile)
2. Lemhi County
     (Leesburg, Shoup, Salmon, Yellowjacket, Blackbird)
3. Custer County
     (Yankee Fork)
4. Boise Basin
     (Centerville)
5. Elmore, Camas, Blaine Co.
     (Big Wood River)

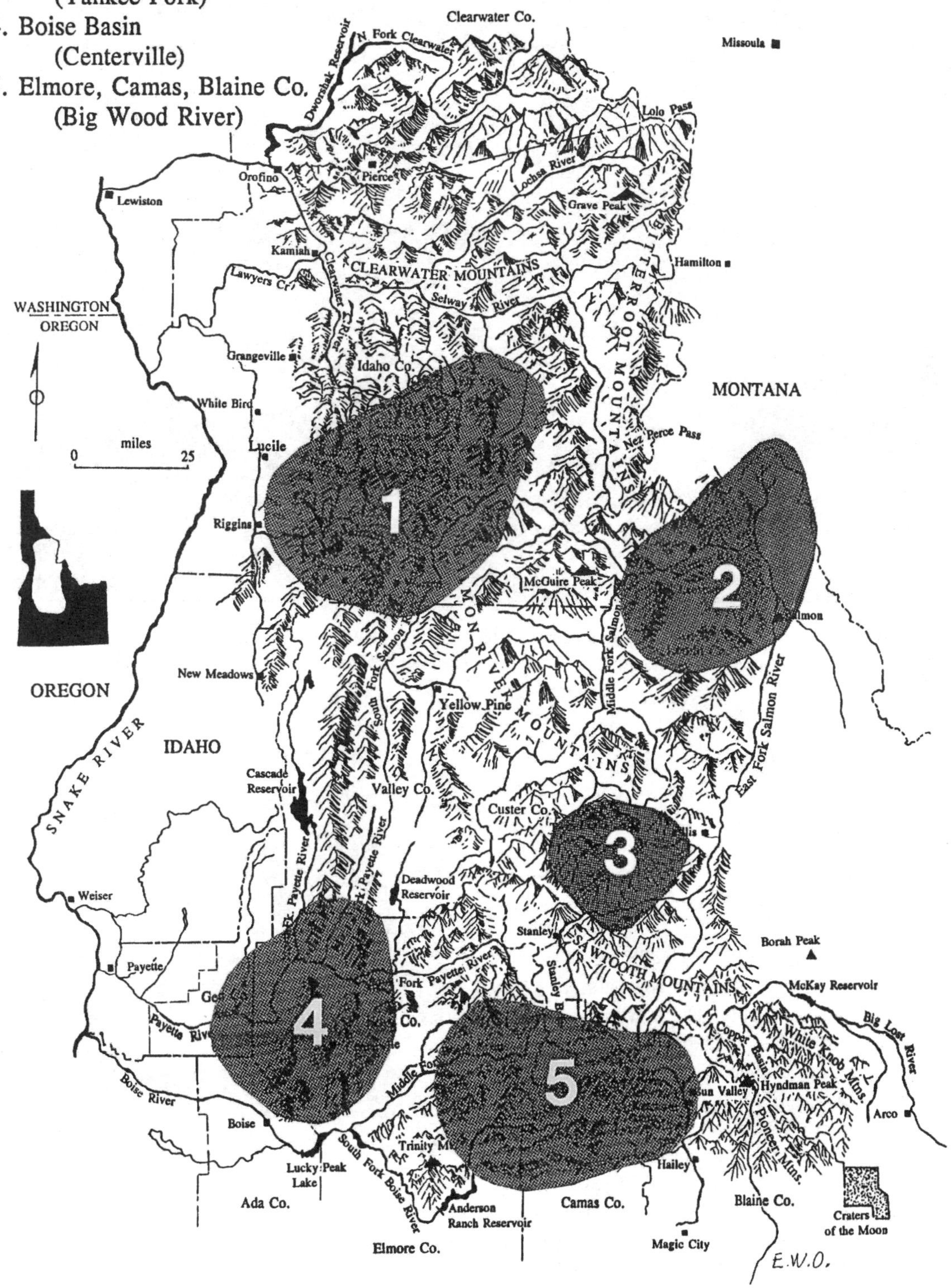

Figure 7.19 Primary mining districts of central Idaho.

Figure 7.20 By employing a continuous conveyor belt of 71 buckets, each with a capacity of 8 cubic feet, the massive dredge on Yankee Fork mechanized the process of extracting gold from stream deposits (Idaho Dept. of Transportation).

Placers at Leesburg, Yellowjacket, Blackbird, and on adjacent creeks in northcentral Lemhi County totalled 720,000 ounces of gold through 1959.

Exploration programs are currently underway in this county. Heap-leaching facilities are planned, and a new surface mine and mill have been opened at Yellowjacket.

IN CENTRAL CUSTER COUNTY news of rich finds at Yankee Fork, a tributary of the Salmon River, kept miners working feverishly. The search for gold on Yankee Fork got off to a slow start after gold resources had dried up elsewhere. Limited but steady profits of $10 per day of panning in and around the creek justified continued exploration, but even when $60,000 was extracted from one small vein in the lucrative General Custer mine in 1877, there was no overwhelming gold fever. It was not until 1879 that the traditional rush came to Bonanza City in Custer County, where the Salt Lake Tribune reported the roads "lined with stampeders. . . ." Operating steadily for a period of years utilizing both gold and silver ores from a network of veins, the Custer Mine produced approximately $8 million in 1880, half of amount that was mined at Yankee Fork. Total output in this county reached 266,600 ounces. Fractured Late Paleo-

Figure 7.21 The Custer Mill was at the center of mining on the Yankee Fork of the Salmon River (Idaho State Historical Society).

zoic Wood River Formation was the host rock for most of these minerals.

ANOTHER MAJOR GOLD-PRODUCING DISTRICT to the southeast was the Boise Basin, where shear zones through rocks adjacent to the Idaho batholith have

Figure 7.22 The Bullion Mine in the Wood River gold, silver, and lead district (Idaho State Historical Society).

been filled with lode deposits of gold and silver. After Bannock Indians revealed the presence of gold to prospector Moses Splawn, he led a party of men to the area in the summer of 1862 camping near the present site of Centerville. Constant Indian attacks and lack of provisions forced the miners to withdraw only to return the next year well-supplied and heavily armed. In the inevitable rush that followed the discovery of gold, a horde of 20,000 prospectors thronged to Boise County where $10 to $100 per day was the average extracted from sluice boxes. In this district, placers were the most profitable, and later placer mining was augmented by powerful streams of water hydraulically washing gold out of gravel banks. Lode deposits contributed just a small portion to the total of 2,900,000 ounces of gold from this county.

ELMORE, CAMAS, AND BLAINE COUNTIES, along the southern part of the Idaho batholith, accounted for more than 650,000 ounces of lode gold, however, silver and lead production predominated over gold throughout the districts. The Big Wood River area at the northern tip of Blaine County, with more than 212,000 ounces of gold as well as additional profits from silver and lead, made this a particularly prosperous region.

Ores were extracted from veins and faults within sediments of the Paleozoic Milligen, Wood River, Carrietown, and Dollarhide formations. The intrusion of small Cretaceous plutons redistributed minerals, which

Figure 7.23 The abandoned Triumph silver mine southeast of Ketchum in the Wood River valley operated in the late 1800s and again in the middle 1900s (Idaho Dept. of Transportation).

had been precipitated from deep-sea ridges onto the ocean floor during the Precambrian. In a second stage of mineralization, metals were emplaced in veins at the same time pluton intrusion took place.

With the Indians disposed of in the Bannock War of 1878, prospectors were quick to locate lead and silver deposits. The inaccessibility of Big Wood River and the lack of an ore-processing facility held up the fabulous profits, which came with the construction of a stage line, railroad, and local smelters. Aided by a rail-road capable of shipping ore to the midwest, profits soared to over $2 million a year in the middle 1800s. The years from 1880 to 1887 were the most prosperous, but after that period the price of silver dropped. The exploitation of the Triumph Mine at $30 million between 1936 and 1957 far outstripped all other Wood River mines combined. Beautiful large crystals of ruby silver, a red silver sulfide mineral, from Sawtooth City in this district accounted for the incredible output of silver tonnage.

ADDITIONAL READINGS

Armstrong, R.L., 1975

Ballard, W.W., Bluemle, J.P., and Gerhard L.C., coord., 1983

Breckenridge, R.M., et al., 1988

Conley, C., 1982

Dingler, C.M., and Breckenridge, R.M., 1982

Evenson, E.B., Cotter, J.F.P., and Clinch, J.M., 1982

Foley, D., and Street, L., 1988

Hintze, L.F., coord., 1985

Howard, K.A., Shervais, J.W., and McKee, E.H., 1982

Johnson, K.M., et al., 1988

Kaysing, B., 1984

Ketner, K.B., 1977

Kiilsgaard, T.H., Fisher, F.S., and Bennett, E.H., 1986

Lewis, G.C., and Fosberg, M.A., 1982

Lewis, R.S., et al., 1987

Maley, T., 1987

McIntyre, D.H., Ekren, E.B., and Hardyman, R.F., 1982

Moye, F.J., et al., 1988

Nilsen, T.H., 1977

Poole, F.G., and Sandberg, C.A., 1977

Richmond, G.M., 1986

Ross, S.H., 1971

Stephens, G.C., 1988

Vallier, T.L., and Brooks, H.C., eds., 1987

Vallier, T.L., and Brooks, H.C., eds., 1994

Wells, M.W., 1963

Wells, M.W., 1983

Wust, S.L., and Link, P.K., 1988

8

SNAKE RIVER PLAIN AND OWYHEE UPLANDS

Even 50 years ago, the extraordinary rough lava surface, cinder cones, and buttes at Craters of the Moon National Monument were a popular tourist spot (Idaho Dept. of Tourism).

T he Snake River physiographic province extends 400 miles across southern Idaho from Island Park in the northeast to Payette in the southwest. This broadly curved lowland is draped over the raised block mountains of the Basin and Range on the southeast and the edge of the Idaho batholith to the north.

The eastern margin of the province blends with the Yellowstone volcanic center in Wyoming, while the western portion widens to encompass the raised highlands of the Owyhee plateau in Oregon and Idaho.

Although it has a different topography, the Owyhee uplands are geologically part of the Snake River province. From near Twin Falls, Idaho, the highland expands into Oregon where it takes in much of Malheur County. This region may be characterized as a high desert that averages 5000 feet above sea level. Steep gradients of the plateau have given rise to a net-

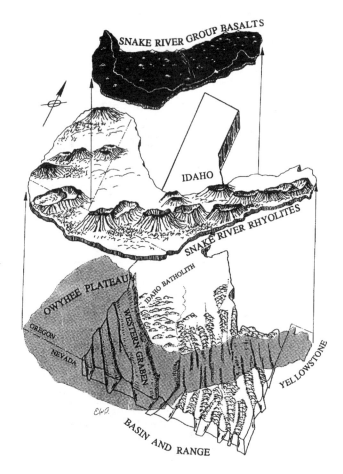

Figure 8.1 Three stages of Snake River geology: the youngest lava of the Pleistocene and Recent is the lop layer; large calderas from the explosive volcanic episodes of the Miocene and Pliocene are in the middle; and on the bottom are the broken linear mountain and valley structures of the Basin and Range.

Figure 8.2 Area and landmarks of the Snake River Plain and Owyhee uplands.

work of meandering, narrow canyons cut into the landscape by the Owyhee, Bruneau, and Snake rivers, and their tributaries.

The entire province is dominated by the Snake River that originates near the continental divide in Wyoming to flow south and west across Idaho before turning abruptly northward at the Oregon-Idaho border. In southern Washington the Snake cuts a deep channel westward to join the Columbia River at the great bend near Pasco.

INTRODUCTION

Contrasting geologic features divide the Snake River province into western and easterns sections at a point near Twin Falls. While the western portion is a pronounced depression filled with stream and lake sediments, the eastern valley is a broad deeply incised plain of thick Tertiary volcanic flows covered by thin Quaternary lavas.

Fresh-looking lavas interlain with sedimentary rocks containing plant and animal fossils tell the geologic story of this province. Since the Miocene, the opposing geologic agents of water and volcanism have been at work here. The energy of the streams was spent

cutting channels and widening valleys. At the same time volcanic episodes contrived to cover and smooth over the topography as well as fill in and plug the canyons, periodically impounding water in large natural reservoirs. As the lava dams were breached, the lakes were drained to start the process anew.

Volcanic activity began with catastrophic rhyolitic eruptions that created enormous calderas across southern Oregon and Idaho. As eruptive sites propagated over the Snake River Plain from west to east, they left 70-mile wide track of thick Miocene ash and lava that culminated in the Island Park and Yellowstone caldera complexes. During a second volcanic phase of the Pleistocene and Holocene, quiet basaltic flows issued from hundreds of long fissures and cracks in the eastern Snake River Plain, covering much of the older rhyolites to produce a flat even surface.

Within the faulted depression of the western Snake River valley, Pliocene-Pleistocene lakes, empounded when rivers were dammed by lavas, carpeted the basin with fossil-rich layers of sediments. A final veneer of sand and gravel was flushed into the province by the Bonneville Flood that swept down the valley 14,500 years ago. With this event and smaller Holocene eruptions of lava, the Snake River province assumed its present appearance.

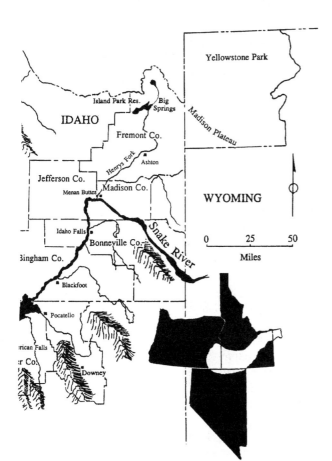

Figures 8.3 Just north of Twin Falls, Idaho, three small Blue Lakes with spring-fed sapphire-colored water from the Snake River aquifer lie in a deep canyon carved into basalts (Idaho Dept. of Transportation).

Figure 8.4 The Snake River canyon near Twin Falls (Idaho Division of Tourism).

GEOLOGY OF THE SNAKE RIVER PLAIN AND OWYHEE UPLANDS

The oldest rocks within the Snake River and Owyhee region are limited to two small packages of annexed terranes. On the Oregon-Idaho border, 15 miles south of Jordan Valley, a small exposure of Late Paleozoic schists, quartzites, and marble is known as the Owyhee terrane. Near the southern margin of the Owyhee uplands on the Idaho-Nevada border and into Nevada, deepwater shales, turbidites, pillow basalts, and cherts of the Golconda terrane are similar in age. A much larger section of the Golconda terrane extends across the center of Nevada. Other occurrences of exotic terrane rocks throughout this province are doubtless obscured by thick Tertiary lavas and fluvial sediments.

A CHANGING VOLCANIC STYLE IS ONE OF THE HALLMARKS of Cenozoic geology in the Snake River province. A variable history of bimodal eruptions, taking place in two distinct phases, begins with violent eruptions of quartz-rich rhyolitic lavas and tuffs from large calderas then moves to quiet flows of fluid black basaltic lavas from lengthy cracks and fissures.

The great bulk of volcanic material in the Snake River Plain is rhyolitic or quartz-rich in composition. The lower temperatures of this lava, as compared to that of basalt, make it stiff and viscous. Although intermittent, rhyolitic eruptions are particularly explosive and tend to be localized to one area, forming domes rather than broad flows. Accompanying the rhyolitic events, great wind-borne clouds of ash eventually settled and consolidated to produce a cohesive rock mixture called tuff. The

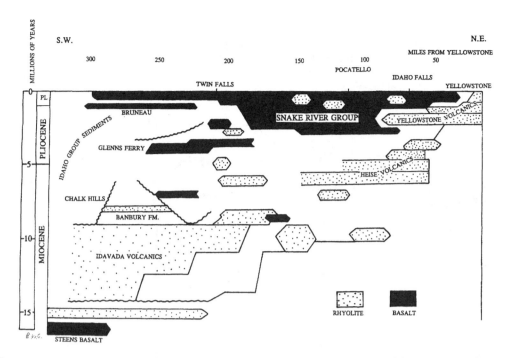

Figure 8.5 From the Miocene to Pliocene, bimodal volcanism and eastward propagating rhyolitic centers moved across the Snake River Plain from Steens Mountain in Oregon to Yellowstone, Wyoming (after Armstrong, et al., 1975; Hackett and Morgan, 1988; Leeman, 1982).

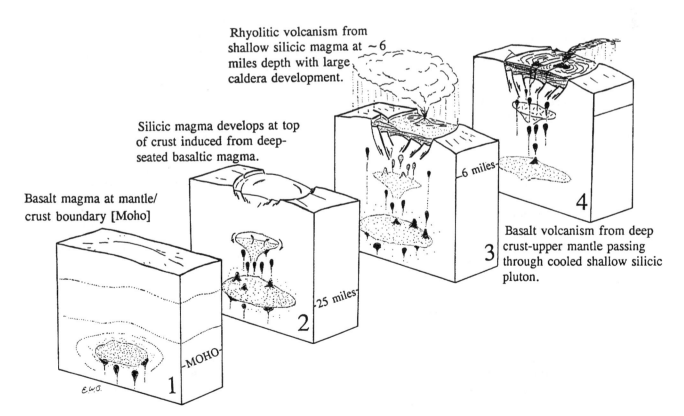

Figure 8.6 Eruption model for Snake River volcanics explains the transition from rhyolitic to basaltic activity, two distinct bimodal phases (after Leeman, 1982).

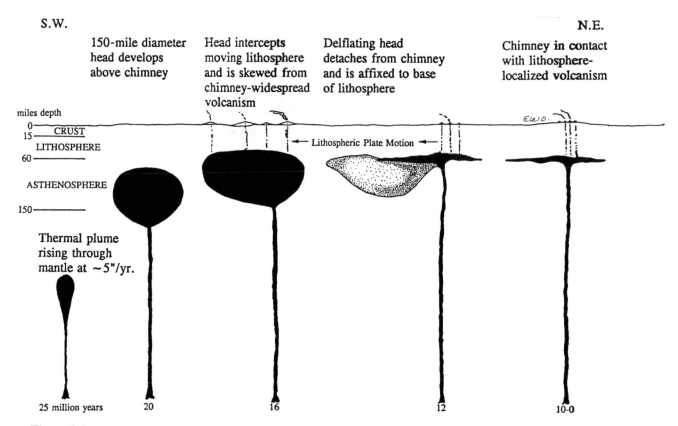

S.W.

150-mile diameter
head develops
above chimney

Head intercepts
moving lithosphere
and is skewed from
chimney-widespread
volcanism

Deflating head
detaches from chimney
and is affixed to base
of lithosphere

N.E.

Chimney in contact
with lithosphere-
localized volcanism

miles depth
0
15   CRUST
LITHOSPHERE
60
ASTHENOSPHERE
150

← Lithospheric Plate Motion ←

Thermal plume
rising through
mantle at ~5"/yr.

25 million years   20   16   12   10-0

Figure 8.7 Geometry and evolution of the Yellowstone mantle plume (after Pierce and Morgan, 1992; Richards, et al., 1989).

rapid removal of enormous amounts of molten rock material from underground chambers resulted in failure or collapse of the surface rocks during the final stage of the episode to form immense calderas. Measuring tens of miles across, hundreds of these calderas cover the entire width of the Snake River Plain, overlapping each other in intricate volcanic fields. Although thin layers of younger basalts obscure much of the older calderas and lavas, the rhyolitic flows had a much greater volume.

With the progression of time from the Middle Miocene, the focus of these rhyolitic eruptions steadily shifted across the Snake River Plain from the southwest to the northeast, beginning in the Owyhee region of Oregon and terminating in the Yellowstone volcanic province in Wyoming. This activity is interpreted as an example of tectonic plate movement over a mantle plume or "hot spot." As the North American Plate moved southwestward, a volcanic source known as the Yellowstone hot spot buried deep in the mantle periodically released molten material called magma. While the volcanic source remained stationary, the movement of the plate left a trail of eruptions like a smoke plume. Since the most recent rhyolitic eruptions along this chain are at Yellowstone Park, the hot spot lies beneath that plateau at present.

The Snake River Plain hot spot allows geologists to gauge accurately the direction and speed of the North

American Plate over the past 16 million years. Passing over the stationary mantle plume between 16 and 10 million years ago, North America moved west-southwest at 2½ inches per year. For the past 10 million years, however, the plate has shifted direction to a more south-southwesterly course and reduced its speed to about 1 inch per year.

The changing geometry of the mantle plume is believed to be responsible for the pattern of volcanism across the Snake River and Owyhee areas. Experimental research suggests that a rising mantle plume develops an inflated "head" above the "chimney" with a configuration like a balloon on a string. As the North American plate moved westward over the buried Yellowstone plume, two events occurred. First the head of the plume, well-imbedded in the crustal rocks below southeast Oregon at that time, began to move and shear away from the chimney. Then as the two separated the head began to deflate and shrink in size. In the Middle Miocene interval, 16 million years ago, the source of the widespread volcanic fields in southeast Oregon was the head, which was estimated at over 150 miles in diameter. By Late Miocene time and into the Pliocene, as volcanic activity progressed eastward, the narrow chimney provided lava and ash to an eruptive zone only 5 to 10 miles wide near the Idaho-Wyoming border.

It has been suggested that the Yellowstone mantle plume-hot spot may be much older than Middle Miocene and that it might be the source for Early Tertiary volcanic activity beneath the Coast Range of western Oregon and Washington.

Mantle plumes emerge as hot spots at more than 40 sites around the globe. Perhaps the most famous hot spot in the world lies beneath Hawaii where a cluster of huge coalescing shield cones on the Big Island are situated over a mantle plume. Off to the northwest a chain of progressively older volcanic islands string out as Maui, Molokai, Oahu, and Kauai. Below sea level, the trail of cones continues all the way to the margin of the Pacific Ocean as the Emperor seamount chain of volcanic mountains on the sea floor.

In spite of the myriad of hot spots, their origin is unclear. One passing notion was that these plumes might mark old sites of direct meteor hits. Such massive meteor strikes were thought to have penetrated deep into the mantle, miles below the crust, to generate tremendous heat with subsequent large volumes of magma. While this idea is creative, there is no hard evidence in the rock record to support it.

THE McDERMITT CALDERA, regarded as the starting point for the Yellowstone hot spot in its 400-mile passage, lies in the southeast corner of Oregon. Between 15 and 16 million years ago a remarkable volume and

Figure 8.8 Distribution of rhyolitic calderas and volcanic centers in the Snake River province shows both an eastern progression of activity and increased narrowing of eruptions (after Hackett and Morgan, 1988; Pierce and Morgan, 1992; Rytuba, et al., 1991).

Owyhee caldera complex
1. Castle Peak caldera
2. Three Fingers caldera
3. Mahogany Mountain caldera
4. Saddle Butte caldera

McDermitt caldera complex
5. Hoppin Peak caldera
6. Jordan Meadow caldera
7. Calavera caldera
8. Long Ridge caldera
9. Pueblo caldera
10. Whitehorse caldera
11. Washburn caldera

12. Juniper Mountain volcanic center
13. Bruneau-Jarbidge volcanic center
14. Magic Reservoir caldera
15. Twin Falls volcanic field
16. Taber caldera
17. Picabo volcanic center

Heise volcanic field
18. Blue Creek caldera
19. Kilgore caldera
20. Blacktail caldera

21. Rexburg caldera
22. Henrys Fork caldera
23. Island Park caldera
24. Yellowstone caldera

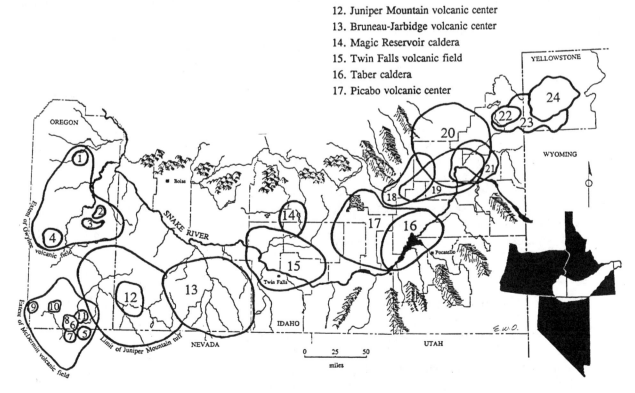

Figure 8.9 Bizarre shapes have been eroded into rhyolitic lavas at the City of Rocks near Gooding (Idaho Dept. of Transportation).

variety of explosive activity began in the McDermitt and Lake Owyhee fields. Quartz-rich rhyolitic eruptions in this broad volcanic field produced a series of a dozen gigantic overlapping calderas that straddle the Oregon-Nevada border. Less than 100 miles to the north in the Lake Owyhee region Mahogany Mountain, Three Fingers, Castle Peak, and Saddle Butte volcanoes erupted ash and lavas about 15.5 million years ago before collapsing into similar vast depressions from 5 to 15 miles in diameter.

Rhyolitic flows and ash were not the only event in this vicinity during the Middle Miocene. The Steens basalt, with an astonishing total volume of close to 3000 cubic miles, poured out over 6000 square miles. Extending well beyond both the Lake Owyhee and McDermitt fields, the Steens lavas covered most of southeast Oregon, spilling into Idaho and Nevada. The proximity and similarity in age of the Lake Owyhee rhyolites to the Steens basalt suggest that both events relate to the early history of the mantle plume.

THE BRUNEAU-JARBIDGE EVENT BEGAN at least 12 million years ago as continuing eruptions of the Yellowstone mantle plume progressed eastward. Focused on the Owyhee plateau in southwest Idaho, voluminous quantities of ash and lava were released before the central cone of the volcano collapsed into an enormous oval

crater 30 by 60 miles across. Rhyolitic flows from the Bruneau-Jarbidge volcano typically are 300 feet thick, and the largest exceeds 800 feet with almost 50 cubic miles of lava. Today the only visible remnant of the splendid buried caldera is the Grasmere escarpment along the west side of the eruptive center. At least 40 smaller buttes, which dot the floor of the older caldera, represent basaltic cones from a later phase of activity.

Also in Owyhee County, Juniper Mountain to the southwest of Bruneau River canyon is a low dome above the plateau that issued sheets of rhyolitic lava about 14 million years ago. Because of the age and great depth of the magma chamber, clear-cut evidence of a caldera is lacking. Tuffs from Juniper Mountain are spread over the southwest corner of Idaho and northern Nevada, producing the region known as the Badlands, where rapid erosion of the soft ash has created a distinctive topography.

THE SERIES OF VOLCANIC EVENTS AT MAGIC RESERVOIR, midway along the track of the hot spot plume, began 3 to 6 million years ago with tremendous amounts of rhyolitic lavas and ash smothering an extensive area of what is now Blaine and Camas counties. Although this early episode probably produced a caldera, a veneer of younger sediments obscures the exact margins of the structure. Having an approximate diameter of 15 miles, the Magic Reservoir caldera is similar in size to

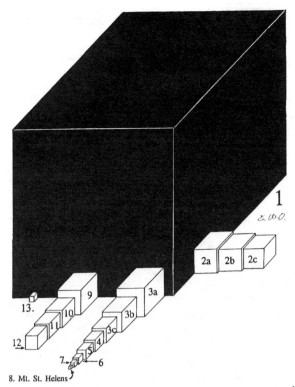

1. Columbia River plateau basalts 6-15 million, 55,680 miles^3
2. Heise volcanic field
    2a. Blacktail tuff, 6.5 million, 366 miles^3
    2b. Blue Creek tuff, 6.0 million, 342 miles^3
    2c. Kilgore tuff, 4.3 million, 244 miles^3
3. Yellowstone-Island Park
    3a. Huckleberry Ridge tuff, 2.0 million, 610 miles^3
    3b. Lava Creek tuff, .6 million, 244 miles^3
    3c. Mesa Falls tuff, 1.2 million, 68 miles^3
4. Tambora, 1815, 37 miles^3
5. Mazama, 6,900, 18.3 miles^3
6. Krakatoa, Indonesia, 1883, 4.3 miles^3
7. Katmai, Alaska, 1912, 1.7 miles^3
8. Mt. St. Helens, Wash., 1980, .25-.5 miles^3
9. Bruneau-Jarbidge volcanic field, 11-12 million, 488 miles^3
10. McDermitt volcanic field, 15-16 million, 404 miles^3
11. Owyhee-Humboldt volcanic field, 14 million, 349 miles^3
12. Picabo volcanic field, 10.3 million, 293 miles^3
13. Mount Pinatubo, Philippines, 1991-92, 2.5 mi.^{3}

Figure 8.10 Comparison of the volumes of Snake River volcanic material to those of the Pacific Northwest and other historic eruptions (after Christiansen, 1982; Hackett and Morgan, 1988; Pierce and Morgan, 1992).

others on the Snake River Plain. Inside the rim of the caldera, smaller amounts of basaltic lava were extruded to build domes as recently as 3 million years ago.

South of Magic Reservoir, the City of Rocks above Gooding displays bizarre erosional hoodoos—vertical eroded columns—carved into the older 9 million-year old rhyolitic tuffs that preceded those of Magic Reservoir.

USED AS A GUNNERY RANGE during World War II, Big Southern Butte in Butte County, at 2500 feet

in elevation, is the largest of three buttes which rise above the flat basalt plain here. Approximately 4 miles across at the base, this prominent 300,000-year old rhyolitic butte began as a pocket of underground magma that broke through to the surface as viscous lava that accumulated into a dome. When the center of activity moved northwesterly, construction of a second overlapping dome pushed up and tilted a 200-foot thick basalt slab, part of a former surface flow, on the northern slope of the butte. Middle Butte and East Butte are aligned to the northeast of Big Southern Butte. East Butte is a small 600,000-year old rhyolitic center, while Middle Butte is a raised block of older basalt. At the intersection of two structural trends running northwest by southeast, these three buttes may obscure an older buried caldera.

THE HEISE VOLCANIC FIELD in the northeast region of the Snake River province lines up with the Yellowstone plume. In this vicinity lava and ash from three overlapping calderas originated 6.5 million years ago, spanning the Miocene-Pliocene boundary. Erupting in three episodes during a 2 million-year interval, the Heise activity began with the explosion and collapse of the massive 70-mile wide Blacktail caldera. Reaching entirely across the Snake River Plain, the caldera produced over 350 cubic miles of rhyolitic lava and related volcanic material, an amount nearly 100 times the volume of the catastrophic 1883 Krakatoa eruption in Indonesia.

Shortly after the Blacktail event, the Late Miocene Blue Creek caldera of slightly smaller size formed, overlapping the older caldera on the western flank. After 1.5 million years, the Kilgore caldera coalesced with the southern margin of Blacktail. The Kilgore episode was close to the Blacktail in size but only erupted two-thirds the volume of ash and lava. Of the three tuffs, the Kilgore, at almost 250 cubic miles, is the smallest, but even that amount is almost 1000 times the volume released by Mount St. Helens in 1980. In comparing these volcanic occurrences, it is evident that such an eruption today would be of unprecedented destructive proportions.

ISLAND PARK VOLCANIC CENTER, at the northeastern margin of the Snake River Plain, is a component of the Yellowstone plateau. Beginning 2 million years ago during the Late Pliocene and continuing up to 70,000 years ago, three volcanic episodes with extensive ash flows formed two calderas. Small initial amounts of lava were followed by voluminous rhyolites before the central roof over the magma chamber collapsed to form a large crater. Much of the eastern boundary of this caldera is buried under the Yellowstone plateau, although a southwest section along Bishop Mountain is visible as Big Bend Ridge.

A brief second cycle of volcanism 1.3 million years ago produced the unique formation, craters, buttes, and

Figure 8.11 Middle Butte and East Butte near Arco, Idaho, are smaller than Big Southern Butte, but the three are part of the same volcanic episode thousands of years ago (Idaho Dept. of Transportation).

Figure 8.12 A geyser at Yellowstone Park reflects ongoing volcanic activity from the Yellowstone hot spot (courtesy J.K. Hillers, U.S. Geological Survey).

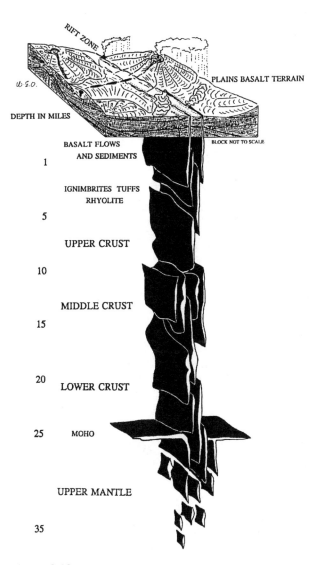

Figure 8.13 Model for eastern Snake River volcanic plumbing systems. Basalts, derived from magmas deep in the upper mantle and lower crust, erupted onto the surface as flows (after Greeley, 1982; Kuntz, 1992).

Rhyolitic volcanism subsided here as activity moved eastward with time toward Yellowstone Park where the third cycle began about 1.2 million years ago. After erupting over a span of 600,000 years, the circular fracture system sent several flows over into Island Park before the collapse of the magma chamber. The youngest ash flows were exceptionally thick and voluminous, producing the wide Madison, Pitchstone, and Central plateaus of Yellowstone National Park and nearby regions.

FLUID BASALT FLOWS CHARACTERIZE THE SECOND PHASE of bimodal volcanic activity on the Snake River Plain. During these comparatively quiet eruptions, dark quartz-deficient lavas welled up from cracks like hot syrup to fill low spots in the landscape and create vast flat volcanic plateaus. Spreading lava fields and low-profile shield volcanoes typify the topography produced by these basalt eruptions in contrast to the preceeding explosive rhyolitic dome and caldera-forming phase. Many of the basalt flows look remarkably recent, although most of them date back over 2000 years.

Along the eastern Snake River Plain, fissure eruptions coalesced as low shield cones, where lavas spread as much as 20 miles from a central vent to average 1.2 cubic miles of basalt per flow. There may have been as many as 8000 individual cones, with an eruption every 1000 years during a 10 million-year period. The total volume of basalt on the eastern Snake River Plain is estimated at 10,000 cubic miles.

Basalts issued from regularly spaced fissures or cracks oriented northwest by southeast and parallel to the underlying structural trend of the Basin and Range. With crustal stretching and thinning, Basin and Range-style fractures and cracks below the Snake River Plain were transmitted upward to appear as fissures. More than a half dozen major rift zones extend northwest by southeast across the axis of the Snake River Plain at intervals of 10 to 20 miles. As zones of weakness and linear fractures, the open rifts are natural avenues for volcanic materials, particularly basalts, that follow these cracks upward to spread out as surface flows.

One of the largest and best-exposed of these multiple open rifts is the appropriately named Great Rift that extends for 62 miles in a curved northwest by southeast trend from the Pioneer Mountains west of Arco to the Sublett Mountain range. Throughout its length, the zone is defined by a complex series of parallel faults, depressions, tensional cracks, fissures, and volcanic cones and vents. The northern half of the Great Rift zone in the Craters of the Moon volcanic field is narrow, but toward the south it expands from 1 to 10 miles in width as fissures of the Open Crack, Kings Bowl, and Wapi lava

other features common to Island Park. Again, when the magma chamber emptied, the roof collapsed inward to create Henrys Fork caldera situated within the southwest limits of the older volcanic depression. Ash-flow sheets from this eruption are a distinctive pinkish color with feldspar crystals up to an inch long. After caldera development, six small steep-sided rhyolite domes, Moonshine Mountain, Silver Lake, Osborne Butte, Elk Butte, Lookout Butte, and Warm River Butte, formed. Except for Warm River Butte to the south, these lava domes project above the floor of the younger second caldera. Today remnants of the initial crater are approximately 60 miles long and 20 miles wide, while the second smaller Henrys Fork caldera, nestled within the older one, is approximately 12 miles across.

Figure 8.14 Buttes, fissures, and volcanic cones of the Craters of the Moon lava field.

fields. Within the Open Crack system, fractures are up to 300 feet long and 800 feet deep. Volcanic activity, triggered by the opening of the Great Rift, is relatively recent, and it is estimated that the Craters of the Moon eruptions originated about 15,000 years ago and wound down about 2100 years ago. The Great Rift has been designated as a National Landmark.

Basaltic lavas of the Pleistocene and Holocene created the flat landscape seen today in the eastern Snake River Plain. Eight separate lava fields here include the Shoshone, Craters of the Moon, Wapi, Kings Bowl, Cerro Grande, North and South Robbers field, and Hells Half Acre.

THE SHOSHONE LAVA FIELD, from Gooding through Lincoln and Blaine counties, was active as late as 10,100 years ago, Black Butte Crater at the northern tip of the field exuded a 35-mile arcuate strip of lava that filled Big Wood and Little Wood river valleys. The streams were forced to change their channel so that today they are situated 25 miles west of their original junction. Extending southward from the summit of the crater is a 3-mile long lava tube that has partially collapsed. Within this tunnel system, Shoshone Ice Cave has been commercially developed. Water seeping into the cave through cracks and crevices freezes during the wintertime into curious delicate ice crystals.

Figure 8.15 In the Devils Orchard at Craters of the Moon National Monument, trees were frequently engulfed in hot flowing lava (Idaho Dept. of Transportation).

CRATERS OF THE MOON VOLCANIC FIELD, one of the larger basalt wastelands in North America, began with eruptions 15,000 years ago from a 25-mile long fissure across Blaine and Butte counties of south-central Idaho. Lava within the field issued from cracks at the north end of the Great Rift system in several episodes that produced over 25 cones and related volcanic features. Individual eruptions lasted about 100 years, and it is possible that additional activity could occur in the fu-ture. Craters of the Moon was declared a National Monument in 1924.

Most of the volcanic activity here was a quiet release of lava accompanied by steam and frothy gasses. Billowy, ropy pahoehoe lava flows, which resemble a solidified river, make up most of the fresh-looking surface, whereas blocky aa lava here cooled in thick broken slabs. Both the pahoehoe and aa lavas are dark brown from oxidized iron, but some of the pahoehoe displays a blue-green iridescence present at the Blue Dragon flow.

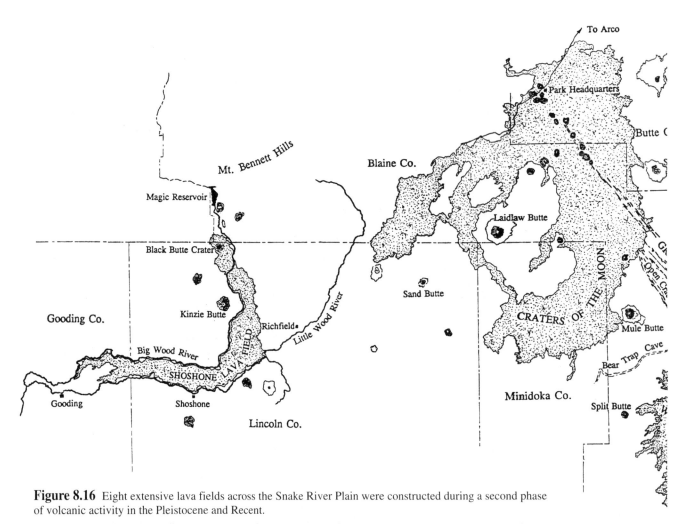

Figure 8.16 Eight extensive lava fields across the Snake River Plain were constructed during a second phase of volcanic activity in the Pleistocene and Recent.

Figure 8.17 The Great Rift, volcanic cones, and flat lava plain of Craters of the Moon National Monument are typical of the eastern Snake River landscape. The view is from the top of Big Cinder Butte looking southeast (courtesy H.T. Stearns, U.S. Geological Survey).

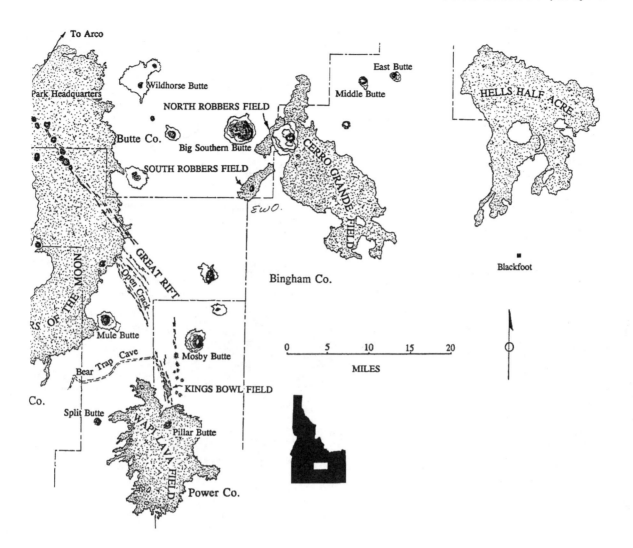

Figure 8.16 Continued.

Cinder cones, spatter cones, and shield volcanoes developed in neat lines along the rift and fissure system. Cinder cones, with a central crater, have smooth conical profiles covered with black, loose debris. Big Cinder Butte, which provides a wonderful view from its summit at 6516 feet, is an outstanding example of a pure basalt cinder cone. Spatter cones up to 50 feet high were built by fluid blobs of lava thrown tens of feet into the air and then annealed to the steep slopes. Good examples of cinder cones are Paisley, Inferno, and The Watchman. Smallest of all, lava cones or domes, some less than 20 feet above the plain, are broad and flat. These diminutive features developed when pahoehoe lava welled up quietly to the surface from underground tubes. Indian Tunnel, Great Owl Cavern, and Needles Cave are located near typical domes.

Other volcanic structures are lava bombs, tree molds, lava stalactites, chasms, and lava tube channels. The distinctive "bombs" lie scattered about on the surface. Ranging from ½ inch to more than 13 feet, bombs form when molten lava hurled through the air spins in flight and hardens into long twisted shapes. Because they cool and expand while airborne, many bombs have a cracked bread crust-like surface. Where lava engulfed a forest, standing and fallen trees, which were encased in a rind of basalt, formed casts. Because the trees burned, bark impressions are not normally preserved.

During eruptions, underground lava tubes or tunnels developed when the upper surface of a molten stream formed into a crust over a moving channel. After the liquid interior of the flow had drained away and the inner roof and sides of the tunnel cooled, hot lava continued to drip down to create volcanic "stalactites." Where sections of the roof of the lava tube have fallen through, inner caves are revealed, and in some cases isolated arches may be all that are left of large tube systems.

Figure 8.18 Aerial view of Kings Bowl, the large circular crater on the Great Rift at the center of the dark lava flow. The light area east of the crater is ash from one of the eruptions (U.S. National Archives CYN-7B-103).

One of the best preserved and longest lava tube systems is Bear Trap, which extends northeastward from the Great Rift into Minidoka County for 13 miles. Collapsed in several places, the walls of the tube experienced multiple flows as lava advanced miles from its source.

With the passing of time, thin soils supporting vegetation have developed in Craters of the Moon field. Because they have a high moisture content, cinders and cinder cones are especially susceptible to plants taking hold and growing.

THE 2222-YEAR OLD KINGS BOWL FIELD near American Falls is immediately north of the slightly older, 2270-year old Wapi lavas. Both are situated along the Great Rift fissure system. The eruptions from these two fields were almost simultaneous, but the two differ in that lava from the central fissure at Kings Bowl ceased after possibly only 6 hours, while the Wapi vent was active for months. The larger volume of basalt from the Wapi fracture created a low volcanic cone and covered a wider area.

Figure 8.19 Spatter cones near Big Craters in the Craters of the Moon lava field (Idaho Dept. of Transportation).

Kings Bowl is unique because of its variety of volcanic features such as explosion pits, basalt mounds, squeeze-ups where stiff lava was forced upward through cracks, and thick flows where lava lakes ponded up and hardened. Eruptions were explosive when molten lava encountered groundwater, and the 100-foot wide crack here became obstructed with blocky volcanic debris. Ice sometimes formed in a number of deep sections along extensive fissure systems. One locality, Crystal Ice Cave, was opened commercially until development destroyed the unusual ice creations.

The Wapi field displays a number of vents, basins, and lava tubes, along with the prominent landmark, Pillar Butte. This knob of aa lava, seen for some distance on the plain, is a spatter cone rising 50 feet above the south side of the main shield cone.

AT CERRO GRANDE, a large lava field roughly 30 miles southeast of Arco in Butte County, basalts were expelled from a small shield volcano that developed 13,400 years ago. Inside of the present 1/2-mile wide basin, the central vent is obscured by basalt that welled up in a lava lake. Adjacent to Cerro Grande, the small North and South Robbers fields resulted from lava extruded through long cracks 12,000 years ago. Rows of small spatter cones and ridges mark the fissures.

THE HELLS HALF ACRE VOLCANIC FIELD, adjacent to Idaho Falls on the eastern Snake River Plain, is second only to Craters of the Moon in size. A shield volcano, circular pit craters, spatter cones, rough broken slabs of lava, and collapsed lava tubes lie along a fissure over 2 miles long. The center of the shield structure was

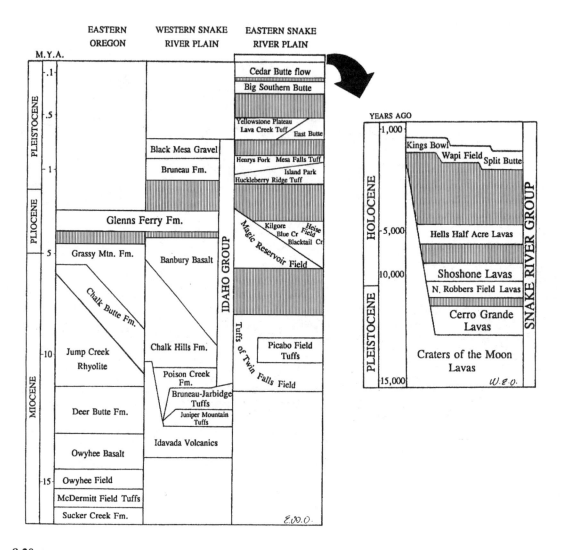

Figure 8.20 Stratigraphy of the Snake River Plain and southeast Oregon (after Leeman, 1982; Morgan, 1992; Rytuba, et al., 1990; Walker, 1979).

filled with large volumes of molten material 5200 years ago producing a lava lake. Pahoehoe lava, overflowing from the lake, spread out for almost a mile.

Among the numerous isolated Late Pleistocene buttes that are located within the eastern Snake River Plain, Sand Butte, Split Butte, and Menan Buttes are distinct from other volcanic cones because they are maar craters, shallow basins that resulted when hot rising basaltic lava came into contact with groundwater. The ensuing explosions forcefully ejected shattered glassy debris that settled back as a high ring around the opening. The symmetrical tuff ring surrounding the central crater at Sand Butte southwest of Arco sits astride a long deep fissure. Here airborne tephra and basaltic lavas interacted to create the cone and surrounding lava flows.

Near the west side of the Wapi lava field, the ash ring at Split Butte grew with each eruption until it reached the present-day size close to a mile across. At a later phase of development, a lake of lava filled the cen-

tral depression, and small cinder cones sprouted along the inner wall of the crater. The two large rings and three smaller ones at Menan Buttes, about 20 miles north of Idaho Falls in Jefferson County, are composed of basalt cinders and broken debris, which have accumulated in mounds up to 800 feet high. Prevailing winds here as at the other buttes have carried the light ejected tephra to the north and east, elongating the rings in that direction.

LAVA FLOWS ARE PRONE TO INVADE AND PLUG RIVER valleys, while streams cut down and flush them out, so that these two geologic agents are in perpetual conflict. Within the Snake River province, these opposing forces have interacted from the Miocene through Pleistocene to produce large lakes and deep river canyons.

From the middle Miocene, 13 million years ago, lava flows began to alter the course of the Snake River, damming the water at the narrows of Hells Canyon on

Figure 8.21 The otter *Satherium* inhabited the Snake River Plain during the Middle Miocene, approximately 8 million years ago (skull from C.L. Gazin, U.S. National Museum).

the Oregon-Idaho border to produce ancient Lake Idaho. In actuality there were two successive large lakes separated by an interval of several million years.

This natural reservoir filled the western Snake River graben in Idaho as well as major river tributaries in eastern Oregon. With a length of 150 miles and a width of 50 miles, Lake Idaho rivalled present-day lake Ontario in New York state. At its highest level, the waters reached 3800 feet in elevation. Although the shorelines of the lake are difficult to follow, they can be traced by floodplain silts, beach sands, fossil algae deposits, and distinctive pillow basalts that occur along the lake margins where lava flowed directly into the waters. Lake levels rose and fell several times, finally to recede as the water drained during the Early Pleistocene.

Stream, floodplain, marsh, and lake deposits of the Glenns Ferry and Chalk Butte formations from ancient Lake Idaho date back to the Middle Miocene and Pliocene epochs and contain rich accumulations of fossil plants and animals. Fish bones, of a size and species that reflect a very large body of water, have been collected from close to 200 sites between Hagerman, Idaho, and Adrian, Oregon, where they litter the old lake floor. Thirty-nine separate species of fish have been identified. The most common are whitefish, trout, salmon, minnows, and members of the sucker family. Following the draining of the lake, most of these fish became extinct.

At the celebrated Hagerman quarry, fossil horse remains along with those of rabbits, peccaries, gophers, beaver, small reptiles, and birds have been uncovered. Herds of the rare ancestral horse, *Plesippus*, resembling a small pony, roamed here during the Pliocene, 3 million years ago. Rodents such as squirrels, rats, mice, and muskrats supported a group of predators that included cats, weasels, badgers, coyotes, and a bone-cracking scavenger dog. The most frequent fossil carnivore was the otter, *Satherium*, which is curiously rare today throughout North America. As indicated by plant remains of hickory, alder, poplar, oak, and fir, the climate was much wetter and warmer than at present. The

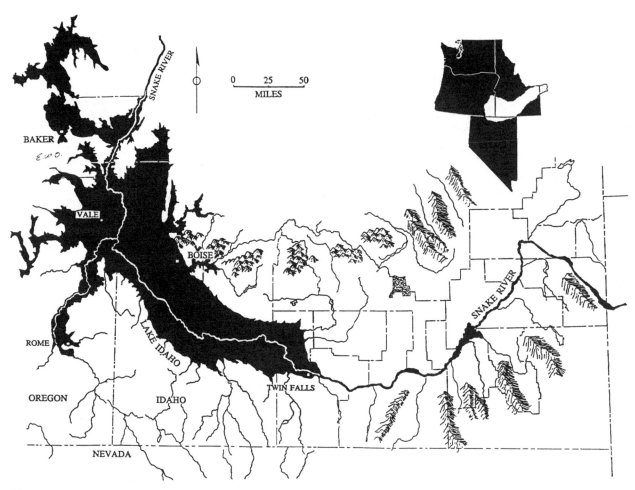

Figure 8.22 Bounded on its margins by large-scale faults, Lake Idaho [shaded], which stretched into eastern Oregon, was one of the most extensive lava-dammed lakes of the Miocene (after Middleton, Porter, and Kimmel, 1985; Smith, et al., 1982).

Figure 8.23 Tranquil Hagerman Valley in Gooding County, the site of an immense lake dating back 3 million years ago, is famous today for its abundant fossil remains of animals that lived here during that time (Idaho Dept. of Transportation).

Figure 8.24 Lavas, which covered close to 2000 square miles of southwest Idaho millions of years ago, are now layered in the canyon walls of the Bruneau River (Idaho Dept. of Transportation).

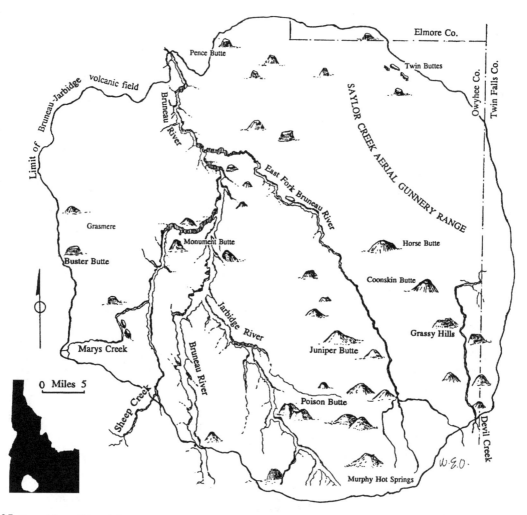

Figure 8.25 Over 40 small basalt buttes are spread across the floor of the older Bruneau-Jarbidge caldera in southwest Idaho.

Hagerman fauna site in Twin Falls County has been declared a National Monument.

To the southeast of Lake Idaho, a somewhat smaller Late Pleistocene lake formed when a lava dam backed up waters of the Snake River into Lake McKinney. Located in Gooding County, basalt from McKinney Butte impounded lake waters to depths up to 600 feet. At its maximum, the body of water extended upstream along the Snake River channel through Hagerman Valley and Thousand Springs to Melon Valley before being drained. Yahoo clay deposits are evidence of the old lake bed.

Lava from the eruption of Cedar Butte, a basaltic shield cone in Power County, blocked the Snake River at Eagle Rock around 70,000 years ago to create American Falls Lake. This impoundment of water extended for 40 miles upstream with a maximum water level at an elevation of 4500 feet. A variety of pollen, leaves, mollusc shells, and vertebrate fossils from these lakebed sediments testify to the cooler, rainy Late Pleistocene environment around American Falls. Conifers, sagebrush, and members of the goosefoot plant family lived on a high plateau surrounding the lake, while water-loving cattails crowded the shore. Enormous mammoth, bison, and modern-sized horses frequented the plateau.

The end of American Falls Lake came in a spectacular deluge as Bonneville flood waters poured into the basin. With this immense and rapid influx of water the lake spilled over its natural dam to downcut and destroy the obstruction. Once the Snake River had receded to its present level, the 50-foot high American Falls was left to cascade down over basalt blocks.

AS REPEATED FLOWS OF LAVA TOOK OVER STREAM VALLEYS, rivers reestablished their drainage by flushing out the basalt or cutting new pathways. Working to erode the blocked channels, waters left a network of deep gorges, sheer walls, and dramatic

Figure 8.26 Owyhee canyon was cut by the river through thick ash and lava of the eastern Oregon high plateau (Oregon State Highway Commission).

precipices in the Owyhee high desert plateau where the Bruneau, Jarbidge, and Owyhee rivers cut through successive layers to reveal a history that began with volcanic eruptions 15 million years ago.

The Jarbidge River and Sheep and Clover creeks of southeast Idaho, with canyons well over 800 feet deep, merge with the spectacular Bruneau River to flow between 1300-foot wide steep walls that narrow to only 30 feet in places. Iron-stained volcanic layers, visible in the canyon, are shades of brown, hence the name Bruneau, derived from the French "brun" used by early trappers. The 240 miles of canyonlands cut by the Owyhee River and its tributaries extend over 450,000 acres in Nevada, Idaho, and Oregon. Successive accumulations of ash and lava, which delineate individual flows, have decomposed and oxidized to beautiful hues of yellow, brown, and red. Fossils of ancestral horse, deer, camels, rhinoceroses, and a terrestrial beaver that was equal to a bear in size are entombed in sedimentary layers between the lava. These fossils attest to long intervals with lush grasslands between violent episodes of volcanism.

ICE AGE LAKES were yet another phenomena that left an array of deposits on the Snake River Plain. When Lake Bonneville in Utah spilled northward through Red Rock pass in the southeast corner of Idaho 14,500 years ago, the Snake River Plain experienced a flood of biblical proportions. Pouring along the Marsh Creek and Portneuf valleys to Pocatello, the water swept down the Snake River canyon eventually joining the Columbia River in southern Washington. Unlike the 40 to 100 floods that came down to the Columbia River from glacial lakes impounded near Missoula, Montana, the Bonneville occurrence was apparently a single catastropic event. Although the maximum discharge of the flood was 20 times the flow of the Columbia River, it did not approach the size of some of the glacial floods from Montana.

At its height during the Ice Ages, Lake Bonneville covered over 20,000 square miles in western Utah, parts of Nevada, and Idaho, an area almost equal to the state of West Virginia. A rise in the water level in Lake Bonneville began around 600,000 years ago when basalt eruptions diverted the flow of the ancestral Bear River southward into the Bonneville basin. Onset of the wetter Pleistocene climate 50,000 years ago and reduced river drainage combined to raise the lake surface to 5090 feet. After maintaining a constant level for hundreds of years, erosion of the natural Red Rock pass brought about a dramatic and sudden lowering in the water. A short time later the lake levels subsided close to those of the present day.

On both sides of the Snake River channel, the Bonneville flood waters spread out for several miles, scouring basalt layers and cutting deep scablands. By

Figure 8.27 The Bruneau Sand Dunes State Park on the Snake River south of Mountain Home has crescent-shaped barchan dunes hundreds of feet in length that are composed of fine grains of basalt eroded from the surrounding plateau. These features, with horns pointed downwind, indicate a predominate southwest wind (courtesy G.K. Gilbert, U.S. Geological Survey).

hydraulically vacuuming up millions of tons of rock, the cascade left unmistakable signs of erosion and deposition along its route. Among these are the many high sand bars, vast beds of coarse gravel, and thick layers of clay, silt, and sand collected in basins. Well-rounded mellon-sized boulders and smaller pebbles, transported by the turbulent flood waters, were released and strewn over sand bars as the force slackened. Huge boulders of basalt were torn loose and tumbled along by the water to their present locality downstream from American Falls Reservoir, where they may be seen today at Massacre Rocks. The area is so named because of the role the rocks played in Indian attacks along the Oregon Trail.

Figure 8.28 Pleistocene Bonneville Lake [in black] and the path of the devastating flood from Utah through the Snake River valley in Idaho, joining the Columbia River in Washington (after O'Connor, 1993).

Figure 8.29 Beaches and wave-cut shorelines of ancient Lake Bonneville show up clearly around the present Salt Lake basin (courtesy C.D. Walcott, U.S. Geological Survey).

The stripping of covering sediments from basalt surfaces resulted in a scabland topography, with deep potholes at Eden in Jerome County as well as the cataracts and falls along the river from Twin Falls and Shoshone Falls to Crane Falls and Swan Falls downriver. The picturesque Shoshone Falls was a tourist attraction as early as 1883, but stream diversion for power plants has drastically altered the scenery since those early days. The same is true of Twin Falls where water diversion has choked off much of the flow, greatly diminishing its previous grandeur.

THE SNAKE RIVER AQUIFER is the largest source of underground water in Idaho and one of the most productive anywhere in the world. Underlying an area over 10,000 square miles across the Snake River Plain, permeable volcanic rocks saturated with groundwater make up the aquifer. The aquifer is estimated to have twice the volume and several times the water quality of that in Lake Erie. In the western region, water moves very slowly through buried sand and gravel

Figure 8.30 Moving through fractures and cracks in basalt, water from the Snake River aquifer emerges as springs near Twin Falls, Idaho (Idaho Dept. of Commerce and Development).

Figure 8.31 Water tumbles over a massive basalt lip at Twin Falls on the Snake River (Idaho State Historical Society).

layers, whereas under the eastern plain it easily flows at three to five feet per day through open spaces, shrinkage cracks, and lava tube systems in basalt. The level of the Snake River aquifer drops 3000 feet in less than 200 miles. At the surface, water emerges as springs, many of which are artesian. Heavy private and commercial withdrawal from the aquifer is causing water levels to fall slowly, and, in places, livestock and agricultural operations severely threaten the quality of this resource. With 3 million acres of land fed from the aquifer, Idaho has the highest per capita use of water in the United States.

Water, as at Thousand Springs, commonly flows from basalt bluffs on the Snake River. In southern Gooding County, Thousand Springs, partly made up of water from the Snake and Lost rivers, exits from underground conduits at a rate of nearly 600 cubic feet per second. As with most other water-related features along the Snake River valley, this spring has been significantly altered by hydroelectric dams.

ALTHOUGH HUNDREDS OF THERMAL SPRINGS and water wells are scattered throughout Oregon and Idaho, the rapid flow of cold ground water through the Snake River aquifer virtually eliminates warm water springs in the plains section of this province. The border with Yellowstone National Park was considered to be a prime target for geothermal potential, but thermal waters, as at Big Springs in Fremont County, only reach temperatures of 56 degrees Fahrenheit.

In contrast, hot water springs commonly arise on the Owyhee plateau, and drilled wells frequently yield water with high temperatures. Water reaches over 150 degrees Fahrenheit near Bruneau and Mountain Home and a pleasant 90 degrees at Murphy Hot Springs in Idaho. At Vale along the Oregon border, the potential for geothermal power is exhibited by the significant number of surface springs and wells with temperatures over 200 degrees.

In Blaine County, at the northern margin of Magic Reservoir, hot well water temperatures of 160 degrees are a result of its being located at the edge of an older rhyolitic eruptive center.

Figure 8.32 Massacre rocks, 12 miles southwest of American Falls, marks the site of ill-fated Indian attacks on travelers along the Oregon trail. Scattered among the rocks are massive boulders left behind by dynamic Bonneville flood waters (Idaho Dept. of Transportation).

Figure 8.33 The Silver City district of Owyhee County experienced a brief boom for about 10 years before the ore was mined out, even though the silver-gold veins were extremely rich (Idaho State Historical Society).

MINERAL PRODUCTION

While Idaho produces a wide variety of metals and minerals, industrial mining products have less of a "romantic" appeal than gold and silver, both of which have played an important historically role in Idaho.

The Owyhee uplands was the scene of a spectacular boom from 1863 until 1875 with the discovery of silver on Jordan Creek. As news of the riches spread, hundreds of prospectors rushed to the district only to find over 300 claims already staked out, and Silver City on its way to becoming a full-fledged town with six general stores, eight saloons, and a hospital. Further investigation led to the exploitation of the fabulously rich Orofino and Morning Star veins. Mines tunneled into these veins on War Eagle Mountain yielded $13,000,000 in both gold and silver over an 11-year period. Higher production costs, the presence of mostly low-grade ores, and the collapse of the banks in 1875 brought about closure of the War Eagle mines. In the 1990s new targets on War Eagle Mountain were being reexamined for their potential.

A mining resurgence between 1889 and 1914 was triggered by the discovery of new ore bodies on Florida Mountain in the same district, where approximately $30,000,000 in silver was mined during this 25-year interval. Dredges continued to operate on Jordan Creek long after the lode operations ceased. Following renewed interest in this area, a gold and silver mine south of Silver City was the foremost gold producer in Idaho during 1991. A mill and heap-leaching facility at this locality processed ores from nearby sites, and there are plans to reopen the Florida Mountain mine.

Veins of silver, lead, and antimony ores are restricted to fractures in the Silver City granite where hydrothermal emplacement took place prior to the Miocene.

THE ECONOMIC POTENTIAL for uranium and mercury in Tertiary lavas of the McDermitt field in southeast Oregon is the highest in the United States. Here mineral ores associated with rhyolitic lavas have been mined sporadically for almost 40 years. In the McDermitt calderas, faults have allowed mineral-laden hot water to percolate throughout, precipitating the ores in veins. Severe power shortages in both the summer and winter of the year 2000 renewed interest in nuclear power. In turn, McDermitt uranium reserves are being reaccessed and evaluated for additional mining of these ores.

Figure 8.34 Three Island crossing near Glenns Ferry was a common spot for wagons on the Oregon Trail to ford the Snake River. The islands have been built up by sand deposited at the sharp bend in the river (Glenns Ferry Chamber of Commerce).

Figure 8.35 Sheep Trail Mountain cinder cone from the northeast at Craters of the Moon in Idaho (H.T. Stearns, U.S. Geological Survey).

ADDITIONAL READINGS

Asher, R.R., 1968

Baker, V.R., et al., 1987

Bjork, P.R., 1970

Bonnichsen, B., 1982

Bonnichsen, B., et al., 1988

Bright, R.C., 1982

Christiansen, R., 1982

Conley, C., 1982

Ekren, E.B., et al., 1982

Gidley, J.W., 1930

Greeley, R., 1982

Greeley, R., and King, J.S., 1975

Hackett, W.R., and Morgan L.A., 1988

Houser, B.B., 1992

Hughes, S.S., et al., 1999

Jenks, M.D., and Bonnichsen, B., 1989

Kimmel, P.G., 1984

King, J. S., 1982

Kuntz, M.A., 1992

Kuntz, M.A., Covington, H.R., and Schorr, L.J., 1992

Kuntz, M.A., et al., 1982

Link, P.K., and Phoenix E.C., 1996

Leeman, W.P., 1982

Mabey, D.R., 1982

Malde, H.E., and Powers, H.A., 1962

Maley, T., 1987

Middleton, L.T., Porter, M.L., and Kimmel, P.G., 1985

Morgan, L.A., 1992

O'Connor, J.E., 1993

Pierce, K.L., and Morgan, L.A., 1992

Rytuba, J.J., Vander Meulen, D.B., and Barlock, V.E., 1991

Scott, W.E., et al., 1982

Shotwell, J.A., 1970

Smith, G.R., et al., 1982

Stearns, H.T., 1928

Stephens, G.C., 1988

9

BASIN AND RANGE

Composed of folded Precambrian and Paleozoic rocks, Mt. Borah in central Idaho is the highest point in the state at 12,662 feet (Idaho Dept. of Transportation).

The vast Basin and Range reaches into Utah, Nevada, Idaho, Oregon, Arizona, California, New Mexico, and Mexico. Covering an area of 300,000 square miles, this single province comprises almost 1/10 of the United States. Although the northern Basin, which takes in parts of Idaho, Oregon, and California, includes areas of diverse surface topography, the underlying north-south trending mountain ranges and alternating valleys tie the entire region together.

In eastern Idaho, this distinctive topography appears on both sides of the Snake River Plain where the north-east portion encompasses the Beaverhead Mountains, Lemhi Range, and Lost River Range interposed with broad basins. Southeast of the Snake River, the Caribou Range, the Blackfoot and Chesterfield Mountains, the Portneuf Range, Deep Creek, and Sublett mountains follow the same trend. In the Lost River Range, Borah Peak is the highest point in Idaho at 12,662 feet, while a short distance east of these mountains, the Beaverhead Range marks the Continental Divide as well as the border between Idaho and Montana.

Across central and southern Oregon, the Basin and Range takes in the volcanic region of the High Lava Plains. Except for recent small-scale fault scarps and

Figure 9.1 Lemhi Pass, on the Continental Divide between Idaho and Montana, is a National Historic Landmark and one of the highest points on the Lewis and Clark Trail at 7373 feet in elevation. The pass was cut by glaciers working back-to-back (Idaho Dept. of Commerce and Development).

isolated buttes, the topography of the Lava Plains is smooth with little relief. On the western margin, Paulina Peak at Newberry Crater rises to 7984 feet above sea level while Harney Basin on the east sits in a broad depression at an elevation of 4080 feet.

That section of the Basin and Range that extends into northeast California includes the volcanic terrain of the Modoc Plateau and Medicine Lake Highland. Averaging several thousand feet in elevation, the Modoc Plateau is a flat volcanic tableland broken up by a pattern of sharp linear fault scarps and shallow valleys. The broad basins of Surprise Valley, Goose Lake, Tule Lake, and Big Valley lie between the Warner Range, the

ridges near Adin, and the Big Valley Mountains. The Warner Mountains at the Nevada border reach 9883 feet at Eagle Peak, the highest point.

On its western border the Modoc Plateau merges with the Medicine Lake Highland, which is an upland constructed of recent lava flows. Medicine Lake, an old caldera at the center of the highland, has an elevation of 6700 feet. Lava Beds National Monument, which occupies the northeast corner of the region, presents a singularly impressive tableau of volcanic caves, buttes, and flows.

Rivers within the northern Great Basin have no direct outlet to the sea, and only the Klamath, which originates in Upper Klamath Lake, Oregon, enters the Pacific

Figure 9.2 Features of the Basin and Range province in southeastern Oregon and northern California.

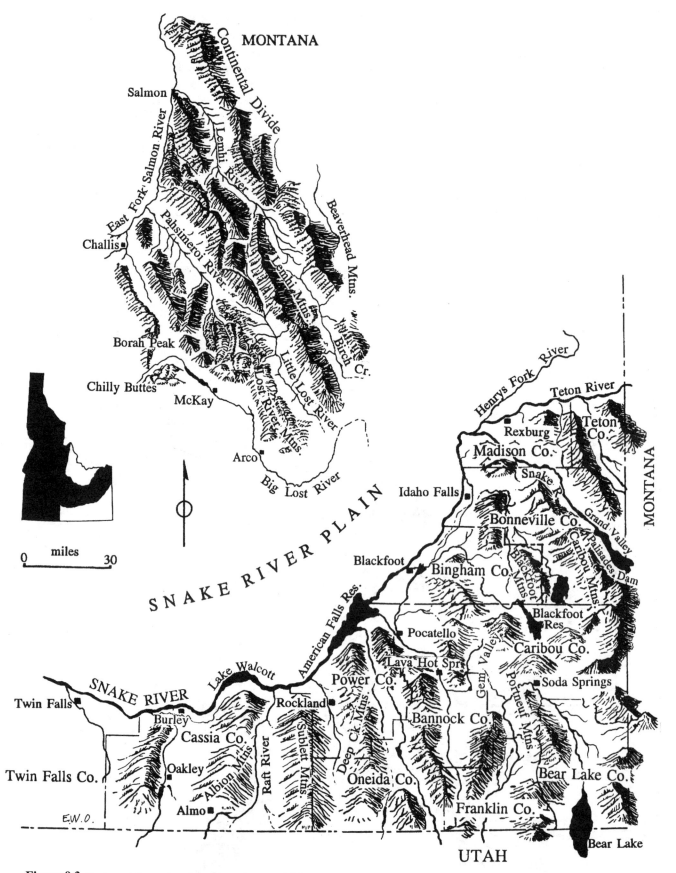

Figure 9.3 Basin and Range locations in southeastern Idaho.

Figure 9.4 Pits and fractures of this flow in Devils Garden, Lake County, Oregon, are typical of recent lava fields in the northern Basin and Range (Oregon Dept. of Geology and Mineral Industries).

Ocean after a torturous route through northern California. The lengthy Deschutes River follows the northwestern border of the province to the Columbia while the Silvies and Donner und Blitzen rivers end in the Harney Basin. Meandering through the lava of the Modoc Plateau, the Pit River flows from northeast to southwest before draining into Lake Shasta.

In southeast Idaho, rivers tend to have short runs. Those that drain the mountains north of the Snake River flow less than 25 miles before sinking into the basalt of the Snake River Plain. South of the plain, rivers, which have their headwaters in Utah, Idaho, or Wyoming, enter the Snake River after travelling 70 miles or less. The exception is Bear River, which flows north from Utah to Soda Springs, Idaho, then back south to the Great Salt Lake, a distance of 450 miles.

Most of the northern Basin and Range is semiarid with less than 10 to 15 inches annual rainfall. The climate has imposed a typical desert topography of enclosed lake playas and broad alluvial fans washing out into widening basins.

INTRODUCTION

The geologic signature of crustal stretching that characterizes the Basin and Range extends well beyond its conspicuous physiographic boundaries. Marked by elongated valleys and eroded chains of hills, this province displays the scars of intensive extension that began 17 million years ago and continues even today. As the earth's semibrittle crust was slowly pulled apart, it thinned considerably before the rocks failed, and faulting broke up the surface. This attenuated crust and pattern of long tensional faults give the province its distinctive tilted block mountains, wide valleys, high heat flow, and volcanic activity.

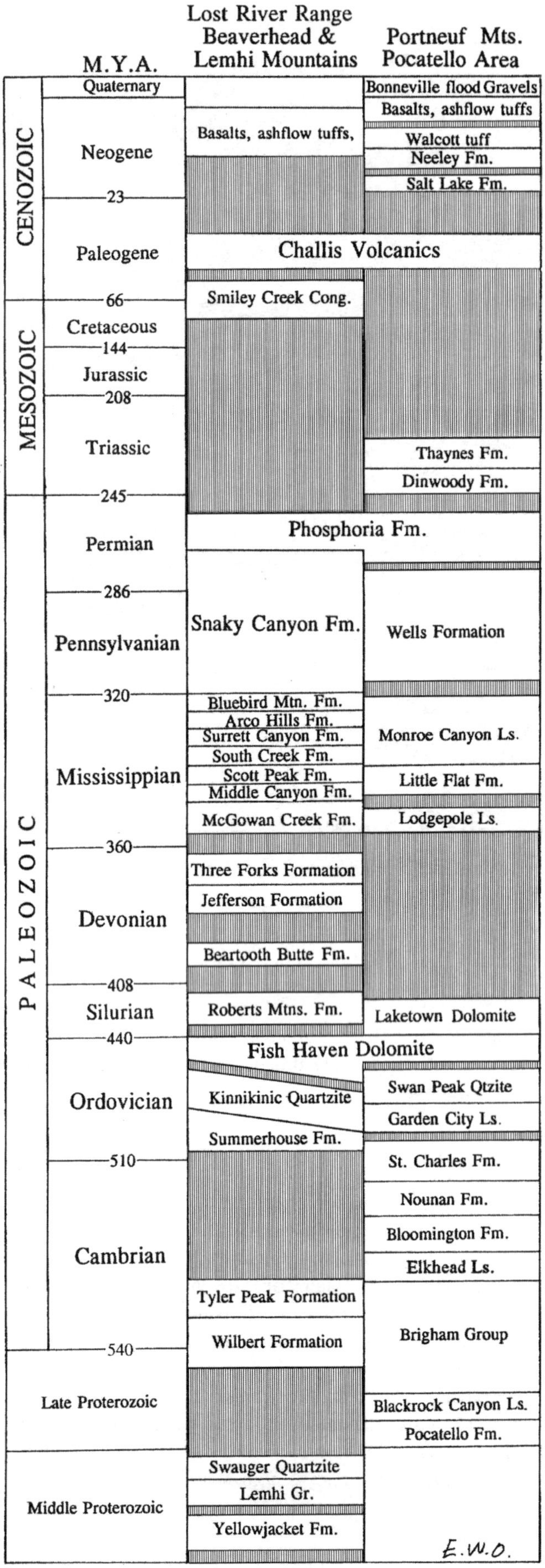

Figure 9.5 Stratigraphy of central and southeast Idaho (after Hintze, 1985; Link, et al., 1988; McFaddan, Measures, and Isaacson, 1988).

Era	M.Y.A. / Period	Lost River Range Beaverhead & Lemhi Mountains	Portneuf Mts. Pocatello Area
CENOZOIC	Quaternary		Bonneville flood Gravels
CENOZOIC	Neogene	Basalts, ashflow tuffs,	Basalts, ashflow tuffs / Walcott tuff / Neeley Fm. / Salt Lake Fm.
CENOZOIC	—23—		
CENOZOIC	Paleogene	Challis Volcanics	Challis Volcanics
	—66—	Smiley Creek Cong.	
MESOZOIC	Cretaceous		
MESOZOIC	—144—		
MESOZOIC	Jurassic		
MESOZOIC	—208—		
MESOZOIC	Triassic		Thaynes Fm. / Dinwoody Fm.
	—245—		
PALEOZOIC	Permian	Phosphoria Fm.	Phosphoria Fm.
PALEOZOIC	—286—		
PALEOZOIC	Pennsylvanian	Snaky Canyon Fm.	Wells Formation
PALEOZOIC	—320—	Bluebird Mtn. Fm. / Arco Hills Fm. / Surrett Canyon Fm. / South Creek Fm. / Scott Peak Fm. / Middle Canyon Fm.	Monroe Canyon Ls.
PALEOZOIC	Mississippian		Little Flat Fm.
PALEOZOIC		McGowan Creek Fm.	Lodgepole Ls.
PALEOZOIC	—360—	Three Forks Formation	
PALEOZOIC	Devonian	Jefferson Formation / Beartooth Butte Fm.	
PALEOZOIC	—408—		
PALEOZOIC	Silurian	Roberts Mtns. Fm.	Laketown Dolomite
PALEOZOIC	—440—	Fish Haven Dolomite	Fish Haven Dolomite
PALEOZOIC	Ordovician	Kinnikinic Quartzite / Summerhouse Fm.	Swan Peak Qtzite / Garden City Ls.
PALEOZOIC	—510—		St. Charles Fm. / Nounan Fm.
PALEOZOIC	Cambrian	Tyler Peak Formation / Wilbert Formation	Bloomington Fm. / Elkhead Ls. / Brigham Group
PALEOZOIC	—540—		
	Late Proterozoic		Blackrock Canyon Ls. / Pocatello Fm.
	Middle Proterozoic	Swauger Quartzite / Lemhi Gr. / Yellowjacket Fm.	

E.W.O.

Long before the Basin and Range was distorted by stretching and thinning, it was subjected to severe compression that began during the Late Mesozoic 100 million years ago. Triggered by the docking and suturing of exotic terranes to the ancient western margin of the North American landmass, a strong eastward pressure crumpled and severely shortened the rocks of this province.

The Late Cenozoic era was a time when fairly continuous lava flows filled valleys and built vast elevated platforms across the northern margin of the province. As one eruption followed the other, thick lavas, oozing to the surface from cracks and fissures, obscured and smoothed over much of the earlier topography. Hundreds of buttes and larger stratocones, dark sheets of cinders, and blocky lava highlight these episodes.

Following this volcanic phase, the cooler Pleistocene epoch saw heavy rainfall that filled wide depressions in the landscape with pluvial lakes. Fish, birds, and mammals crowded the waters and banks of the lakes until a worldwide warming trend diminished these aquatic environments to relatively modest bodies of water or dry playas.

GEOLOGY OF THE NORTHERN BASIN AND RANGE

During the Early Paleozoic up to 400 million years ago, western North America was covered by a broad seaway with shorelines across what is now Colorado and Wyoming. To the west of the strand, a volcanic island archipelago separated this seaway from the proto-Pacific Ocean. Paleozoic sediments, derived from the North American continent, were carried into the ocean basin over long periods of time. These strata are dominantly marine and reflect alternating periods of advancing and regressive seas interspersed by intervals of deposition and erosion. Variable oceanic environments hosted faunas of corals, brachiopods, molluscs, crinoids, and trilobites.

With few breaks, deposition was ongoing into the early Pennsylvanian when it was followed by a Late Pennsylvanian period of faulting and subsiding basins that continued into the Permian. Sandstones and limestones of the Wells Formation in southeast Idaho, which preserve a variety of microfossils, corals, and crinoids, reflect these disruptions. The shallow marine shelf of the Permian was punctuated by deep basins. The vast Phosphoria basin of Idaho, Wyoming, and Montana that accumulated mudstone, chert, and phosphorus-rich deposits is

Figure 9.6 Developed in Paleozoic limestone, Minnetonka Cave near Bear Lake, Idaho, is the largest in the state (Idaho Dept. of Transportation).

noted for the build-up of coral banks, a characteristic of the Permian. Bioherms of this seaway supported dense communities of brachiopods, crinoids, byrozoa, and algae. The Phosphoria Formation covers 175,000 square miles at an average thickness of 700 feet; the region is known for its economically valuable deposits of phosphatic shale.

IN THE LATE MESOZOIC THE BASIN AND RANGE was altered by intense compressional forces from the west that signaled the arrival, collision, and accretion of exotic terranes. Carried on crustal slabs beneath the Pacific basin, a succession of well-traveled plate fragments moved slowly eastward to merge with the ancient North American continent.

Exotic terranes predominate in the central and southern Great Basin, but within the northern portion, they are largely obscured by the volcanic overprint. While the entire subsurface basement of the Basin and Range was originally thought to be a mosaic of accreted terranes, it has been shown that the crust here is very thin and that Late Tertiary lavas probably form most of the foundation rocks. The small exposures of exotic terrane rocks within this province may be isolated scraps left behind after the severe crustal stretching that took place less than 50 million years ago.

At Elko, Nevada, the Basin and Range can be divided into eastern and western portions, with rocks native to North America on the east and accreted rocks on the west. In Oregon, the Late Triassic Jackson terrane is composed of oceanic sediments, making up part of the Trout Creek Mountains. South in Nevada a fault separates rocks of this terrane from those of the Late Paleozoic to Middle Triassic Black Rock terrane of island arc volcanics associated with shales, cherts, and lavas. Near Winnemucca, a curved exposure of the Black Rock terrane is again separated by faults from deep-water ocean deposits of the Jurassic Jungo terrane.

Following Mesozoic collision of the terranes, compressional forces were transmitted into Idaho where thin sedimentary layers of the Paleozoic ocean on the edge of ancestral North America slowly began to bend and fold. As pressure increased, rocks failed or faulted and were thrust eastward over each other in a repeating series of domino-like overlapping plates. Prior to compression, the Wyoming-Idaho thrust belt, which runs along western Idaho, Wyoming, and the Rocky Mountain front range, may have been twice as wide as its present 75 mile width. Even as the front range was shrinking, the western margin of North America was growing significantly by the acquisition of layer upon layer of exotic terranes. Inland, however, along the Rockies there was a considerable loss in width as crustal pieces were now severely telescoped together.

After the final stages of collision, the mainland had been elevated well above sea level, and marine sedimentation ceased. Cretaceous deposits in Idaho are limited to the southeast corner of the state where 14,000 feet of the Gannett Group and Wayan Formation represent freshwater lakes and streams. Fossil conifers, cycads, and the fern-like tree *Tempskya* in the Wayan indicate a subtropical environment. In addition, this formation yields fragmentary bones of crocodiles, turtles, and at least three species of herbivorous dinosaurs.

EARLY IN THE CENOZOIC, opposing tectonic plate interactions imparted a shearing action with acute tensional stresses to a belt across Idaho and into Nevada. In addition to widespread faulting, these forces raised ancient metamorphic rocks called core complexes from intervals deep in the crust to sites near the surface.

A second phase of extensional tectonics in the Basin and Range began in the Miocene, 17 million years ago, when conflicting plate movements as well as back-arc spreading on the North American slab was imposed on a broad area. With continued plate adjustments, the North American slab was pulled northward by the Farallon Plate, creating tears and faults throughout the continental mainland. These faults run diagonally northwest by southeast for hundreds of miles across the Basin and Range.

At the same time back-arc spreading complicated this picture of tension across western North America.

Figure 9.7 Looking north, Abert Rim and Lake Abert in central Oregon display the typical Basin and Range elevated block structure (Oregon Dept. of Geology and Mineral Industries).

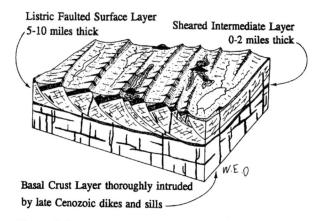

Figure 9.8 The multilayered crust of the Great Basin (after Eaton, 1979).

During this process, a local spreading center was formed in the basin between the volcanic island chain and the continent. Mantle heat and magmas rose in the center where the crustal rocks began to stretch and thin. This process is believed to be the main contributor to the tensional force that produced severe thinning of the crust of the Basin and Range.

Since the late 1970s estimates of the amount of expansion in the Great Basin seem to have increased by about 5% per year. Current calculations are that the Basin and Range has been stretched or distorted over 100%, effectively doubling its original width. Crustal rocks become thinner as they are pulled apart; thus, one way of determining the amount of extension is to measure the thickness of the crust. Beneath the Basin and Range, the distance between the crust and the mantle varies between 10 and 20 miles, whereas under the Rocky Mountains crustal thickness is twice that. Measuring only the crust, the amount of extension is computed at 40%, but the picture is additionally complicated by volcanic intrusions that have thickened the lower crust. By eliminating intrusions, the corrected figures yield a much higher total for stretching.

The amount of extension in the Basin and Range is also difficult to access because of the nature of the faults that create the mountain and valley fabric of the province. Faults here are "normal" with a 60 degree to nearly vertical orientation where they intersect the surface. Clearly faults that maintain this configuration into the subsurface do not allow for much expansion. The normal faults, which make up the margins of the valleys, are spaced about 5 to 10 miles apart, yielding a distance of 10 to 20 miles between mountain divides and as much as 3 miles between the tops of the mountain blocks and the valley floors. Many of the major faults can be traced for hundreds of miles even though they may be broken into multiple segments.

Figure 9.9 The linear east margin of Bear Lake is formed by the Bear Lake fault, which runs from Idaho south into Utah (Idaho Dept. of Commerce and Development).

Figure 9.11 Near Whiskey Springs in Custer County, the vertical tension cracks created by the Borah Peak earthquake in 1983 are dramatic evidence of the force of earth movement (courtesy H.E. Malde, U.S. Geological Survey).

Figure 9.10 Spectacular fault scarp generated by the 1983 Borah Peak earthquake in Custer County, Idaho (courtesy H.E. Malde, U.S. Geological Survey).

In the central and northeastern Basin and Range, faults appear to be "listric." These faults display a 60 degree angle near the surface but curve at depths until they are virtually horizontal. As listric faults move, the blocks between them rotate so the surface tilts away from a central axis of extension. In much of this province, fault blocks tend to tilt to the east on the eastern side of the province and to the west on the western side.

ON BOTH SIDES OF THE EASTERN SNAKE RIVER PLAIN, Late Cenozoic northwest trending faults trace the pattern of the Basin and Range beneath this area. Many of these faults show evidence of movement within the last 15,000 years. Crossing into Wyoming south of the Snake River, individual segments of the 90-mile long Grand Valley fault scarps are dated at 11,000 years. The Grand Valley and adjacent Snake River faults form a narrow trough or graben that traps the Snake River near Rexburg. North of the river, the Beaverhead, Lemhi, and Lost River faults run the length of steep valleys, and movement on the Lost River fault was responsible for the powerful Borah Peak quake of 1983.

Earthquakes in the Pocatello Valley in 1975 registering 6.1 in magnitude and at Hebgen Lake, Montana, in 1959 at 7.1 demonstrate that extension and faulting are active today in this province. One of the largest earthquakes in the western United States in recent years, the 1983 Borah Peak event registered 7.3 in magnitude. Two children were killed in Challis by falling debris, and the damage to building foundations and chimneys was estimated at $15 million. Groundwater levels rose and fell after the quake, and even the Old Faithful geyser at Yellowstone was affected. Immediately after the quake, groundwater spewed from the flank of Chilly Buttes and the surrounding plains 6 miles west of Borah Peak, shooting 20 feet into the air and creating over 50 craters.

One curious aspect of the early spring Pocatello Valley earthquake was that a mile-long fracture appeared in the crust of snow and ice on top, even though there was no discernable displacement of surface rock or soil below.

THE 130-MILE LONG BROTHERS FAULT ZONE, the Eugene Denio fault system, and the Mt. McLoughlin zone extend southeast by northwest across central Oregon as large-scale fracture systems. Induced by crustal shearing action, these belts of overlapping smaller faults are relatively recent and are readily visible on aerial photographs. Although over 100

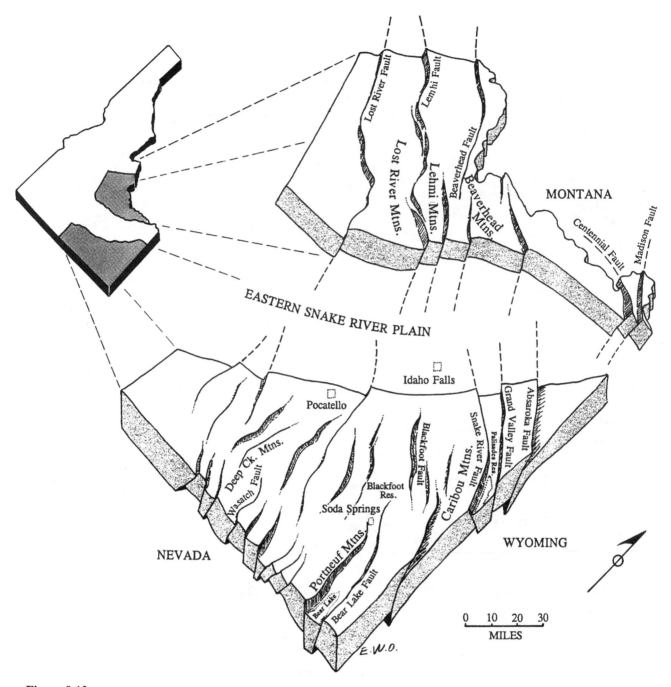

Figure 9.12 Faults within the Basin and Range extend from southeast Idaho beneath the Snake River Plain and well into eastern Idaho (after Piety, Sullivan, and Anders, 1992; Boyer and Hossack, 1992).

Holocene volcanic centers mark the Brothers Fault zone from Steens Mountain to Bend, Oregon, earthquake activity in the High Lava Plains during historic time has been minimal. Along a parallel fault system, the 1993 Klamath Falls 6.0 quake, which was the most recent, collapsed brick walls, broke windows, and damaged building foundations.

IN CALIFORNIA AN EXTENSIVE FAULT SYSTEM between Mt. Lassen and Medicine Lake Highland is over 50 miles long and 20 miles wide. Small volcanic cones erupted along the fault zone, and high scarps are frequent, the longest being the Hat Creek Rim south of Fall River Mills. Although movement is relatively infrequent, an earthquake swarm at Medicine Lake Highland in September, 1988, knocked over dead trees and created waves in Little Medicine Lake.

VOLCANISM DOMINATED the northern Basin and Range during the Late Cenozoic. With tension and stretching, the thin semibrittle crust cracked and faulted, allowing the molten material to reach the surface

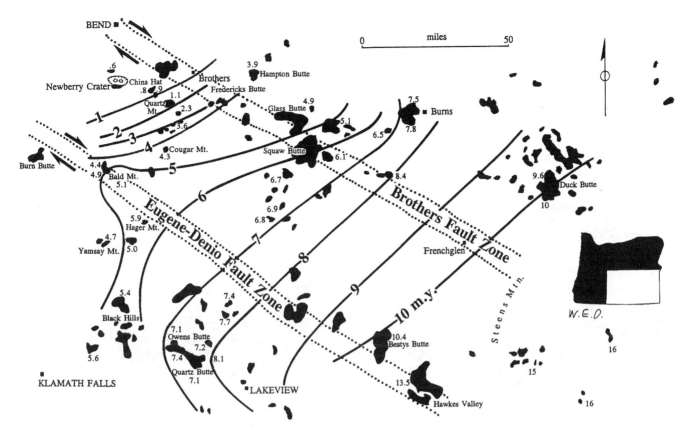

Figure 9.13 The dates of silicic eruptions and rhyolitic domes show a progressive movement of activity northwestward across Oregon of about 1.2 inches per year in the Miocene slowing to about .4 inch per year in the Pliocene and Quaternary (after Walker, 1979; Walker and Nolf, 1981).

through open conduits. Eruptions followed a set sequence beginning with ash and lava pumped out of an elongated crack or fissure before a localized vent formed small cones or a larger stratovolcano. Crevices and chasms, caves, and basalt ridges were created as part of a dark lava field surrounding the volcanic edifice.

Caves in lava begin when a hardened shell forms over the surface of a liquid flow sheet, allowing the material underneath to drain away and leave a diversity of tubes. Once the roof collapses an entrance is provided into the lava cave system, which often meanders for several miles underground.

VOLCANIC ACTIVITY IN SOUTHEAST IDAHO was not as pervasive as elsewhere in this province even though the Blackfoot, Willow Creek, and Gem Valley lava fields covered parts of Caribou County with basaltic and rhyolitic lava from the Pliocene to Pleistocene. This volcanism is closely related to the fault system that runs beneath the area, and cinder cones, like those at Pelican Ridge and Soda Springs Hills, are commonly located at the base of pronounced scarps.

At Blackfoot Reservoir rhyolitic domes cover fault traces, but sections of the volcanic cones south and west of the reservoir at China Hat and Reservoir Mountain

have been truncated by faulting that postdated the volcanism. The two islands within Blackfoot Reservoir and the prominent hills at the southern end are volcanic domes. Close to the reservoir, Bear River probably drained northward about 2 million years ago until basalt from isolated eruptive centers blocked and diverted it southward into ancient Lake Bonneville. A heavy layer of Pleistocene loess, fine windblown glacial silt, covers most of the lava flows and cones here.

ON THE HIGH LAVA PLAINS OF OREGON, the earliest volcanic activity was in Steens Mountain around 16 million years ago. Successive, voluminous outpourings of basaltic lava from broad shield cones filled and smoothed the landscape. Because it was later elevated by mountain-block faulting and subjected to intensive glaciation, the average thickness of the Steens flows can only be estimated at 3000 feet over a 6000-square mile area. Near Alvord Creek in Malheur County, a single flow is up to 1000 feet thick. Even though a copious amount of lava was produced, the entire Steens eruption lasted only 50,000 years.

A peculiarity of volcanic history in the High Lava Plains is the systematic progression of a wave of eruptive sites of rhyolitic or silicic lava. Moving in a westerly

Figure 9.14 Looking northwest across Newberry Crater in central Oregon shows East Lake in the foreground separated from the distant Paulina Lake by the Central Pumice Cone. Even though the interior of the crater is carpeted by a pine forest, lava and obsidian flows are easy to identify (Oregon Dept. of Geology and Mineral Industries).

direction, lavas are oldest on the east and younger to the west. The oldest in this sequence of cones at Beatys Butte and Duck Butte in Malheur County date back 10 million years. Squaw Butte, Hager Mountain, and Black Hills in the central plains were active around 5 million years ago, while in Deschutes County Newberry volcano on the western edge erupted during the last 10,000 years. This remarkably uniform westward procession of parallel volcanic centers is intriguing because its trend and direction rule out a hot spot. Present ideas as to its origin include clockwise rotation of the crust or a slowing of the subducting plate and subsequent steepening of the slab angle, which, in turn, would affect surface volcanic patterns.

In central Oregon south of Bend, Newberry volcano is a great caldera occupied by East and Paulina lakes with numerous obsidian flows and volcanic domes. The eruptive life of this volcano began 500,000 years ago when successive layers of basalt, rhyolite, ash, and dust built a low profile shield cone 4000 feet above the surrounding plain. The crest of the cone collapsed during several eruptions creating a caldera 5 miles in diameter. As recently as 2000 years ago, the Big Obsidian lava

flow emerged from a crack in the south rim of the caldera to spread for 1½ miles toward the center.

The gentle slopes of Mt. Newberry are dotted with over 400 smaller parasitic cones. Most notable is Lava Butte, straddling a fissure system on the north face. Around 6000 years ago basaltic lava from these fissures, that reached as far as the Deschutes River, engulfed whole forests. Today curious hollow cylinders mark casts of trees left standing above the rugged surface. Below the crust of the flows, a maze of lava tunnels and caves meanders for miles. Lava River Cave, 15 miles south of Bend, is one of the most lengthy volcanic tunnel systems in Oregon. The 1-mile-long public section of the cave is decorated with delicate formations, ledges, and "sand castles" carved by groundwater, which give it a unique quality.

Newberry Crater, Lava Butte, the Lava Cast Forest, and Lava River cave were dedicated as the Newberry Volcanoes National Monument in 1991. This monument offers a panoramic view from the rim of the crater with fresh volcanic topography in a compact area easily accessible by car.

Figure 9.15 At Lava River Caves State Park south of Bend, Oregon, the tunnels have a typical keyhole shape in cross-section, which formed when the lava drained out of the tube (Oregon Dept. of Geology and Mineral Industries).

Figure 9.16 The explosion that results when molten lava encounters groundwater produces clouds of debris that fall to the ground in tuff rings.

Other volcanic highlights in central Oregon are tuff rings and bowl-like maars, eruptive features that originated during the Pleistocene. Scattered widely throughout the High Lava Plains, tuff rings developed when lava, traveling upward through the crust, encountered groundwater. Catastrophic explosions that ensued sent rocks, ash, and blocky debris high into the air. As this material settled, it constructed a distinctive high ridge of shattered breccia around the central crater. Today one of the better-known and picturesque tuff rings in the world is Diamond Head that rises just east of Honolulu, Hawaii.

Three famous ring-shaped volcanic features in Lake County are Fort Rock, Hole-in-the-Ground, and Big Hole. At Fort Rock, the south rim has been eroded and breached by waves from a former pluvial lake that occupied the basin. About 1/3 mile wide and over 300 feet

Figure 9.17 Hole-in-the-Ground in central Oregon is a circular explosion crater (Oregon Dept. of Geology and Mineral Industries).

above the surrounding plain, Fort Rock is more visible than Big Hole, whose overwhelming size is now largely hidden by forests. Just across the Oregon border at Tule Lake, California, Prisoners Rock and The Peninsula are also cones and tuff rings.

IN NORTHERN CALIFORNIA, volcanism in the Modoc Plateau and the Medicine Lake Highland began in the Oligocene 32 million years ago and continued sporadically to within the last 1000 years. In Siskiyou and Modoc counties, the Modoc platform is underlain by Oligocene to Miocene Cedarville andesitic lavas and rhyolitic ash flows interspersed with lake bed deposits. Stretching in the Basin and Range broke the Cedarville lava surface into mountain scarps separating enclosed basins.

Ash and sheets of very fluid Warner Basalt, up to 100-feet thick, poured into these basins during the Late Miocene and Pliocene, and the Devils Garden, a continuous plateau of basalts, extends north and west from Alturas for 700 square miles. Pliocene faulting and tilting imposed ridges and broad basins atop the Warner lava surface.

Pliocene lakes that filled many of the depressions were rich in diatoms along with thin layers of ash, lava,

and mudflows. In Modoc and Shasta counties, intact leaves of cottonwood, willow, and other deciduous plants are found near Alturas, Lake Britton, and in the Pit River valley. These fossils of the Alturas Formation suggest a semiarid climate of cold winters and hot summers in contrast to floras of the earlier Cedarville period that indicate a high rainfall and moderate Miocene climate.

Pleistocene and Holocene lavas issuing from cracks and fissures were the last to erupt, spreading a thin basalt veneer over older flows. Basaltic lavas, cinder cones, and lengthy lava tube systems at Lava Beds National Monument cover about 75 square miles near the California-Oregon border. More than a dozen cones, projecting several hundred feet, are scattered throughout the park, but most of the basalt flows can be traced back to Mammoth Crater.

The most striking aspect of the Lava Beds is the broken maze of intricate fissures and basalt ramparts that scar the surface. This irregular landscape was created when a thick crust congealed on top of a moving front of a lava flow. A new surge of molten material, carried underground through the tunnel system, lifted and broke apart the hardened crust, allowing long tongues of the still-fluid lava to leak out. As flows con-

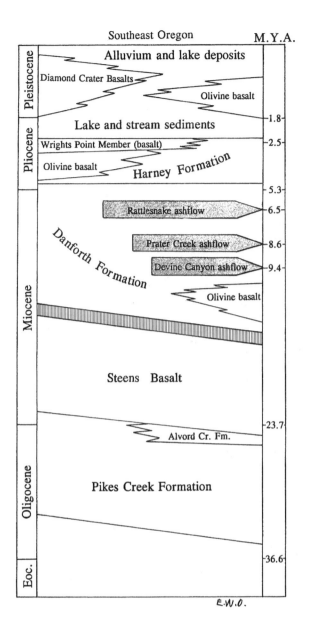

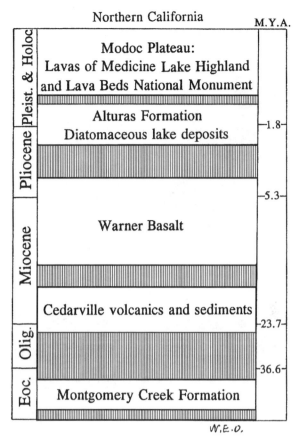

Figure 9.18 Tertiary stratigraphy of northeast California and southeast Oregon (after Bishop and Davis, coord., 1984; Walker, 1979).

tinued, large pieces of cooled crust separated and were carried along on the lava river. Eventually they were scattered as large detached blocks over the landscape.

Over 300 separate tunnels perforate the basalts in an extensive cave system that runs deep within the lava covering. From 1917 to 1933, an early settler, J.D. Howard, explored, mapped, and named most of the caves. A devotee of ancient history and mythology, Howard gave such names to the caves as Catacombs and Cleopatra. Some of the caves, like Mushpot, are lighted and easily accessible, but Post Office Cave, among others, is more remote and requires permission to enter. Crystal Cave is of special interest because of its delicate ice formations.

The uneven rugged volcanic topography here was to play a critical role in the war between the Modoc

Figure 9.19 Captain Jack, Modoc leader during the war of 1872 to 1873 (courtesy J. Orr-Mooney).

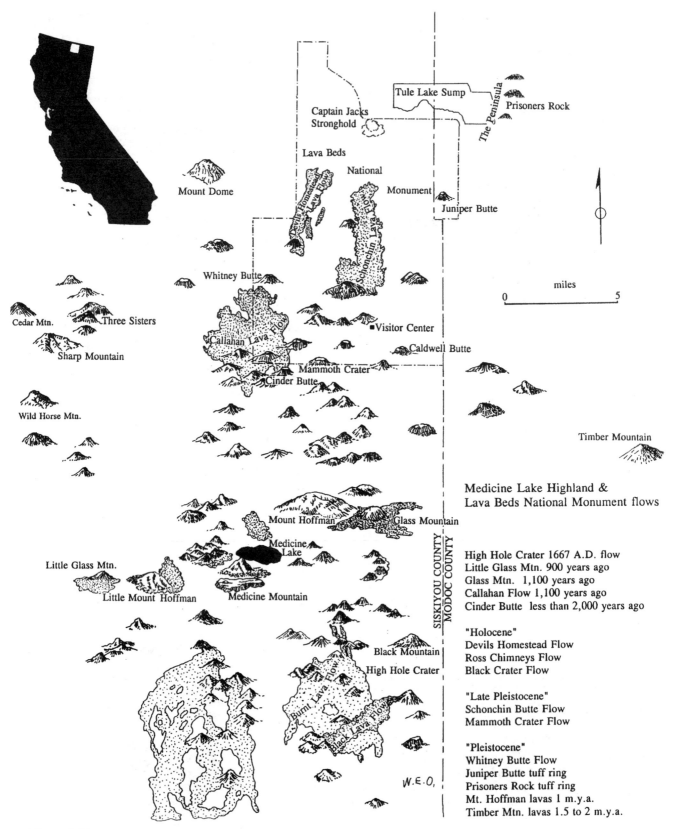

Figure 9.20 Buttes, volcanic cones, and lava flows of the Modoc Plateau and Medicine Lake Highland.

Mount Dome

Tule Lake Sump

The Peninsula

Prisoners Rock

Captain Jacks Stronghold

Lava Beds

National

Monument

Juniper Butte

Whitney Butte

Devils Homestead Lava Flow

Schonchin Lava Flow

Cedar Mtn.

Three Sisters

Sharp Mountain

Callahan Lava Flow

Visitor Center

Caldwell Butte

Mammoth Crater

Cinder Butte

Wild Horse Mtn.

Timber Mountain

Mount Hoffman

Glass Mountain

Medicine Lake

Little Glass Mtn.

Little Mount Hoffman

Medicine Mountain

Black Mountain

High Hole Crater

Burnt Lava Flow

Black Lava Flow

W.E.O.

Medicine Lake Highland & Lava Beds National Monument flows

High Hole Crater 1667 A.D. flow
Little Glass Mtn. 900 years ago
Glass Mtn. 1,100 years ago
Callahan Flow 1,100 years ago
Cinder Butte less than 2,000 years ago

"Holocene"
Devils Homestead Flow
Ross Chimneys Flow
Black Crater Flow

"Late Pleistocene"
Schonchin Butte Flow
Mammoth Crater Flow

"Pleistocene"
Whitney Butte Flow
Juniper Butte tuff ring
Prisoners Rock tuff ring
Mt. Hoffman lavas 1 m.y.a.
Timber Mtn. lavas 1.5 to 2 m.y.a.

SISKIYOU COUNTY
MODOC COUNTY

miles
0   5

Figure 9.22 Eruptions at Glass Mountain (in the foreground) were among the latest to occur on the Medicine Lake Highland 1000 years ago. Mt. Shasta is in the background (California Division of Mines and Geology).

Figure 9.21 Mushpot cave, one of the many in the Lava Beds National Monument (courtesy C. McKillip).

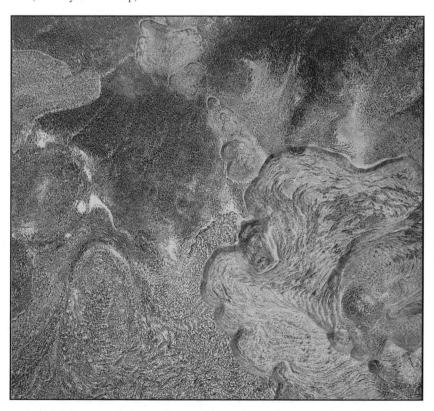

Figure 9.23 Although many of the Medicine Lake Highland lava flows date back over 2 million years, in aerial view they appear to be only weeks old (California Division of Mines and Geology).

Figure 9.24 Lava flows and the extensive cave system (inset) of Lava Beds National Monument in Siskiyou County, California (modified from Waters, Donnelly-Nolan, and Rogers, 1990).

Indians and the United States cavalry that took place for 8 months from November, 1872, to June of 1873. The plateau provided a stronghold and refuge for approximately 130 Modocs who held off assaults from 650 heavily armed troops. The Indians, led by Captain Jack and Hooker Jim, were thoroughly familiar with the area and took full advantage of the tunnels and other natural rock fortifications to evade capture and deliver punishing defeats to the soldiers. In one memorable skirmish, a handfull of warriors, led by Scar-Faced Charley, ambushed and killed 15 of the soldiers. The army transmitted a report that read, "We have sickening news again from the lava fields." Ultimately betrayed by other Indians, Cap-

tain Jack's band was captured, and he and the other leaders hanged.

THIRTY MILES NORTHEAST OF MT. SHASTA, Medicine Lake is a 4000-foot high plateau built of Pliocene to Pleistocene tuff, lava, and stream sediments capped by Holocene volcanic debris. Spanning the last 2 million years, eruptive activity began with basalt lava from the Timber Mountain cone immediately east of the Medicine Lake platform. This phase continued when the Mt. Hoffman shield volcano issued an immense amount of andesitic, dacitic, and rhyolitic lavas less than 1 million years ago before the center collapsed into a caldera.

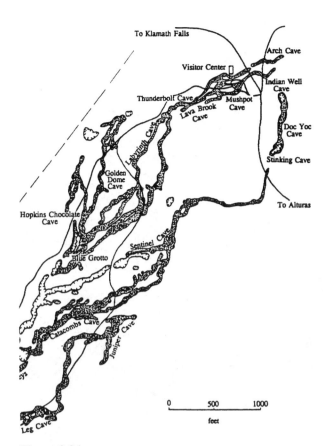

Figure 9.24 Continued.

Containing a small lake, the caldera is 7 miles across and up to 500 feet deep. The Medicine Lake eruption, with an estimated volume of 150 cubic miles, was significantly larger than that of Mt. Shasta, the most voluminous of the Cascade volcanoes. Close to one hundred smaller andesitic domes, like Sharp Mountain, Wild Horse Mountain, Three Sisters Mountain, and Cedar Mountain west of the Highland caldera are between 300,000 to 1.3 million years old. Lava from these later eruptions buried the rim of the older Mt. Hoffman.

Through the Ice Ages, the Medicine Lake peak was glaciated, and a thick soil cover developed prior to the final volcanic episode. Andesitic rocks on the flanks of the mountain display the unmistakable striations and polish of glacial activity.

The latest events here were on the slopes of the older Medicine Lake cone when Glass Mountain and Little Glass Mountain erupted basalt followed by rhyolitic lavas only 1000 years ago. Limited erosion and weathering on Little Glass Mountain give obsidian flows the appearance of having just cooled.

Figure 9.26 An aerial view of Medicine Lake Highland in northern California showing the volcanic cones in the Burnt Lava flow. High Hole Crater is to the left (California Division of Mines and Geology).

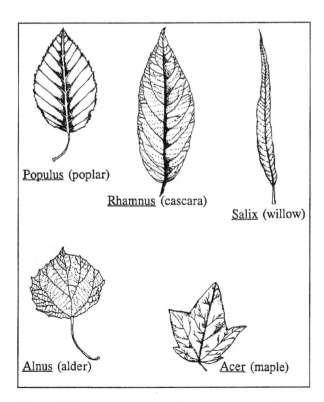

Figure 9.25 A variety of plants of the Alturas Formation flourished across the northern Great Basin during the Late Cenozoic.

THE PLEISTOCENE EPOCH BEGAN just under 2 million years ago, initiating increasingly cooler climates and higher precipitation in the northern Basin and Range. Well beyond the reach of the Canadian continental ice lobes, Steens Mountain in Oregon along with mountain ranges on both sides of the Snake River Plain experienced limited valley glaciers. Long U-shaped valleys were scooped out of the subdued western slope of Steens Mountain, while the steep eastern fault scarp was scalloped by ice that traveled only short distances down from the peaks. Along the crest of the Steens, moving masses of ice left thin sharp ridges separating the canyons.

Figure 9.27 Glacial valleys have been deeply cut into the gentle western face of Steens Mountain of southeastern Oregon. Just over the rim to the east the dry playa of the Alvord Desert is visible (Oregon State Highway Dept.).

In Idaho, the upper reaches of the Lost River and Lemhi Ranges were subjected to several episodes of Late Pleistocene alpine glacial advance and retreat 15,000 years ago. Although no enveloping ice sheets covered the peaks, evidence of glaciation is visible in sharp crests and accumulations of gravel and sand distributed by the diminishing ice masses. Large deposits of gravel reflect previous periods flooding. Augmented by melting ice and snow, stream discharge was as much as 10 times higher than at present. The extremely coarse material in these gravel deposits and lack of finer debris attest to the power of moving water. The last of the floods in the Big Lost River was about 16,000 years ago, at the same time as the Lake Missoula floods in Montana.

Throughout the Late Pleistocene and Holocene, several thousand square miles of southeast Idaho were blanketed with wind-borne loess deposits averaging up to 3 feet thick but reaching accumulations of 100 feet south of the Snake River Plain. This fine buff-colored material, which originated as glacial rock flour, was, for the most part, transported from the flood plains and alluvial fans by the Snake River and its tributaries.

During the cool, moist Pleistocene epoch, broad pluvial lakes occupied old playas in the Basin and Range of southcentral Oregon and northeast California. These shallow lakes diminished rapidly in size with the onset of warm climates about 10,000 years ago. Today, in the midst of a dry interglacial period, most of the smaller

Figure 9.28 In the northern Great Basin, a comparison of Pleistocene pluvial lakes (shown in pattern) to modern lakes (black) shows how much these bodies of water have diminished (after Dicken, 1980; Snyder, Hardman, and Zdenek, 1964).

bodies have dried up, and the remainder are only a fraction of their original size. Across the northern Great Basin, the main bodies of water were Lake Modoc, Lake Alturas, Fort Rock Lake, Lake Chewaucan, Warner Lake, Malheur, and Alvord lakes.

Ancient Lake Modoc, in Oregon and California, reached 75 miles in length and covered 1096 square miles. As the waters began to evaporate, the lake diminished and individual basins formed. Each varied considerably in size, and the smaller ones shrank even faster.

Today Upper and Lower Klamath lakes and Tule Lake are the largest of the eight remaining bodies of water that were once part of the broader Lake Modoc. Even as late as the turn of the century, shorelines of Tule Lake stretched to the Lava Beds National Monument, whereas today its waters are almost 10 miles north of the lava fields.

At its maximum extent, ancient Lake Alturas, that straddled the Oregon-California border, spread across the floor of the down-dropped basin west of the Warner

Figure 9.29 View east of Lost River Range beyond McKay, Idaho, shows Mt. McCaleb (Idaho State Historical Society).

mountains. A remnant of the larger Alturas, Goose Lake occupies only the northern end of the basin. Once covering 368 square miles, Goose Lake is still shrinking, and at least twice over the past 130 years the water has dried up completely.

In northern Lake County, Oregon, the dry sands of Silver Lake, Christmas Lake, and Fossil Lake are all pieces of the former Fort Rock Lake, which reached depths of over 200 feet. Fossilized bones of anadromous salmon found here indicate the Pleistocene lake had an outlet to the Columbia River and Pacific Ocean.

Sprawling over 461 square miles in central Lake County during the Ice Ages, pluvial Lake Chewaucan has shrunk to Summer Lake, Lake Abert, and the Chewaucan Marshes. The basin of the ancient lake bed is enclosed by fault scarps of Winter Rim to the west and Abert Rim to the east. Ice Age bison, camels, horses, and even elephants lived around the lake. Fossil bones of waterbirds and fish, plentiful here during the Pleistocene, are scattered over the now sandy playa.

Small lakes persist today in Warner and Catlow valleys and the Alvord Desert of eastern Lake, Harney, and Malheur counties. Once filled by much larger pluvial

bodies of water, all that remain today are Bluejoint, Campbell, Flagstaff, Anderson, Hart, Crump, and Alvord lakes. At Alvord Lake the water is rimmed with a white alkalai crust, the residue left by evaporating waters. It is unfortunate that the natural lake bed here has been defaced by graffiti cut by an "artist" using a rototiller, which, because of slow erosion processes, will remain for some time.

Once covering much of what is now Malheur County, shallow Malheur and Harney lakes continue to fluctuate dramatically with heavy seasonal rains. As recently as 1984 these lakes expanded from 125,000 to 175,000 acres, although in previous years the basin was virtually dry. Lake waters here serve as a wildlife refuge attracting thousands of birds.

GEOTHERMAL RESOURCES are indicative of areas with thin crust and a history of volcanic activity. In the northern Great Basin a remarkably high heat flow and elevated crustal temperatures are manifest on the surface by numerous thermal springs. Because of the fault-block structure, hot springs are typically located along the valley margins.

Figure 9.30 Ring-shaped Ft. Rock in central Oregon has been eroded by waves of the Pleistocene lake, which once filled this basin (Oregon State Highway Dept.).

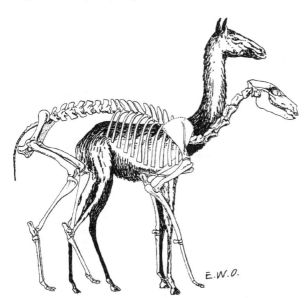

Figure 9.31 Camels and a number of other exotic animals lived around Pleistocene lakes in the Basin and Range (after Scott, 1937).

Figure 9.32 Unlike the dry conditions today in the northern Basin, the rainy Pleistocene here several million years ago was the habitat to a wide range of animals.

Figure 9.33 An 1877 drawing of striking calcium carbonate features deposited by spring waters at Soda Springs, Idaho (U.S. Geological Survey).

Abundant hot springs in the Great Basin are south of the Snake River at Heise Hot Spring, Lava Hot Springs, Steamboat Springs, and those in the Raft River valley. Within the deep Portneuf River canyon of Caribou County, multiple vents at Lava Hot Springs discharge 1500 gallons of water per minute averaging 120 degrees Fahrenheit, although waters from nearby Steamboat Springs only reach 80 degrees. In the same region, Albion Basin, Raft River valley, and Rockland Valley all have high geothermal potential. Even though springs and water wells in the Raft River basin maintain an average temperature of 60 degrees, several localities have spectacular boiling water wells. Two springs supply 90 degree Fahrenheit water to a natatorium in Rockland Valley.

Throughout southcentral Oregon, hot springs are prevalent. In the Harney, Alvord, and Warner basins, thermal waters frequently range in excess of 100 degrees. Borax Lake, a large pool in the Alvord basin, discharges water at a temperature of 97 degrees Fahrenheit. The water is laden with borax salts that precipitated in long low domes. During the late 1800s the borax was mined and removed by mule-drawn wagons to the railhead at Winnemucca, Nevada. Several of the springs, as the 180 degree Hunter Hot Spring near Lakeview, have been exploited commercially. Radium Hot Springs in the Harney Basin, which opened to the public at the turn of the century, today is the largest warm water pool in the state with temperatures hovering around 70 degrees.

In northern California hot springs are related to recent volcanic activity as well as to faulting. Percolating

Figure 9.34 North of Adel in Warner Valley, Oregon, Crump Geyser first erupted explosively in 1959 from a drilled well (Oregon Dept. of Geology and Mineral Industries).

through conduits in the basalts, over 30 known springs register temperatures above 90 degrees, and near Mt. Hoffman steam emerges from a vent at a temperature of 200 degrees.

Although not geothermal in nature, one of the largest springs in the United States is to be found near Fall River Mills in Shasta County, California, 50 miles south of Lava Beds National Monument. Moving though cracks and underground channels in an older basalt flow, the water discharges at the astonishing rate of 1290 million gallons daily at Fall River Springs.

The water system feeding nearby Burney Falls disappears into underground passages through old basalt layers above the falls, travels less than a mile, and reemerges as combined stream and spring waters. The stream drops 120 feet over a basalt lip in an impressive falls with a curtain of smaller cascades behind.

THE GREAT BASIN is not known for its economic minerals. In southeast Idaho the Late Permian Phosphoria Formation provides 10% to 15% of the phosphate produced in the United States, much of which is ex-ported for fertilizers. Easily the most important geologic resource in this area, the phosphate ore, apatite, is mined in open pits that are up to 300 feet deep.

Another economic product, diatomite deposits of microscopic glassy algal remains are mined from several dry lake beds in this province. Although the soft light-weight material finds use to a natural filter, it is most familiar as the absorbent chips in kitty litter.

ECONOMIC MINERAL DEPOSITS in this province of California have been minimal although three small gold districts at Hayden Hill in Lassen County and the High Grade and Winters in Modoc County yielded $3,500,000 during peak operation in the early 1900s. Veins ran as deep as 835 feet at Hayden Hill, but at High Grade rich lodes were found less than 100 feet below the surface. These mines had long been inactive until Hayden Hill was opened within the past 10 years. Present operations include a facility for extracting gold by cyanide heap-leaching. In this locality both gold and silver were deposited during the Miocene in fractures by hot spring waters saturated with dissolved minerals.

Figure 9.35 Near Lakeview in Lake County, Oregon, Hunter Hot Springs began erupting in October, 1923, when a well was being drilled at this site (Oregon Dept. of Geology and Mineral Industries).

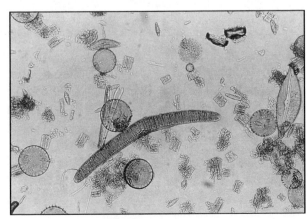

Figure 9.36 Millions of diatoms, the glassy remains of single-celled microscopic plants, accumulated in lake beds to form the soft light-weight rock, diatomite (Oregon Dept. of Geology and Mineral Industries).

Figure 9.37 The immense Dry Falls is one of the many geologic features in the Grande Coulee of Washington. Here flood waters cascaded over basalt during the Pleistocene carving the falls that is almost 4 miles wide and 300 feet deep (Washington Dept. of Natural Resources).

ADDITIONAL READINGS

Allmendinger, R.W., 1982

Axelrod, D.I., 1944

Ballard, W.W., Bluemle, J.P., and
    Gerhard, L.C. coord., 1983

Coogan, J.C., 1992

Craddock, J.P., 1992

Davis, G.A., 1979

Dicken, S.N., 1980

Donnelly-Nolan, J.M., 1992

Eaton, G.P., 1979

Fields, R.W., et al., 1985

Fiero, B., 1986

Fiesinger, D.W., Perkins, W.D., and
    Puchy, B.J., 1982

Hughes, S.S., and Thackray, G.D., eds.,
    1999

Hunt, C.B., 1979

Johnston, D.A., and Donnelly-Nolan, J.,
    eds., 1981

LaMotte, R.S., 1936

Lavine, A., 1994

Lewis, G.C., and Fosberg, M.A., 1982

Link, P.K., 1988

Link, P.K., and Phoenix, E.C., 1996

Figure 9.37 Continued.

McDonald, G.A., 1968

McFaddan, M.D., Measures, E.A., and Isaacson, P.E., 1988

McNutt, S., 1989

Mertzman, S.A., 1981

Pease, R., 1965

Pierce, K.L., and Scott, W.E., 1982

Rodgers, D.W., and Janecke, S.U., 1992

Ross, S.H., 1971

Ruppel, E.T., 1982

Skipp, B., and Link, P.K., 1992

Skipp, B., Link, P.K., Sando, W.J., and Hall, W.E., 1979

Snyder, C.T., Hardman, G., and Zdenek, F.F., 1964

Stevens, C.H., 1977

Stewart, J.H., and Suczek, C.A., 1977

Suek, D.H., and Knaup, W.W., 1979

Walker, G.W., 1979

Walker, G.W., and Nolf, B., 1981

Waters, A.C., Donnelly-Nolan, J.M., and Rogers, B.W., 1990

Wills, C.J., 1991

10 COLUMBIA RIVER PLATEAU

Looking downriver at the Columbia River Gorge just west of The Dalles, Oregon, multiple basalt flows are exposed in the canyon walls (Washington Dept. of Natural Resources).

The Columbia River province covers much of northeast Oregon and southeast Washington with smaller extensions into Idaho near Coeur d'Alene, Lewiston, and Weiser. The irregular outline of the province was produced by successive flows of lava, which spread for hundreds of miles to give the plateau its level rolling topography of projecting higher ridges and buttes.

The basalts that make up the province are confined to a basin surrounded by mountain ranges. The Okanogan Highlands are to the north, the Cascade Mountains are on the west, the Blue Mountains are south, and the Clearwater Mountains and foothills of the Rocky Mountains rise to the east. Within the plateau, low ridges that trend east-southeast are the Frenchman Hills, Saddle Mountains, the Yakima Ridge, Toppenish Ridge, the Rattlesnake Hills, and Horse Heaven Hill in Washington.

Watersheds of the Columbia and Snake rivers have the greatest impact on the plateau. Originating in the Rocky Mountains of British Columbia, the Columbia River watershed is equal in size to the state of Texas, covering a total of 260,000 square miles in Canada, Washington, and Oregon. Flowing into Washington from the north, the river curves around the margin of the province before looping east at Wallula Gateway where it heads west to the Pacific Ocean. The Snake River merges with the Columbia at the Wallula narrows, having come along the eastern margin of the plateau before turning sharply westward at Lewiston.

The Columbia Plateau province takes in the adjacent Deschutes basin of Oregon, an area lying between the Cascades and Ochoco mountains, which is drained by the Deschutes River.

INTRODUCTION

This region has been largely shaped by the combined effects of lava flows and glacial flood waters. East of the emerging Cascade Mountains a vast area of dull black volcanic rock blanketed the landscape in much of northeast Oregon, southeast Washington, and western Idaho to construct a rugged basalt platform. For over 11 million years, during the Miocene epoch, flow upon flow of Columbia River basalts poured across an ever-widening area following stream valleys much like water, eventually reaching the Pacific Ocean. Because they were extremely fluid, these lavas are called flood basalts.

The eruption and spread of lavas were intermittent, and lapses of thousands of years between events allowed soil layers to develop atop flows. Lakes formed in shallow basins around the margins of the plateau where streams had been blocked, attracting a diverse group of animals to the region. Occasionally invaded by lava or

filled with lake and stream sediments, the basins retain a remarkably complete history of this province.

Following the extended volcanic stage, the basalt plateau was repeatedly scoured by great Ice Age floods. Late in the Pleistocene epoch, an enormous lake in Montana periodically sent torrents of water across the province in a wide path of destruction. Stripping away the soil and cutting deep into the bedrock, floodwaters left behind a distinctive scabland topography in eastern Washington before exiting through the Columbia gorge to the Pacific Ocean. Like the thick platform of basalts, the action of these torrential waters was responsible for today's topography.

GEOLOGY OF THE COLUMBIA RIVER PLATEAU

The Columbia River basalts spread widely over parts of three states to conceal the underlying terrain so thoroughly that exposures of rocks below the Miocene lavas are limited. The geologic history of events prior to the flood basalts can only be surmised from an occasional glimpse of the older strata at the bottom of deep canyons or along raised and eroded areas. Small outcroppings of the Wallowa and Baker exotic accreted terranes in Washington and Oregon indicate these rock sheets extend far to the north and west beneath thick volcanic flows. Similarly lavas and sediments of the Eocene through Miocene Clarno and John Day formations of central Oregon disappear under layers of the dense black basalts.

In the Clearwater lava embayment of northwest Idaho, Mason Butte, Cottonwood Butte, and Woodland Butte are granites of the 70 million-year old Idaho batholith that stand out in sharp contrast above the sea of younger basalts. Within the same lava plain between Spokane and Pullman, Tekoa, Dunlop, Granite, Steptoe, and Kamiak, along with assorted smaller buttes along the eastern border of Washington, are knobs of resistant Precambrian sedimentary Belt rocks over a billion years old.

Kamiah Butte, projecting 1000 feet above the Columbia River flows in northern Idaho County, records an Eocene volcanic episode 50 million years ago that was simultaneous with eruptions of the Challis lavas. The Columbia River basalt lapped up against these older buttes and covered most of the earlier flows.

Ancestral highlands of older terranes at Peck Mountain, Cuddy Mountain, and Sturgill Peak in southwest Idaho emerge like islands or "steptoes" from the flat cover of Weiser lavas. Older terrane rocks are also displayed in the steep walls of the adjacent Snake River canyon.

THE MIDDLE MIOCENE, dating back more than 17 million years, opened with massive outpourings of lavas that would eventually flood the entire Columbia

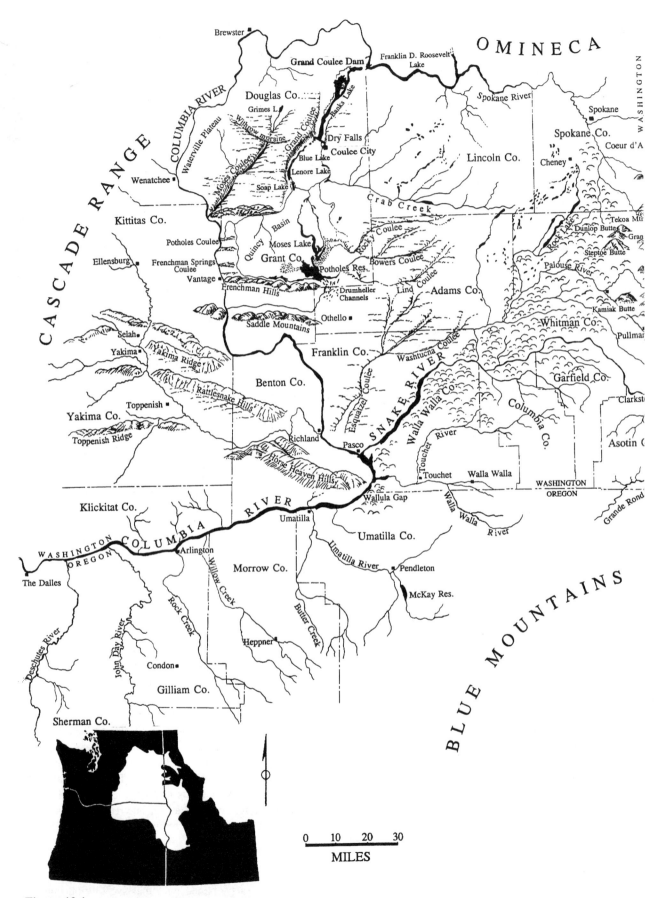

Figure 10.1 Main localities of the Columbia Plateau.

Figure 10.1 Continued.

Figure 10.2 An oil painting by Dee Molenaar of Dry Falls along the Grand Coulee during Pleistocene floods. In the distance the Okanogan lobe of the vast continental ice sheet from British Columbia is visible (Washington Dept. of Natural Resources).

basin. Successive flows filled a natural broad, spoon-shaped depression that was steep toward the east but thinned along its margins. Near Yakima in southcentral Washington, almost 3 miles of the dense basalts accumulated in the deepest part of the basin. Even as the lavas were erupting, the entire plateau was being gently tilted to the west and north by progressive uplift of the Idaho batholith. Accompanying the tilting, the flows extended further west and north, while the basin itself gradually deepened with the immense weight of the cooling rock.

Issuing from zones of fissures and cracks called dike swarms, the lavas originated as the Monument group within the John Day basin of Oregon, the Grande Ronde dike system that cut the eastern Oregon-Washington border, and the Cornucopia north of Weiser, Idaho. Although some of the fissures were obscured by later flows, many can be traced for up to 100 miles. Each flow lasted only a few days or weeks, while thousands of years often lapsed between eruptions. Individual flows are estimated to have been several stories high along their front edge as the lava moved up to 3 miles per hour.

Despite intensive study of this basalt mass over the past several decades, its origin is still not clear. However, the composition and remarkably homogeneous chemistry of the lavas suggests that they rose from deep in the crust too rapidly for extraneous rocks to mix in from adjacent strata. Early ideas that the flows might be derived from the continental landmass passing over the Yellowstone hot spot below the crust are difficult to

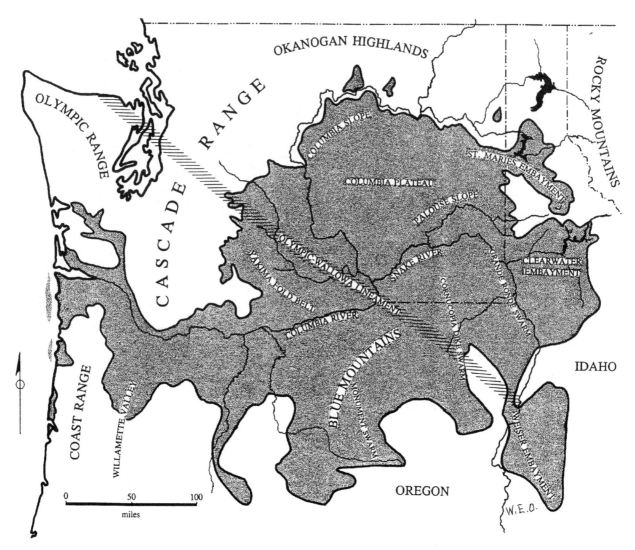

Figure 10.3 Maximum extent of Columbia River lavas that covered parts of three states (Anderson, et al., 1987; Hooper and Swanson, 1987).

support because these fissure eruptions were widely spaced unlike the narrow focused pattern of cones or calderas characteristic of a mantle plume.

More recently it has been proposed that the same Late Tertiary tensional forces, which produced the fault block topography of the Basin and Range, might be responsible for the Columbia River basalts. Back-arc spreading, where stretching east of the Cascade volcanic archipelago severely thinned the crust, induced widespread faulting that released volcanic material upward along fissures and cracks. Even though this explanation is sufficient to account for eruptions from shallow magmas, it fails to address the deep crustal source for the lavas.

The Columbia River Basalt Group has been divided into 6 formations, which are, in turn, subdivided by subtle differences in chemistry and mineralogy into more than a dozen members and over 300 separate flows. The oldest Imnaha basalts, less than 5% by volume of the total Columbia River lavas, spread out into the Snake and Clearwater drainages, where they inundated canyons to depths of 3000 feet. Further west, the Picture Gorge and Prineville basalts, between 16 and 15 million years old, were limited in volume and may not share the same source as the later flows. Prineville basalts, for example, were derived from much shallower magmas than others of the Columbia River Group. The Picture Gorge basalts occupied the upper John Day wa-

Figure 10.4 Within a single basalt flow the lower 1/4, or *colonade,* is characterized by thick upright regular columns that form as the lava cools and contracts. The upper 3/4, or *entablature,* has much smaller columns, which are often at sharp angles to the colonade (Oregon Dept. of Geology and Mineral Industries).

tershed, but Prineville lavas extended south along the Deschutes River in Oregon.

Grande Ronde Basalts make up almost 90% of the total, although the eruptive cycle of this massive formation lasted less than a million years. During the Grande Ronde interval, the basalts of the Columbia Plateau reached a maximum output of 800 cubic feet of lava per second between 16 and 15 million years ago. Composed of over 120 separate flows, several of the more extensive Grande Ronde eruptions emptied from fissures in Oregon and Washington to breach the Cascades barrier by following the ancestral Columbia River gorge. Traversing this canyon, some of the lava spilled into the Willamette Valley of Oregon as far as Salem, while other molten streams proceeded to the mouth of the Columbia River. Here the heavy liquid plunged deep into soft sediments to reappear as dikes and sills along the Oregon coast as far south as Seal Rock. Overflowing to the north, the moving lava covered the Willapa Hills of southwest Washington. In eastern Washington, successive Grande Ronde flows diverted the channel of the Columbia River into a circuitous route around the northern edge of the plateau.

Figure 10.5 Stratigraphy of the Columbia River Basalt Group (Beeson, et al., 1989; Hooper and Swanson, 1987).

SERIES	GROUP	FORMATION	MEMBER	AGE M.Y.
MIOCENE — UPPER	COLUMBIA RIVER BASALT GROUP	SADDLE MOUNTAINS BASALT — 1% of C.R.B.G. total	LOWER MONUMENTAL MEMBER	6
			TAMMANY CREEK FLOW	
			‖‖‖‖ EROSIONAL UNCONFORMITY ‖‖‖‖	
			ICE HARBOR MEMBER	8.5
			GOOSE ISLAND FLOW	
			MARTINDALE FLOW	
			BASIN CITY FLOW	
			‖‖‖‖ EROSIONAL UNCONFORMITY ‖‖‖‖	
			BUFORD MEMBER	
			ELEPHANT MT. MEMBER	10.5
			SWAMP CREEK MEMBER	
			‖‖‖‖ EROSIONAL UNCONFORMITY ‖‖‖‖	
			POMONA MEMBER	12
			‖‖‖‖ EROSIONAL UNCONFORMITY ‖‖‖‖	
			ESQUATZEL MEMBER	
			‖‖‖‖ EROSIONAL UNCONFORMITY ‖‖‖‖	
			WEISSENFELS RIDGE MEMBER	
			SLIPPERY CREEK FLOW	
			TENMILE CREEK FLOW	
			LEWISTON ORCHARDS FLOW	
			CLOVERLAND FLOW	
			ASOTIN MEMBER	13
			HUNTZINGER FLOW	
			‖‖‖‖ EROSIONAL UNCONFORMITY ‖‖‖‖	
			WILBUR CREEK MEMBER	
			LAPWAI FLOW	
			WAHLUKE FLOW	
			‖‖‖‖ EROSIONAL UNCONFORMITY ‖‖‖‖	
			UMATILLA MEMBER	
			SILLUSI FLOW	
			UMATILLA FLOW	
MIDDLE		WANAPUM BASALT — 5-10% of C.R.B.G. total	‖‖‖‖ EROSIONAL UNCONFORMITY ‖‖‖‖	
			PRIEST RAPIDS MEMBER	14.5
			LOLO FLOW	
			ROSALIA FLOW	
			‖‖‖‖ EROSIONAL UNCONFORMITY ‖‖‖‖	
			ROSA MEMBER	
			FRENCHMAN SPRINGS MEMBER	
			LYONS FERRY FLOW	
			SENTINEL GAP FLOW	
			SAND HOLLOW FLOW	15.3
			SILVER FALLS FLOW	
			GINKGO FLOW	15.5
			PALOUSE FALLS FLOW	
			ECKLER MOUNTAIN MEMBER	
			SHUMAKER CREEK FLOW	
			LOOKINGGLASS FLOW	
			DODGE FLOW	
			ROBINETTE MOUNTAIN FLOW	
			‖‖‖‖ EROSIONAL UNCONFORMITY ‖‖‖‖	
LOWER		GRANDE RONDE BASALT — >85% of C.R.B.G.	SENTINEL BLUFFS UNIT	15.6
			MUSEUM FLOW	
			ROCKY COULEE FLOW	
			SLACK CANYON UNIT	
			FIELD SPRINGS UNIT	
			WINTER WATER UNIT	
			UMTANUM UNIT	
		PRINEVILLE BASALT	ORTLEY UNIT	
			ARMSTRONG CANYON UNIT	
			MEYER RIDGE UNIT	
			GROUSE CREEK UNIT	
			WAPSHILLA RIDGE UNIT	
			MT. HORRIBLE UNIT	
		PICTURE GORGE BASALT	CHINA CREEK UNIT	
			DOWNEY GULCH UNIT	
			CENTER CREEK UNIT	
			ROGERSBURG UNIT	
			TEEPEE BUTTE UNIT	
			BUCKHORN SPRINGS UNIT	16.5
			IMNAHA BASALT 5-10% of C.R.B.G. total	17.5

E.W.O.

Figure 10.6 Around 14 million years ago the Priest Rapids lavas resulted during the final great eruptions of Wanapum Basalt from fissures in Idaho. Columns in the basalt are up to 10 feet in diameter (Washington Dept. of Natural Resources).

THREE EXTENSIONS OF BASALT in Idaho were constructed in the St. Maries, Clearwater, and Weiser embayments. The St. Maries lava embayment lies within the watersheds of the St. Maries and St. Joe rivers on the northeast Columbia Plateau. Both Wanapum and Priest Rapids basalts, which are the primary flows in this embayment, originated from local vents. Erupting from fissures in the upper St. Joe River, the Priest Rapids dammed the stream, permitting the basalt to flow down a relatively dry canyon. Because the plateau was fairly level in this region, the Priest Rapids lava was able to advance westward unimpeded by topography, and one of the later Grande Ronde eruptions could spread across to the eastern edge of the platform.

In the lower Clearwater drainage near Lewiston, Idaho, over 4000 square miles of deeply incised mountains were gradually filled by Imnaha, Grande Ronde, and Saddle Mountains basalt. Imnaha and Grande Ronde lavas are thin along the eastern margin of the basin, and successive flows pinch out further west because of tilting of the Columbia Plateau during eruption of the Grande Ronde basalt.

Adjacent to the subsiding western Snake River Plain, the Weiser basin alternately accumulated lava and sediments for 11 million years throughout the Miocene and into the Pliocene epoch. Imnaha, followed by Grand Ronde then Weiser basalt from local vents, created the delta-shaped Weiser embayment that covers 7500 square miles primarily in Washington and Adams counties of Idaho. Dissected by the Snake, Weiser, and Payette rivers, layers of basalt in the embayment are now exposed in steep canyons such as the 800-foot deep gorge of the Weiser River near Cambridge.

Figure 10.7 Tight looping meanders of the Grande Ronde River canyon in southeast Washington expose ribbon-like layers of basalt (Washington Dept. of Natural Resources).

Figure 10.8 Saddle Mountains basalt, intermittently extruded over 8 million years, makes up less than 1% of the total Columbia River basalt flows (Washington Dept. of Natural Resources).

ERUPTIONS TAPERED OFF over the next 10 million years, and succeeding Wanapum and Saddle Mountains formations comprise between 5% and 1% of the total, respectively. In addition to diminished amounts of lava, these later flows, with tens to hundreds of thousands of years between eruptions, are significantly more silica-rich than the older units.

Across the plateau, shifting stress patterns during this interval wrinkled the basalts as they cooled. An initial compressional phase was followed by stretching be-

tween 10 to 12 million years ago that reoriented folds and faults from east-west to north-south trends. The east-west Yakima fold belt in southcentral Washington belongs to the earlier compressional phase; however, prominent northwest-southeast fault systems, which include the Olympia-Wallowa lineament, are due to extension. Best seen on aerial photographs, lineaments are large-scale prominent lines on the surface of the earth that often reflect major structural features such as faults.

Figure 10.9 Evidence for Miocene lakes can be seen in white layers of diatoms within lava flows. Near Vantage in Kittitas County, diatomite, composed of microscopic aquatic plant skeletons, accumulated at the bottom of lake beds where the plants fell once they died (Washington Dept. of Natural Resources).

The Olympic-Wallowa lineament extends from the Strait of Juan de Fuca, through the Pasco basin, and across the Snake River into Idaho. Resulting from this second phase, Basin and Range extensional features may also be present beneath the flows that make up the older part of the basalt platform.

THE FORCES OF RUNNING WATER AND MOLTEN LAVA acted in continual opposition on the Columbia Plateau. In the broad depression between the slowly rising Western Cascades and the Idaho batholith, Columbia River basalts disrupted drainage patterns of streams. With the encroaching lava, waters were pushed from their channels and either cut new pathways or retrenched in the original valley.

The most extensive stream system to be displaced by lava was the Columbia River itself, which was repeatedly blocked by stiffening masses of cooling basalt flows. The ancestral Columbia was able to recover relatively quickly along its lower reaches through the Cascades, but the upper watershed of the river in eastern Oregon and Washington was a different story. Here lavas spread out well beyond the old channels filling the Columbia basin to the brim. To find a new route into the gorge, the Columbia River took a circuitous pathway well to the north of its original channel. Old abandoned channels today are marked by thick layers of coarse gravel lying between layers of basalt.

While the Columbia River was able to circumnavigate the great mass of Grande Ronde basalts, the ancient Snake River drainage was completely diverted. With advancing flows of basalt into the Clearwater embayment of Idaho, the river moved its course well to the southwest as it altered direction. Once the streambed was filled and plugged with successively younger lavas, it was either flushed out or abandoned when the water moved to reestablish its drainage. Intracanyon flows define these old channels. Today the river runs northward around the edge of the basalt embayment in Idaho but takes a direct route through the lava plateau in Washington.

A staggering 42,000 cubic miles of basalt, which buried the older landscape, not only drastically altered drainage patterns of the larger rivers but completely obliterated many smaller networks. Streams unable to cut through the lavas skirted around the margins or backed up as ponds and lakes. Once the impounded waters reached the top of the lava surface, they spilled across creating vast reservoirs in shallow depressions. These bodies of water were natural sediment traps for stream debris eroded from the surrounding higher slopes as well as airborne volcanic ash and dust.

In the thousands of years between eruptions, soil layers developed around the lakes. Here forested slopes and grassy meadows hosted a variety of mammals that would look unusual today. An aquatic rhinoceros similar to an hippopotamus, giant turtles and beaver, large cat

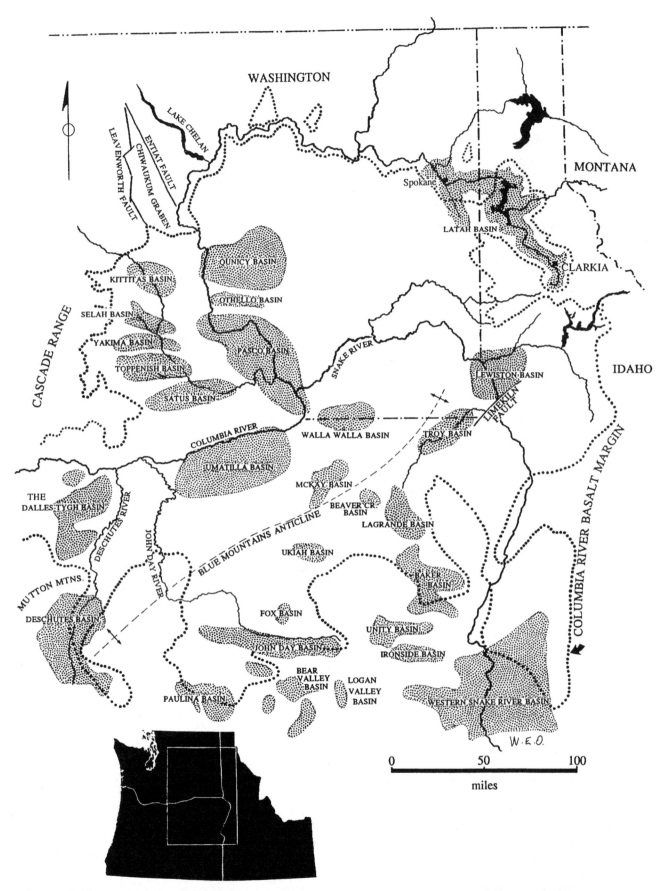

Figure 10.10 Sediment basins across the Columbia Plateau retain a fossil record of Miocene environments (after Smith, Bjornstad, and Fecht, 1989; Fecht, Reidel, and Tallman, 1987).

and dog-like predators, a sheep-like oreodon, diminutive three-toed horses, antelope, deer, bear, camel, and several species of elephants lived on the savannahs of the Early Miocene. During the later Miocene and Early Pliocene, more modern-looking mammals such as horses, raccoons, badgers, and coyotes appeared. Remains of the plants and animals, carried into lake waters and preserved, provide a surprisingly complete picture of environments millions of years ago.

The Early Miocene tended to be rainy, warm, and temperate with short mild winters and humid summers as exists in the southeastern United States today. In this climate, moisture-loving bald cypress, willow, sedges, and cattail were plentiful near lowland streams and lakes, while spruce, fir, oak, maple, and beech adapted to the higher cooler uplands. As the Cascade volcanoes were gradually constructed, much of the ocean-derived atmospheric moisture was trapped west of the range, and the heavy rainfall east of the mountains diminished. In addition, the climate worldwide became cooler and more arid toward the Pliocene.

SEDIMENT BASINS around the eastern edge of the plateau were located near Spokane, Washington, and at Coeur d'Alene, Lewiston, and Weiser, Idaho. A number of small basins scattered in the Blue Mountains, plus those at Umatilla, Pasco, Othello, and Quincy were more central; the small Kittitas, Selah, Yakima,, and Toppenish depressions lay on the western margin of the basalt platform. The Dalles-Tygh and Deschutes basins were developed to the southwest.

Freshwater sediments of the Miocene Latah Formation found from Spokane, Washington, to Coeur d'Alene and Clarkia, Idaho, are the residue of large lakes dammed by volcanic flows. South of Spokane near Cheney, the interruption of west-flowing streams by a ridge of the Wanapum Basalt impounded a lake and created swamps across a wide plain that reached eastward toward Coeur d'Alene. Sluggish meandering streams deposited sand, clay, and volcanic debris of the Latah Formation in the quiet waters. Preserved within the book-like laminations of the Latah are beautiful compressed leaves, microscopic diatoms, and even several species of rare beetles.

In the Idaho panhandle, Miocene lake sediments, identical to the Latah, were deposited within prehistoric Clarkia Lake in Shoshone County when Wanapum Basalts from a local vent dammed the St. Maries River. Surrounded by rolling hills, the quiet lake waters received volcanic ash and floodplain detritus, providing an excellent environment for fossil preservation. At some localities within the lake deposits, a cubic-foot block of sedimentary rock may produce over 1000 separate plant fossils. Most of the 15 million-year old leaves are so well-preserved that the complete cellular tissue remains intact. It is notable that DNA has been extracted from

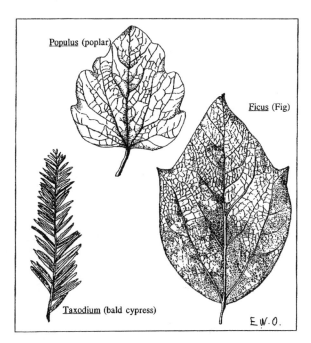

Figure 10.11 Plants of the Latah Formation, living 15 million years ago in eastern Washington, are indicative of a temperate rainy climate.

magnolia and bald cypress leaves from within this anoxic environment. Entire fish skeletons have been recovered, but it is even more remarkable that insects entombed here retain their green, blue, purple, or black iridescence when first uncovered, although the pigment darkens upon exposure to air. Scarabs, click beetles, ants, wasps, and katydids are among the rich fauna of insects removed from Clarkia Lake sediments. Even today the wide flat valley floor south of Clarkia is unmistakable as an ancient lake basin.

FOSSIL PLANTS WITHIN SEDIMENTS OF THE WEISER BASIN in Washington County, Idaho, clearly reflect climate changes for this period. Semitropical plants of the older Miocene Payette Formation give way to those reflecting a drier Poison Creek period, while a mixed flora during deposition of the Pliocene Idaho Group indicates longer winters and cooler summers.

More centrally placed on the plateau, shallow depressions in the Blue Mountains of Oregon at Unity, Ironside, and near Baker collected sediments atop the cooled layers of Picture Gorge and Strawberry lavas. In this region late Middle Miocene lavas from local vents mixed with ash from emerging Cascade volcanoes to trap a wealth of plant and animal remains within the Mascall and Rattlesnake sediments.

Away from drainage systems, the Pasco, Othello, and Quincy basins were largely deprived of sediments, and standing water and swampy conditions were prevalent. The rapidly subsiding Pasco basin of southeast

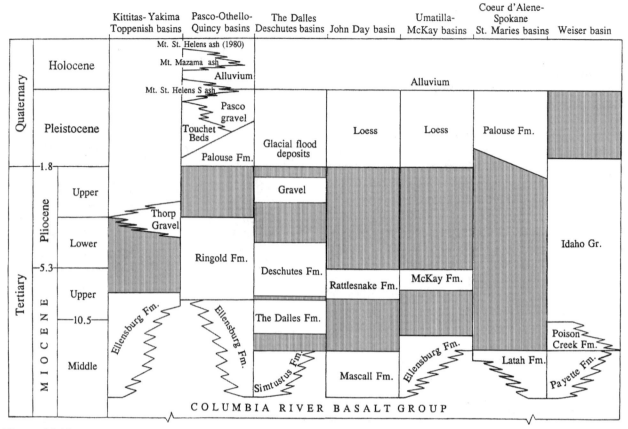

Figure 10.12 Late Tertiary stratigraphy of the Columbia Plateau (Smith, Bjornstad, and Fecht, 1989).

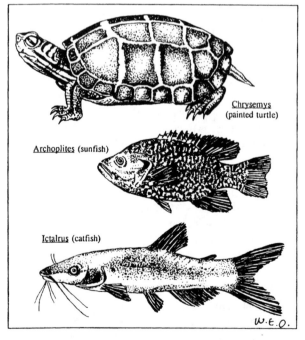

Figure 10.13 Fossil remains of freshwater turtles and fish can be found today where Miocene ponds once filled basins atop the lavas.

Washington is regarded as "sediment starved." Thin deposits of the Ringold Formation, within the settling Pasco trough of Franklin County, are characteristic of a shallow water floodplain with slight depressions, oxbow lakes, and sluggish streams. Molluscs, typically found in freshwater lakes, are rare, and mammal remains are scattered through the sediments, indicating seasonally fluctuating bodies of standing water.

SOUTH AND EAST OF PASCO, THE UMATILLA AND MCKAY BASINS in Oregon are famous for Late Miocene vertebrate fossils recovered from the McKay Formation. In these small local depressions, extensive plant and animal remains reflect an assortment of paleoenvironments that included pond banks, woodlands, and grasslands. The lack of significant sedimentation and presence of swampy conditions in the nearby Troy basin encouraged the development of low-grade coal or lignite beds in the layers between the basalts of this region.

ON THE WESTERN MARGIN OF THE PLATEAU, in southcentral Washington, thick clouds of volcanic ash, blown into the Kittitas, Selah, Yakima, and Toppenish depressions west of the Columbia River,

Figure 10.14 Ice Age shorelines of ancient Lake Missoula are clearly visible in the slopes above Missoula, Montana (courtesy W.H. Nelson, U.S. Geological Survey).

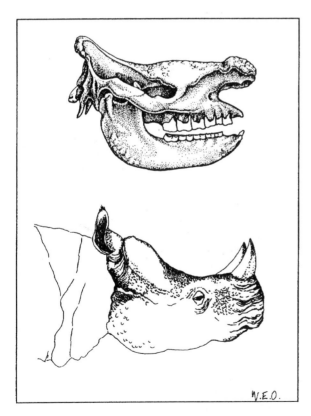

Figure 10.15 Living in eastern Washington 14 million years ago, the ox-sized rhinoceros was smaller than its modern day counterpart and had two horns on its nose (skull after Romer, 1971).

combined to overload streams already choked by lava. Derived from the Cascades to the west as well as from local volcanic sources, most of the sediments were deposited in widespread layers of the Ellensburg Formation by ancient rivers that intersected the basalt platform 12 to 7 million years ago.

Well-developed soil layers in the Ellensburg record lapses of thousands of years between flows of Grande Ronde lavas, when an early Swamp cypress setting was succeeded by deciduous hardwoods followed by a semi-arid forest of white oak. Not only were leaves preserved in the Ellensburg Formation, but, when the land was inundated by the next extrusion of Wanapum basalt, entire trees were overcome and entombed. Often trees in swamps or waterways were protected from burning in the molten lava, only to be buried and preserved intact. Remains of these trees, which also included the less common ginkgo, can be seen at the Ginkgo Petrified Forest State Park near Vantage in Kittitas County. Petrified wood in the Ellensburg Formation has been exposed by ice age floodwaters that stripped away much of the covering layers of lava.

Evidence of animal life during these long spans between basalt flows is strikingly displayed by a rhinoceros encapsulated in the Wanapum Basalts at Blue Lake in Grant County, Washington. The unfortunate animal had fallen into water and become bloated before it was completely encased in pillow basalts that formed when lavas enter water. Preserved as a hollow cast lying

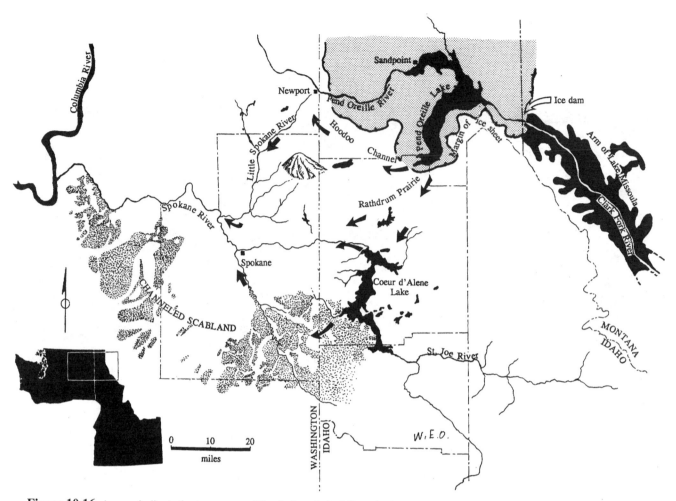

Figure 10.16 Arrows indicate the two routes of floods from Lake Missoula through the Idaho panhandle to the Spokane River (after Breckenridge, 1989; Savage, 1965).

upside down, the animal was held in place by hardened basalt. A partial jaw, teeth, and bones identify the rhinoceros as *Diceratherium,* a curious species, comparable in size to an ox, which had twin horns placed side-by-side at the end of its snout.

ON THE SOUTHWEST MARGIN, the surface of the C-shaped Deschutes and The Dalles basins in Oregon were filled in and smoothed over by the flood basalts. In shallow depressions and low wrinkles on the lava platform, sediment-laden waters spread a thin layer of the Middle Miocene Simtustus mud and sand that was followed by Late Miocene to Pliocene lava, ash, and sediments of The Dalles and Deschutes formations. Camel remains are found from this setting along with the leaves of cool-weather plants. The glassy microscopic skeletons of diatoms, which were abundant in the lake waters, are preserved as clean white layers within the ash.

DURING THE PLEISTOCENE EPOCH, beginning about 2 million years ago, ice sheets from Canada in-

vaded northern Washington and Idaho, plugging river canyons and backing up stream waters to create enormous glacial lakes. As waters rose to overflowing, dams often failed suddenly releasing catastrophic floods that permanently scarred the landscape on the way to the ocean. The hot dry canyons and scablands of eastern Washington today are a sharp contrast to what were probably the world's largest flooding events.

The controversial idea that a scoured topography, known as the channeled scabland of eastern Washington, was due to flooding on a grand scale had been first suggested by J.T. Pardee in 1910 and later confirmed by the painstaking work of J. Harlen Bretz. Despite vigorous opposition by other geologists, Pardee and Bretz concluded that waters of an enormous flood from glacial Lake Missoula sped across Idaho and Washington before entering the Columbia channel near Pasco. Bretz's persistence and research ultimately led to acceptance of the Missoula floods as a reality. Currently evidence suggests that one large cataclysmic flood and scores of smaller releases of water took place.

Figure 10.17 Aerial view showing the dramatic contrast between the Palouse surface (bottom) and pock-marked channeled scabland (top) in northwest Whitman County, Washington (Washington Dept. of Transportation).

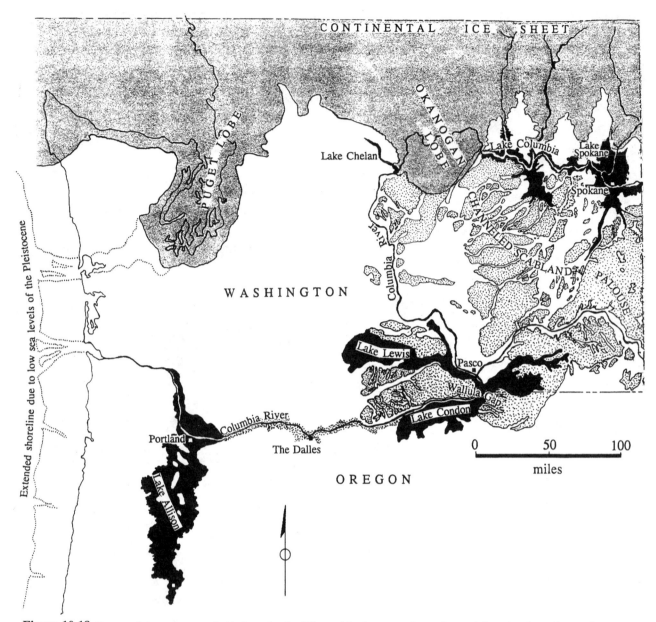

Figure 10.18 By examining a system of old channels, dry falls, scabland topography, and gravel deposits, the pathway of Pleistocene Lake Missoula flooding can be traced from Montana through Idaho, Washington, and Oregon, and on to the Pacific Ocean (after Richmond, 1986; Wright and Frey, 1965).

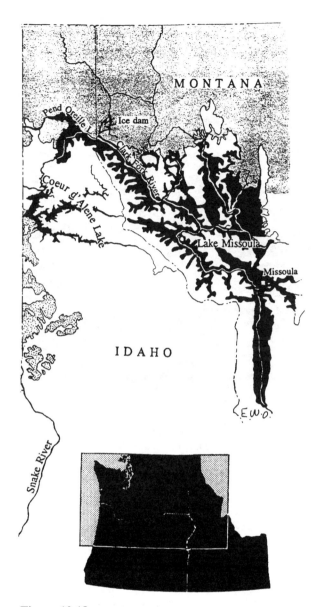

Figure 10.18 Continued.

can be traced. One poured through Pend Oreille Lake, Idaho, then into the Little Spokane River, while a second route emptied directly into the Spokane River valley spreading into a network of separate channels across Washington. The famous channeled scabland of Washington was gouged into the Miocene basalts by each ensuing torrent of water that cut deep, elongated basins and left extensive gravel bars from the Spokane River south to the Snake and westward to the Columbia. Thick sequences of wind-deposited Palouse dust or loess, which originally blanketed southeast Washington, were the first to be stripped away before the layers of Columbia River basalt were removed. Joined by waters that boiled down Crab Creek, the inundations cut Grand Coulee then drained into Quincy basin. Here boulders and gravel, carried in chunks of ice and tumbled along in the stream, were distributed in long mounds hundreds of feet high on both sides of the Moses Lake depression.

Spreading out over Quincy basin in Grant County, the water divided into two streams. One exited south to cut the Drumheller channels, while the rest escaped west across the Potholes and Frenchman Springs coulees, falling over a lip of basalt into the Columbia River. At the Drumheller channels a dramatic view of the erosive force of water and debris is provided by the elongate basins and buttes that parallel the flow.

All floodwater streams flowed into the Pasco basin where the rushing waters paused temporarily at a single outlet through Wallula Gap in Washington to back up as the 1000-foot deep Lake Lewis. Just east of Walla Walla, thinly layered silts and sands of the Touchet Beds were deposited in fluctuating Lake Lewis waters during waning glacial episodes. At one site, 40 separate Touchet layers exposed in the walls of a deep arroyo are regarded as representing an equal number of flood events. Blocks of ice, boulders, sand, and gravel picked up by the ice jammed into the narrows at The Dalles, Oregon, to create Lake Condon before draining and advancing westward. Even though most of the water reached the Pacific Ocean at the mouth of the Columbia, a considerable amount flooded southward to Eugene in the Willamette Valley of Oregon as Lake Allison. With depths of 400 feet at Portland, the lakes deposited thin layers of silt in the valley before the muddy turbulent waters eventually receded back into the Columbia gorge. Glacial erratic stones, many of a substantial size, dot the floor of the Willamette Valley.

These events occurred between 15,000 and 12,500 years ago when ice dams ruptured on great glacial lakes along the Clark Fork River, a branch of the Columbia in western Montana, resulting in over 40 cataclysmic floods. Each is estimated to have borne up to 400 cubic miles of water. Flowing at the rate of 9 cubic miles per hour, it would have taken two days to empty the basin. Two main pathways of the successive floods

Figure 10.19 Thin layers of silt, which settled out of Pleistocene flood waters, were exposed as recently as 1926 in canyon walls near Touchet, Washington, when water broke out of an irrigation canal to cut a deep arroyo (Oregon Dept. of Geology and Mineral Industries).

COULEES, LONG TRENCH-LIKE VALLEYS, are characteristics erosional features in eastern Washington marking the channels of ice age floods. Of the largest coulees, Grand Coulee and Moses Coulee, the Grand Coulee is the most extensive. Stretching for 50 miles south from the Columbia River to Soap Lake in Grant County, Grand Coulee is from 3 to 5 miles wide at the upper end, narrowing to 1½ miles wide in the south before opening up into Quincy basin. This enormous trench was incised 1000 feet into Columbia River flows.

During the Pleistocene, the Okanogan lobe of the continental ice sheet from British Columbia dammed the Columbia River near the north end of the Grand Coulee creating the glacial lakes, Columbia and Spokane. Overflowing southward across the basalt plateau, the lake waters covered the Spokane River valley and its tributaries. Fine laminated lake sediments from this basin are visible around Spokane today. Partially drained when water eroded a rock divide at the northern end of the Coulee, Lake Columbia was ultimately evacuated sometime after 13,000 years ago by a catastrophic flood, which was probably augmented by water from glacial Lake Missoula.

Inside the Grand Coulee, Steamboat Rock is a long basalt mesa level with the surrounding plain. During high water the monolith became an island that divided the floodwaters. Lakes and dry "water" falls dot the floor of the Coulee. At Dry Falls, near Coulee City, a series of five horseshoe-shaped cuts into the basalt drop over 400 feet in 3 miles in what must have been a spectacular cataract during flood events. Dry Falls Lake, Green Lake, Deep Lake, and Castle Lake across the front of the 3-mile wide cataract complex mark ancient plunge pools. Great whirlpools in the narrow chute at Deep Lake rolled enormous boulders around to auger out circular potholes in the lava.

Below Dry Falls, a succession of lakes like Park Lake, Blue Lake, Lenore Lake, and Soap Lake occupy depressions enlarged into fractured basalt. At the bottom of the chain, Soap Lake receives minerals from overflow of the upper lakes. Since it has no outlet other than evaporation, the waters are alkaline. As wind whips across the water, soapy foam collects along the shore giving the lake its name.

Extending from Grimes Lake to the Columbia River, a narrow channel of Moses Coulee in Douglas

County rivals Grand Coulee in size. The wide floor and magnificent walls of basalt have been deepened by successive glacial floods. The north end of Moses Coulee is covered by glacial material marking the terminus of the Okanogan ice lobe. Where the glacier stabilized then retreated, it constructed a hummocky moraine a mile wide and 100 to 200 feet high between the Columbia River and Banks Lake. The impressive size of the Withrow moraine suggests the ice lobe remained in that location for some time.

At the south end of Quincy basin, sand dunes, built by blowing winds, trap water in Moses Lake as well as in the ponds at The Potholes region. Water backed up behind O'Sullivan Dam in southern Grant County has partially covered the dunes of this natural landmark.

ALTHOUGH MULTIPLE FLOODS had a lasting effect on the topography and surficial geology of the Willamette Valley, Columbia gorge, eastern Washington, and the Idaho panhandle, they only had an ephemeral effect on drainage patterns across the plateau. While flooding rivers may leave their existing beds and widen into broad shallow streams, they invariably return to their earlier channels.

Destructive floods from ice dams are not unknown today, and in Alaska and western British Columbia ice often blocks fjords that later release large amounts of water. In 1982 Russell Fjord in upper Yakutat Bay, Alaska, was blocked for a time by Hubbard glacier until the ice broke, and a torrent of brackish water emptied into the ocean.

ROLLING TERRAIN OF THE PALOUSE HILLS covers a 3000-square-mile region across southeast Washington and northern Idaho where wind-blown silty loess sediments over 200 feet thick obscure the undulating upper surface of the Columbia River basalts. Thinner layers extend further into the Pasco basin of Washington and the region between Pendleton and The Dalles in Oregon. Dating back almost 2 million years, the yellowish-brown silt contains ash as well as buried fossil soils (paleosols), which preserve a variety of environmental settings. The remains of large ice age mastodons, mammoths, horses, bison, and sloths have also been recovered from these soft fine-grained deposits. Although loess is regarded as aeolian or wind blown in origin, the fine dust apparently originated as rock flour ground up by glaciers. Dumped on the outwash plain of melting ice sheets, the loose material was collected by strong winds and carried away from the margin of the glacier.

Figure 10.20 At the western end of Frenchman Springs Coulee in Grant County, flood waters, which cut the coulee, merged with the Columbia River. The white patches are diatoms that precipitated to the bottom of Miocene lake beds (Washington Dept. of Natural Resources).

Figure 10.21 The flat basalt butte of Streamboat Rock near Dry Falls in Grand Coulee (Washington Dept. of Natural Resources).

GRAND COULEE DAM, across the main channel of the Columbia River, was built to provide flood control, electric power, irrigation, and recreation. Begun in 1933 and completed in 1942, the water backed up behind the dam into the 151-mile long Franklin D. Roosevelt Lake. Water is pumped from FDR Lake 300 feet up into man-made Banks Lake, which today occupies the northern end of Grand Coulee, Banks Lake, used as a reservoir, has a capacity of 715,000-acre-feet.

In the construction, vast amounts of glacial clays and silts of ancient Lake Columbia were removed to reach bedrock. Excavation oversteepened slopes and precipitated close to 60 landslides between 1934 and 1937. Even with a blanket of rip-rap, as the lake was being filled, 245 additional landslides were triggered, and today they continue to be a problem.

Rising 550 feet above the bedrock at its lowest point, Grand Coulee Dam is over a mile long. The two powerhouses have a total capacity of 2.25 million kilowatts, while the third has a capacity of 3.9 million kilowatts, the largest single power source in the United States.

Figure 10.22 In eastern Washington, Grand Coulee Dam impounds Franklin D. Roosevelt Lake, which fills the deep channel or coulee eroded by Pleistocene floods. Dams such as these effectively limit salmon on the upper reaches of the Columbia River (Washington Dept. of Transportation).

ADDITIONAL READINGS

Allen, J.E., 1979

Allen, J.E., Burns, M., and Sargent, S.C., 1986

Anderson, J.L., 1987

Baker, V.R., 1987

Baker, V.R., and Nummendal, D., 1978

Baker, V.R., et al., 1987

Beck, G.F., 1945

Beeson, M.H., Perttu, R., and Perttu, J., 1979

Behrens, G.W., and Hansen, P.J., 1989

Bond, J.G., 1963

Bretz, J.H., 1923

Bretz, J.H., 1932

Camp, V.E., 1982

Carson, R.J., and Pogue, K.R., 1993

Carson, R.J., Tolan, T.L., and Reidel, S.P., 1987

Chaney, R.W., Condit, C., and Axelrod, D.I., 1959

Fecht, K.R., Reidel, S.P., and Tallman, A.M., 1987

Fitzgerald, J.F., 1982

Gustafson, E.P., 1978

Hooper, P.R., and Conrey, R.M., 1989

Hooper, P.R., and Swanson, D.A., 1987

Hooper, P.R., and Swanson, D.A., 1990

Kaler, K.L., 1988

Knowlton, F.H., 1926

O'Connor, J.E., and Baker, V.R., 1992

Pardee, J.T., and Bryan, K., 1926

Prakash, U., and Barghoorn, E.S., 1961

Reidel, S.P., and Hooper, P.R., eds., 1989

Reidel, S.P., 1994

Richmond, G.M., 1986

Ringe, D., 1970

Smiley, C.J., 1963

Smiley, C.J., 1989

Smiley, C.J., and Rember, W.C., 1979

Smiley, C.J., Shah, S.M.I., and Jones, R.W., 1975

Smith, G.A., 1991

Smith, G.A., Bjornstad, B.N., and Fecht, K.R., 1989

Stradling, D.F., and Kiver, E.P., 1989

Tolan, T.L., and Reidel, S.P., 1989

Waitt, R.B., 1987

Webster, G.D., and Nunez, L., 1982

Williams, I.A., 1991

Wright, H.E., and Frey, D.G., eds., 1965

CHAPTER

11

COAST PROVINCE OF OREGON AND WASHINGTON

Sand dunes South of Florence, Oregon, slowly engulf the channel of the Siletz River (Oregon Dept. of Geology and Mineral Industries).

From the Strait of Juan de Fuca on the north in Washington to the Middle Fork of the Coquille River in southern Oregon, the 150-mile wide Coast Province extends a distance of approximately 400 miles. Bordered to the west by the abyssal plain of the Pacific Ocean and on the east by the Cascade Mountains, the coastal framework encompasses a variety of physiographic areas. The steep-sided Olympic Mountains, at the extreme northwest corner of the continental United States, are separated from the low rounded Willapa Hills by the Chehalis River, both in Washington. South of the Columbia River, which delineates the boundary between Washington and Oregon, a narrow strip of moderately high mountains and prominent headlands characterize the Oregon Coast range. Adjacent to the mountains on the east, the Puget-Willamette basin is a young structural depression, while to the west the continental shelf and slope abut the deep abyssal floor of the Pacific Ocean.

Mountains of the coastal province vary considerably in elevation. Overall the central range is subdued, with the lowest elevations in northern Oregon and southern Washington. In Oregon, Marys Peak near Philomath forms the crest of the southern range at 4097 feet, while Saddle Mountain, Trask Mountain, and Sugarloaf Peak in the northcentral portion of the state are only half as high as the snowy ridges of the Olympics. Most of the Olympic Peninsula of Washington encompasses exceptionally lofty peaks, which lie in a belt that is narrower to the northwest near Cape Flattery but which widens to 50 miles toward Puget Sound. Mt. Olympus, at 7954 feet above sea level, Mt. Deception at 7788 feet, and Mt. Constance at 7743 feet are the highest of the range.

Covered by thick forests and mature soils, the west face of the coastal mountains receives the heaviest precipitation, but watersheds are small. Few streams in the Oregon or southern Washington range cut completely across the mountains, and most flow from the crest westward to reach the Pacific Ocean or eastward to the lowlands. By contrast, rivers in the Olympic Peninsula, with a constant supply of water from melting snow and glaciers, radiate outward in all directions from the high central peaks.

Figure 11.1 Saddle Mountain in the northern Oregon Coast Range is a mass of dark Columbia River basalts that originated in eastern Oregon, Washington, and Idaho. The mountain is centered in a 3000 acre park (Oregon State Highway Dept.).

Figure 11.2 The lighthouse, perched on precipitous cliffs of Eocene volcanic rocks at Cape Disappointment, forms the gateway to the mouth of the Columbia River on the Washington side (Washington Dept. of Transportation).

The broad depression of the Willamette Valley in Oregon extends northward across the Columbia River where it continues as the Puget Lowlands of Washington. The Willamette Valley is a level alluvial plain that averages from 20 to 40 miles wide before it pinches out at the southern end near Cottage Grove. From Eugene with an elevation of 400 feet to Portland near sea level, the valley gradient is less than 4 feet per mile. Beginning at the junction of the Juan de Fuca and Georgia straits,

Puget Sound covers 2000 square miles and reaches a length of 100 miles before terminating south of Olympia. Even the largest ships can enter any part of the Sound, which has an average depth of 600 feet. The overall topography of the lowland is moderate with rolling hills, rivers, lakes, and inlets. Although comparatively limited in area, the Willamette and Puget lowlands support much of the industrial wealth as well as most of the population of both states.

A scenic shoreline that displays an innumerable variety of dunes, sand spits, high cliffs, promontories, and coves typifies the coastal edge of the province. This margin continues offshore as the continental shelf and slope down to the abyssal plain lying 9000 feet below sea level. Scarps, banks, ridges, basins, and canyons give the otherwise level shelf and slope an uneven topography.

INTRODUCTION

Despite the topographic and structural variety of the continental margin, mountains, and lowlands that comprise this area, the similarity of marine rock formations unify the region into a single geologic province. The coastal province was one of the last to form in the Pacific Northwest, and uplift of the range did not begin until long after the significant volcanic and sedimentary events took place.

Rocks deposited within a wide seaway, which covered the area during the Early Tertiary, reflect changing environments and multiple sediment sources against a background of ongoing volcanic activity. Eocene submarine lavas along the coast were rapidly covered with sediments before the sea became increasingly shallow and more constricted during Oligocene and Miocene time. These thick layers of strata were uplifted and folded to become the foundation of the Coast Range, a simple wrinkle in an otherwise much more complicated geologic picture. The Late Miocene uplift pushed the shoreline well to the west near its present location.

East of the coast mountains, development of the Puget Lowland and Willamette Valley by structural processes, and not stream erosion, was directly related to coastal uplift. Driven by continuing subduction, a seesaw effect raised the range as it lowered the interior valleys.

Very late in the Cenozoic era, profound climate changes affected the Pacific Northwest when temperatures began to drop and glaciers formed. An immense Pleistocene ice lobe from Canada sent an arm into the Strait of Juan de Fuca and Puget Sound. Late in this same interval, dozens of catastrophic floods from glacial lakes as far east as Montana surged through the Columbia River gorge and backed up into the Willamette Valley. Thick layers of Ice Age silt covered the Puget-Willamette lowlands, obscuring much of the bedrock today.

GEOLOGY OF THE COAST RANGE

Throughout the early Cenozoic, from 55 to 35 million years ago, a wide shelf, coast plain, and shoreline angled diagonally across Oregon and Washington. During this interval an undersea volcanic chain, carried atop the Kula and Farallon plates, merged with North America to form the volcanic underplate of the Coast Range province.

The Eocene shoreline curved east from Coos Bay to what is now the Western Cascades, crossing the Columbia River along the Puget Lowland and extending northward close to Bellingham. Offshore in the Pacific a discontinuous ridge of volcanoes that occasionally broke the surface as low islands formed a topographic high that would become the foundation of the Coast Range. These underpinnings of the province are Paleocene to Eocene basalts that began as submarine eruptions across the floor of the ocean 63 to 46 million years ago. This sheet of submarine lavas varies considerably in thickness from 20 miles under Oregon to a 3.5 mile veneer beneath southern Vancouver Island.

The origin of these volcanics is not clear, but the Metchosin and Crescent lavas at the northern end of this chain on Vancouver Island and in the Olympic Peninsula, along with the Roseburg and Siletz volcanics in the southern range of Oregon, are significantly older than those erupted in the central portion of the province underlying the Willapa and Black hills of Washington. The apparent age progression of older eruptions at the north and south margins has been cited as evidence that the island chain straddled a mantle plume on an active ocean spreading center that separated the Kula and Farallon plates. Today, moreover, many geologists think that the hot spot beneath Yellowstone was also the source for the coastal volcanics. During the Eocene, eruptions took place as the western edge of the North American Plate passed over that mantle plume.

From the middle to late Eocene, 43 to 38 million years ago, eruptions of the Yachats, Tillamook, Cowlitz, and Northcraft lavas from along the central Oregon coast north to the Willapa Hills, Centralia, and Seattle added to the volcanic pile. Thick accumulations of basalt from submarine vents built up as low shield cones on the offshore platform.

THE EARLY COAST RANGE LAVAS show evidence of having been rotated clockwise after they formed. Before igneous rocks cool and harden, mineral crystals in the melt act as magnets to align themselves with the earth's magnetic polarity. Any later changes in orientation of the rock can then be detected and measured. The rotation of coast basalts took place over a lengthy period of time as older rocks show as much as 90 degrees clockwise rotation, while younger intervals display progressively less. In addition to increased rotation of older rocks, there is a geographic variation in rotation from east to west so that volcanic rocks on the coast display much more intense rotation than those of eastern Oregon and Washington.

Early interpretations of this rotation phenomena suggested that the entire range may have been an exotic

Figure 11.3 Main localities for coastal Washington.

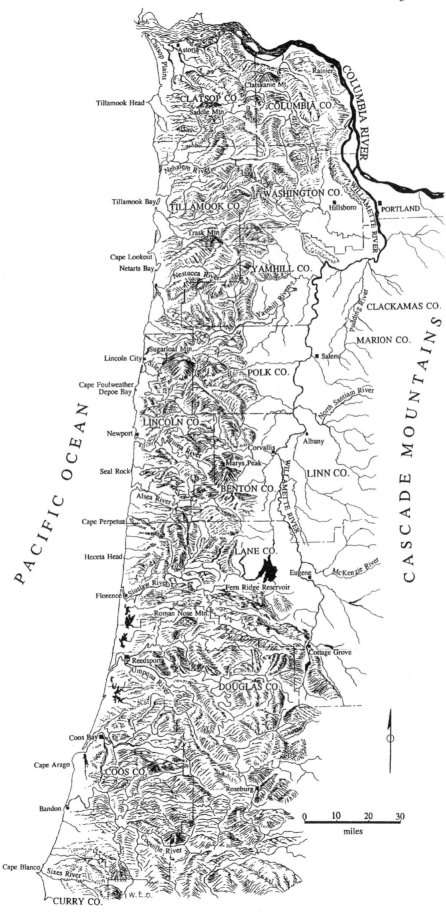

Figure 11.4 Map of locations for the Oregon coast.

Figure 11.5 Eocene Yachats basalts at Cape Perpetua are the foundation of coastal headlands on the southern Oregon Coast (Oregon Dept. of Geology and Mineral Industries).

terrane that swung clockwise as it was docking or ac-creting to the North American continent. More recently it has been shown that rotation could be achieved as a dextral shearing motion with the basalts remaining in place. This explanation for the configuration of these volcanoes is called "continental margin rifting" in which the Coast Range began as a basin that was pulled apart. The shearing action of the north-bound Kula Plate against the west-moving North American Plate produced a series of small spreading ridges separated by short faults. As lavas issued from the undersea rifts or tears, a discontinuous line of Early Tertiary volcanoes began to form on the sea floor.

Elsewhere during the Late Mesozoic to Early Ter-tiary of the Pacific Northwest, this same shearing action is responsible for hundreds of large faults such as the Eocene Fulmer Fault across the shelf of southern Ore-gon. A similar shearing motion is taking place in the Gulf of California today where the Baja Peninsula is being torn away from Mexico, leaving the widening basin of the Sea of Cortez.

AS THE SUBMARINE VOLCANIC PLATFORM SUBSIDED, a long marine basin was created paralleling the margin of North America. River systems delivered sediments, derived from the continent, to cover earlier lavas within this trough. Sediments, which filled the basin, display a variety of ocean environments.

At the southern end of this ocean plateau in Oregon, deltas, submarine fans, and deepwater silt of the Paleo-cene to early Eocene Roseburg, Lookingglass, and Flournoy formations were derived from local undersea volcanoes as well as from the nearby Klamath Moun-tains. Delta and turbidite sand of the later mica-rich Tyee Formation may have been transported by a more mature drainage system that reached much further east.

In shallow seas, construction of the immense middle to late Eocene Coaledo delta near Coos Bay, which ex-tended northward, was in turn supplied by Cascade vol-canic ash that also contributed to the more northerly Cowlitz, Pittsburg Bluff, and Scappoose formations within Columbia County. Shelf conditions of the central basin are seen in silts and sands of the Alsea and

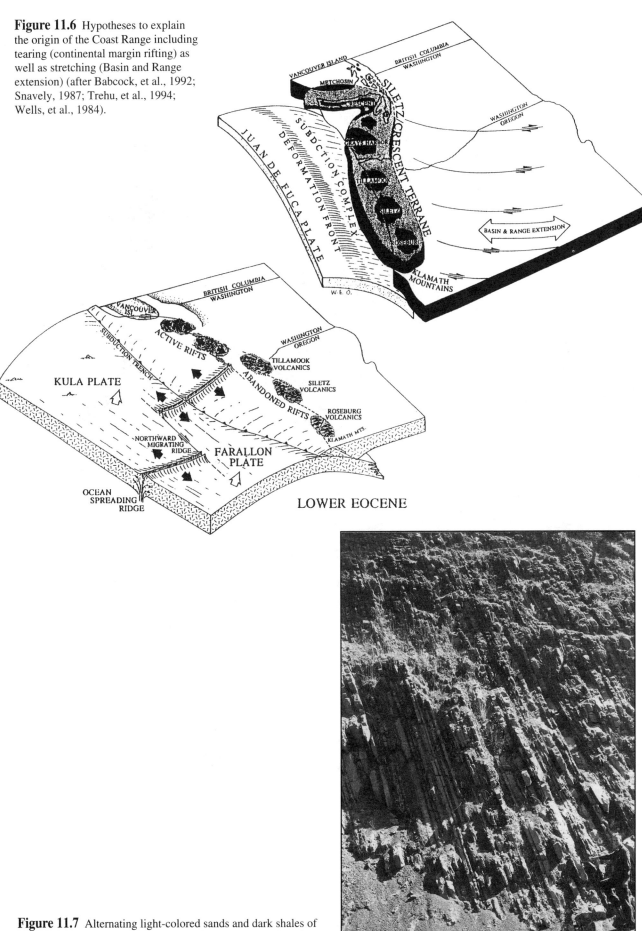

Figure 11.6 Hypotheses to explain the origin of the Coast Range including tearing (continental margin rifting) as well as stretching (Basin and Range extension) (after Babcock, et al., 1992; Snavely, 1987; Trehu, et al., 1994; Wells, et al., 1984).

Figure 11.7 Alternating light-colored sands and dark shales of the upturned Roseburg Formation display successive submarine turbidite flows across a continental slope (E. M. Baldwin).

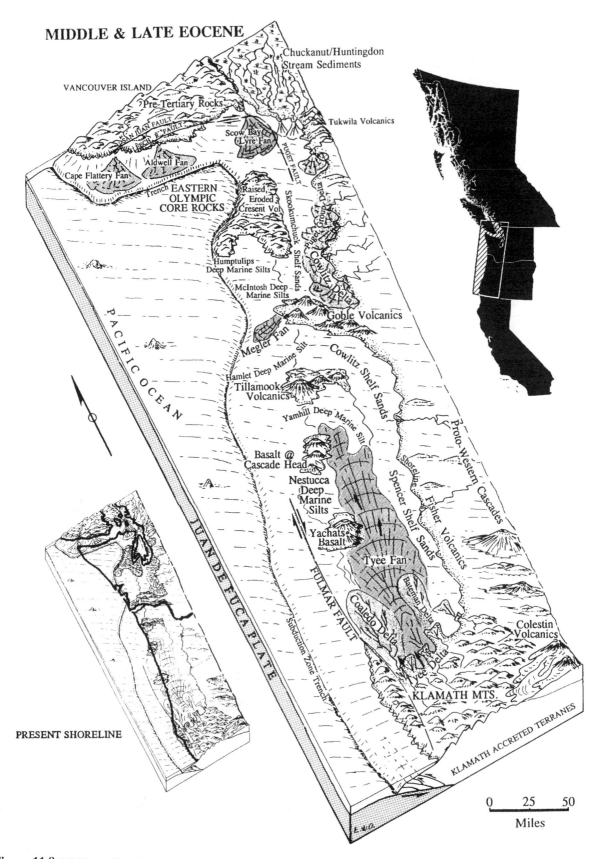

Figure 11.8 Middle and Late Eocene environments along the coast (after Christiansen and Yeats, 1992).

Figure 11.9 Sunset Bay in the upper left and Gregory Point, supporting a lighthouse, are composed of rocks of the Coaledo Formation, a nearshore delta during the Eocene (courtesy W. and M. Robertson).

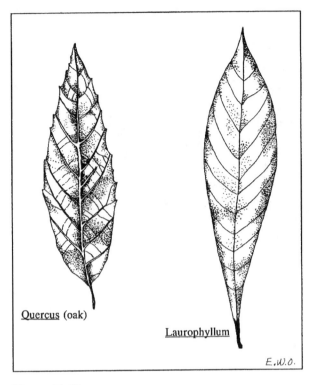

Figure 11.10 Toward the end of the Eocene interval, 36 million years ago, plants along the coastline reflected a mixed temperate-tropical environment.

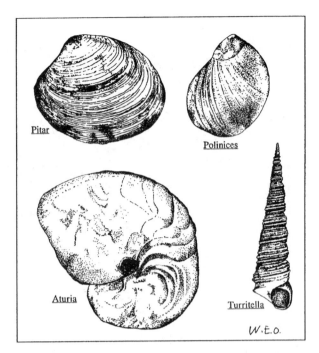

Figure 11.11 Invertebrates from Eocene formations along the West Coast.

Yaquina deposits. Similarly, invertebrates from the Eugene Formation in the southern valley lived in shallow shelf waters. However, a deeper ocean setting is reflected by thin-shelled delicate molluscs and microfossils in the Bastendorff muds adjacent to the south coast and by muds, sands, and silts of the Nestucca in the north. Further west, ocean environments of the Late Eocene Keasey Formation are typical of deeper continental slope water.

Along the western edge of the Willamette Valley, a diverse subtropical broadleaf plant flora is found in volcanic sands of the nonmarine Fisher Formation. Interfingering with the Fisher, shell-rich shelf deposits of the marine Spencer testify to the proximity of the beach here. Ash and lava of proto-Cascade volcanoes were the source for Fisher and Spencer sediments as well as for deepwater muds of the Yamhill Formation in the northern part of the valley. Along the shorelines and out into the deeper waters of the ocean basin, limestone banks of the Rickreall Member within the Yamhill supported an extensive invertebrate fauna.

In Puget Sound, an elaborate stream system built an immense delta that stretched from the marine shoreline east of Centralia and Olympia across a broad inland platform to the vicinity of Mt. Rainier. Within the delta, thick layers of the Middle and Upper Eocene Puget

Group preserve fossil leaves, shelled invertebrates, and coal. Broadleaf oak, palm, and fig indicate humid tropical to subtropical conditions. Puget Group sediments merge westward with shelf sand of the Skookumchuck and deepwater silts of the McIntosh and Humptulips formations. Delta conditions ended with extensive volcanic ash deposits and a lowered sea level during the Oligocene.

DURING THE LATE EOCENE TO OLIGOCENE, 40 to 23 million years ago, local areas of uplift along the axis of the ancestral Coast Range brought increasingly shallow water, and by Late Oligocene time, the seaway only extended as far south as Salem, Oregon. Continued erosion of the older volcanic sediments as well as renewed eruptions in the Cascades added thick layers of debris and ash to the coastal plain and shelf.

The retreating strand can be traced through the northern Willamette Valley by storm deposits and rocky coast conditions of the Scotts Mills Formation. Volcanic ash, which flushed into this nearshore setting from the Western Cascades, entombed a rich fossil collection of molluscs and barnacles along with occasional bones and teeth of sharks and primitive whales. In southwest Washington diverse marine environments of a westward-deepening ocean trough, beginning with the strand and

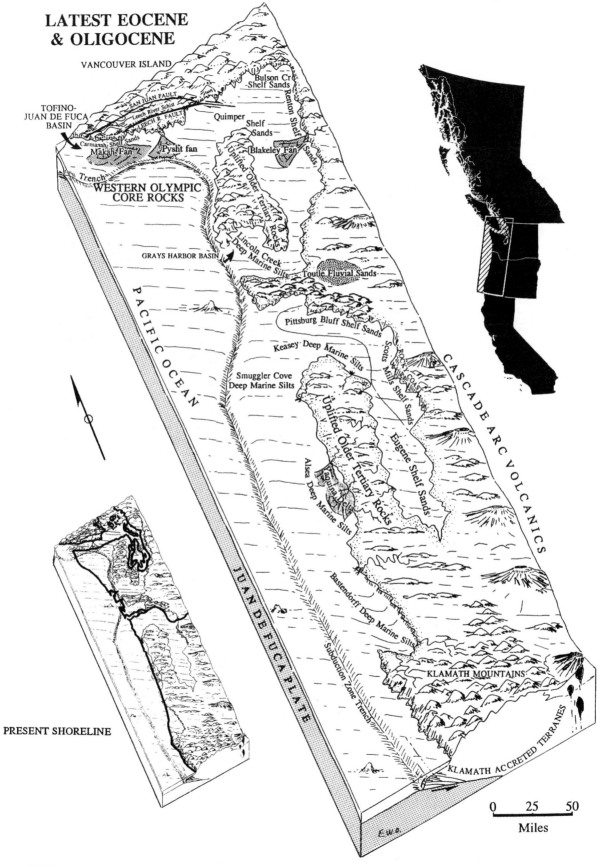

Figure 11.12 Latest Eocene and Oligocene shoreline environments (after Christiansen and Yeats, 1992).

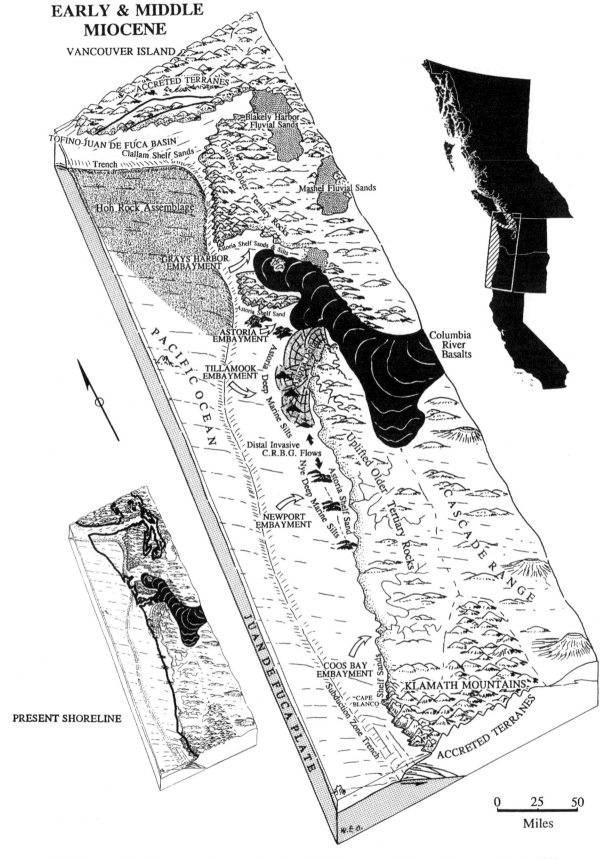

Figure 11.13 Early and Middle Miocene environmental settings in the Coast Range (after Christiansen and Yeats, 1992).

Figure 11.14 A beautiful fossil scallop, *Pectinopecten,* from the Middle Miocene Astoria Formation at Newport, Oregon (photo by H. Howard, courtesy of Oregon Dept. of Geology and Mineral Industries).

Figure 11.15 A reproductive colony of the slipper snail, *Crepidula,* preserved in coastal embayments of the Astoria Formation (photo courtesy John Armentrout).

grading upward into bathyal fine-grained sand and silt, are represented by highly fossiliferous deposits of the Lincoln Creek Formation.

The period from 38 to 29 million years ago saw the emplacement of a number of dikes and sills into rocks of the central coast province from a volcanic source underneath the basin. Following uplift, when erosion peeled off the younger sediments, these intrusive bodies were exposed in the higher elevations of the range. From Roman Nose and Marys Peak in Benton County to Buck Mountain in Tillamook County of Oregon, dominant mountains are composed of these plutons. West of Portland and in adjacent Washington, the Goble lava flows

from this interval were either part of the same volcanic arc or were erupted independently as offshore islands.

FROM THE EARLY TO MIDDLE MIOCENE, 23 to 15 million years ago, continued uplift of the Coast Range and Olympic Mountains saw ocean waters retreat as the shoreline moved to a new position on the west flank of the incipient coastal mountains. The elevation of the range, caused by steady underthrusting of the Juan de Fuca Plate beneath North America, produced a tilting effect that simultaneously depressed the eastern margin along the lowlands of Puget Sound and the Willamette Valley.

Marine deposition during the Miocene was limited to embayments on the edge of the Olympic Peninsula, in the Centralia-Chehalis area of Washington, and on the western fringes of the newly formed coast mountains. Covering the Middle Miocene continental margin of Oregon and Washington, nearshore deltas and shelf sand of the Astoria Formation grade westward into deep water silt beneath the present-day continental shelf. The Astoria is famous for well-preserved fossil molluscs, corals, barnacles, and crabs encapsulated in rounded concretions. Fossil teeth and bones of sharks, whales, seals, and sea lions are equally spectacular.

Tertiary delta and basin deposits formed along the Pacific Coast of Washington. In the northern Olympic Peninsula, sandstones and conglomerates of the Clallam Formation, eroded from Vancouver Island and the North Cascades, built a nearshore delta. South of here a small nonmarine basin near Seattle, choked with over 3000 feet of Early Miocene coal-bearing clays and conglomerates of the Blakely Harbor Formation, was supplied by a mixture of the older eroded Crescent basalts and ash from the Cascades.

On the Oregon coast between Seal Rock and Tillamook Head many of the headlands are formed by invasive Middle Miocene Columbia River flood basalts that originated from vents and fissures over 300 miles away in eastern Oregon, Washington, and western Idaho. These fluid basalts were routed through the ancestral Columbia River gorge where some followed the river to reach the ocean while others crossed the coast range. Here they ponded in deep molten pools of such high density that the lavas sank into softer coastal sediments as local dikes and sills before cooling. Chemically and mineralogically identical to flows of the Columbia River lavas, the Cape Foulweather basalt flow intruded inner shelf sediments north and south of the entrance to the Columbia River. However, the Depoe Bay basalt penetrated rocks as much as 10 miles offshore at depths of 1½ miles. Tongues of lava also spread north into Wahkiakum and Pacific counties of Washington and south into the Willamette Valley as far as Salem.

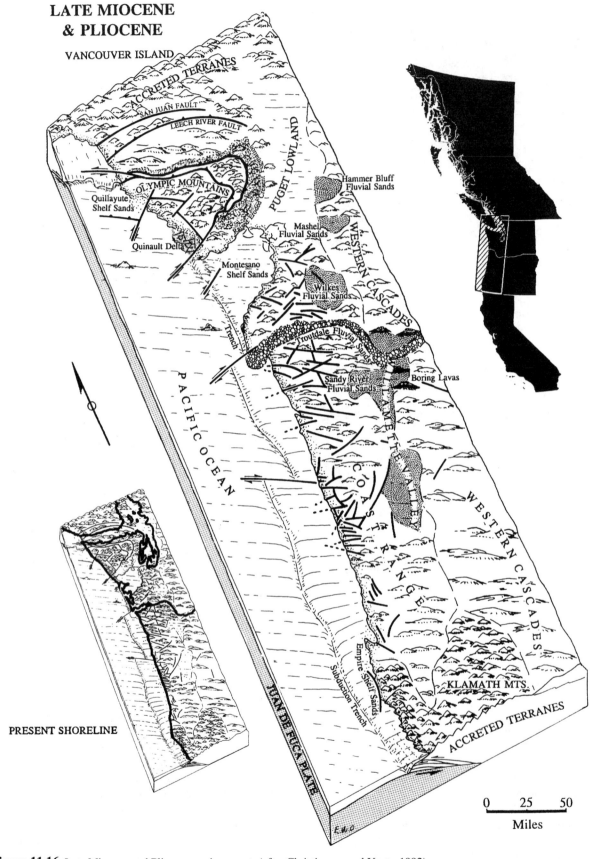

LATE MIOCENE & PLIOCENE

VANCOUVER ISLAND

ACCRETED TERRANES

SAN JUAN FAULT

LEECH RIVER FAULT

OLYMPIC MOUNTAINS

PUGET LOWLAND

Quillayute Shelf Sands

Quinault Delta

Montesano Shelf Sands

Hammer Bluff Fluvial Sands

Mashel Fluvial Sands

Wilkes Fluvial Sands

WESTERN CASCADES

Trench

Troutdale Fluvial Sands

Sandy River Fluvial Sands

Boring Lavas

PACIFIC OCEAN

COAST RANGE

THE VALLEY

WESTERN CASCADES

Empire Shelf Sands

Subduction Trench

KLAMATH MTS.

ACCRETED TERRANES

JUAN DE FUCA PLATE

PRESENT SHORELINE

0   25   50
Miles

Figure 11.16 Late Miocene and Pliocene environments (after Christiansen and Yeats, 1992).

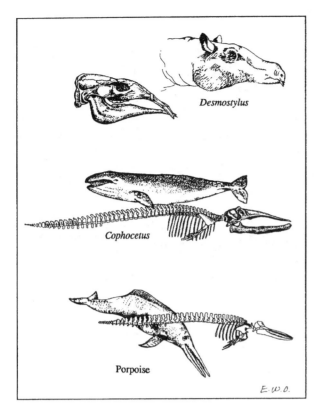

Figure 11.17 Now extinct, the hippopotamus-like *Desmostylus,* the blunt-nosed whale *Cophocetus,* and a long-nosed porpoise inhabited the Pacific Northwest during the Miocene (skeletons modified from Romer, 1971).

THE LATE MIOCENE TO PLIOCENE interval, between 15 to 2 million years ago, was one of continuing uplift, faulting, and deformation of older strata as the Coast and Olympic mountains were folded and raised to their present elevation while the lowlands to the east were depressed. Accelerated erosion increased sedimentation on the coast where rivers transported sand to small embayments and onto the upper shelf.

Invertebrate faunas of the Late Miocene Empire Formation at Coos Bay and Cape Blanco record the shallow warm water of a diminishing basin on the southern Oregon coast. In the same period brackish swamps, lakes, and meandering streams were prevalent during deposition of the Montesano Formation around Grays Harbor, Washington. By Late Pliocene time the shoreline was close to its present location.

ONLY LIMITED PLIOCENE SEDIMENTS, of 5 to 1.8 million years in age, are known from the Coast Range province where most accumulated in structural depressions within the Puget-Willamette lowlands south of Seattle, near Vancouver and Portland, and on the Washington coast. Late Miocene stream and lake sediments of the Mashel, Wilkes, and Hammer Bluff formations, stripped from the Cascades, were piled at the base of the range near Seattle. At the same time, clay and silt of the Sandy River mudstone constructed deltas in the lake waters that filled a bowl-shaped basin from Portland

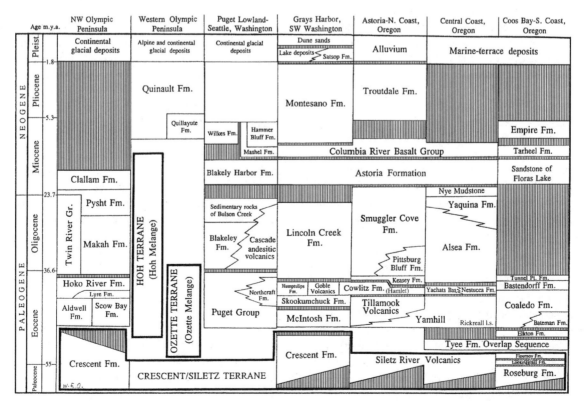

Figure 11.18 Stratigraphy of the Oregon and Washington coastal province (after Armentrout et al., 1988; Christiansen and Yeats, 1992).

Figure 11.19 Volcanic sediments of the Crescent Formation are rich in planktonic foraminiferal microfossils, reflecting open ocean, deepwater conditions (Washington Dept. of Natural Resources).

Figure 11.20 Hoh Head, on the northern Washington coast, is constructed of rocks of the Hoh melange that have been compressed and shattered between two colliding plates. Coarse conglomerate and sand mark this interval of the Hoh Formation (courtesy W. Rau, Washington Dept. of Natural Resources).

to Vancouver. Supplied with sediments carried into the depression by the Columbia River, loosely consolidated gravel and sand of the Troutdale Formation covered the mudstone to depths of 700 feet. Much of this debris was later flushed away by catastrophic Ice Age floods.

Along the northern Washington coast, marine sandstone and silt of the shallow water Upper Miocene to Pliocene Quillayute Formation, as well as thicker littoral and deltaic sediments of the Quinault, were derived from uplifted Eocene Crescent basalts and a shattered mixture of melange rocks at the core of the Olympic Mountains. A particularly rich collection of fossil clams, snails, and foraminiferal microfossils is found within the upper shelf and delta environments of the Quinault.

IN NORTHWEST WASHINGTON, THE OLYMPIC MOUNTAINS, long known for their unequalled scenery, are also geologically unique. In spite of their sharp high peaks, which resemble those of the nearby Cascade Range, the Olympic Mountains are not a volcanic chain. Like the Himalayas, Rockies, Appalachians, and most of the major mountain ranges elsewhere in the world, the Olympics have been trapped and crushed between converging tectonic plates, then turned up on edge and elevated.

The Olympic range provides an excellent opportunity to examine closely the anatomy of an accretionary wedge, a package of sediments or melange of intensely compressed and sheared rocks. Best exposed along Hurricane Ridge only 10 miles south of Port Angeles, pillow basalt, deep ocean clay, and submarine fans are regarded as excellent examples of melange rocks in a subduction zone, which has been annexed to the mainland.

Within the Olympic Mountains, rock formations are arranged in a peculiar horseshoe pattern that has long in-

trigued geologists. Taking in much of the Peninsula, this configuration reflects a sharp eastward curve of the Eocene subduction zone. Opening to the west, the core of the horseshoe is made up of intensely folded Eocene to Miocene melange sediments. Rocks around the outside of this horseshoe include the Eocene Crescent lavas, erupted from a submarine volcanic chain, then covered by marine sediments of about the same age. With ongoing subduction of the Juan de Fuca Plate during the Late Tertiary, loose sediment was scraped off the descending crustal slab and built up into an accretionary wedge along the leading edge of the moving plate. Continued thrusting eventually shoved this thick prism of deformed melange sediments beneath the Crescent lavas.

Successive accretionary packages of sediments within the core of the mountains are composed of folded and faulted Hoh and Ozette melange rocks. Typical of melange mixtures, which have been broken, sheared, and jumbled together by tectonic collision, the Hoh spans the Upper Eocene, Oligocene, and Miocene, while the Ozette is from the Middle Eocene to Oligocene. Extensively exposed in headlands and terraces along the Olympic coast, this resistant sandstone and conglomerate sequence can be found offshore beneath the Washington continental shelf.

ALONG THE NORTHERN MARGIN OF THE OLYMPIC MOUNTAINS and southern Vancouver Island, the Tofino-Fuca Basin extends through the Strait of Juan de Fuca, northwest beneath the Canadian continental shelf and slope, and out to the Pacific Ocean. Bounded on the north by the San Juan and Leech River

Figure 11.21 Looking northeast across the Olympic National Park, Washington, provides a view of spectacular peaks and the 7954-foot high Mt. Olympus in the center of the photograph (Washington Dept. of Transportation).

faults, the basin initially received vast quantities of sands, breccias, and conglomerates of the Eocene Aldwell, Lyre, Hoko River, and Scow Bay formations. These sediments as well as those of the Makah, Pysht, and Oligocene Twin River Group developed as deep-sea turbidite muds and submarine fans atop the earlier Crescent volcanics. As the basin began to fill during Late Miocene time, deltas and swamps of the Clallam Formation, derived from uplifted Vancouver Island and the North Cascades, blanketed older sediments in the trough.

THE PLEISTOCENE BROUGHT ABOUT A WORLDWIDE CHANGE to colder climates almost 2 million years ago. Throughout the Pacific Northwest, lower temperatures and a heavier snowfall that compacted to ice signaled the beginning of profound alterations to the landscape of the Olympic mountains and Puget-Willamette lowlands. Through multiple stages of glaciation, much of the Olympic Peninsula and Puget lowlands was ice-covered leaving a thick glacial outwash over older strata. By contrast, the Willamette Valley to the south was repeatedly inundated by flood waters of immense proportions, which filled the basin then drained after depositing layers of silt.

IN NORTHERN WASHINGTON, the final Late Pleistocene glacial phase began about 28,000 years ago. During glacial cycles, ice lobes from British Columbia advanced into the Puget Lowland and across the northern Cascade Range toward the Columbia Plateau. Sometime before 15,000 years ago, the Puget glacial lobe split into the smaller westward-flowing Juan de Fuca ice sheet that reached the end of the Strait and a main ice tongue that extended south near Olympia.

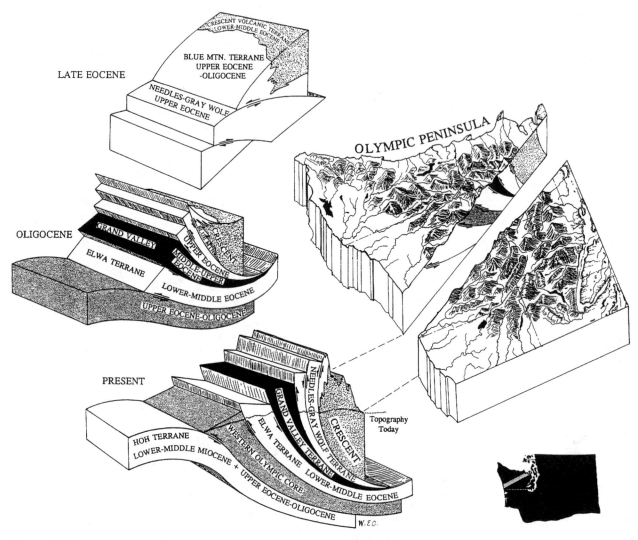

Figure 11.22 Development of the Olympic Mountains of northwest Washington. The sediments were pushed up on end and eroded resulting in the rugged topography seen today (after Christiansen and Yeats, 1992).

During the maximum interval of ice accumulation, the Puget lobe extended approximately 150 miles into the United States and was up to 60 miles wide at the latitude of Seattle. As the mass pushed further south, an ice barricade was confined between the Cascades, Olympic Mountains, and the high divide between Puget Sound and the Chehalis River. The western margin of the Cascade Mountains was covered by ice to elevations of 5300 feet.

Streams, which normally drained from the Cascades and Olympics, ponded up in mountain valleys when their outlets were blocked. Overflowing, the lake waters spilled southward around the margin of the ice, cutting channels along the front of the Cascade Mountains and breaching the divide to reach the ocean through the Chehalis valley. Today the Chehalis River occupies only

a small part of the broad floodplain cut by these waters. Peat bogs, plugged with wood and plant debris, mark these ancient channels, and fossil pollen from the bogs indicates temperatures on the valley floor averaged only 5 to 10 degrees colder than at present.

The most dramatic alteration to the landscape was huge troughs, up to 1200 feet deep, that were carved into the level outwash plain beneath the Puget ice lobe. Throughout the several thousands of years that the ice mass occupied Puget Sound, large amounts of glacial meltwater were discharged from under the glacier carrying away more than 250 cubic miles of sediment and excavating 10 troughs into the bed under the glacier. The passage of ice through the lowlands left the deep channels known today as Admiralty Inlet, Hood Canal, Possession Sound, Puget Sound, the western Strait of Juan

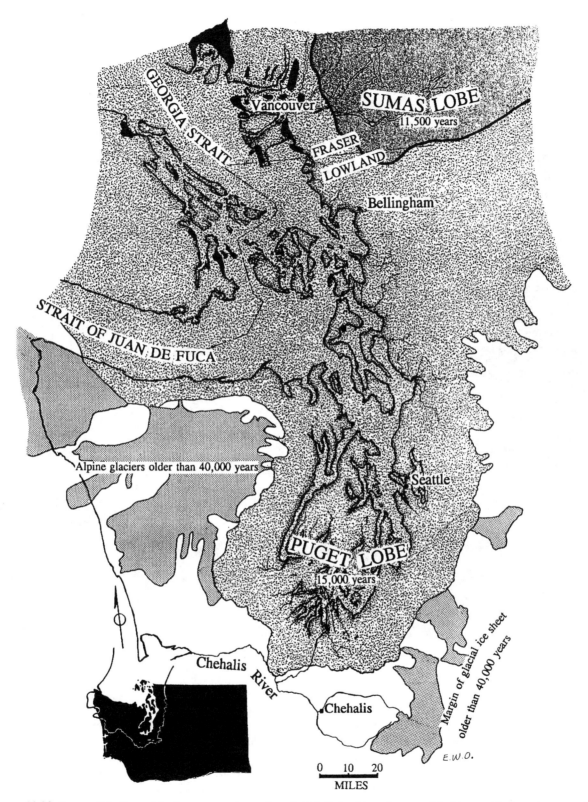

Figure 11.23 Extent of the Puget lobe ice coverage in southern British Columbia and western Washington. Dark patches around Vancouver are Quadra Sand (after Armstrong, 1981; Crandall, 1965; Easterbrook, 1979).

Figure 11.24 Steep canyons and shattered rocks in the Olympic range are by-products of intense pressure from colliding tectonic plates (Washington Dept. of Transportation).

de Fuca, lakes Washington and Sammamish, as well as an intricate network of smaller canals and inlets, all part of an elaborate glacial waterway beneath the ice lobe.

THE WILLAMETTE VALLEY OF OREGON repeatedly experienced tremendous ice-related floods at the height of the glaciation. Far up in the watershed of the Columbia River, glacial ice and debris plugged the Clark Fork River in western Montana. The enormous lake behind this natural dam, covering the site of present-day Missoula, Montana, extended at least 200 miles southeastward. When the dam ruptured, water, ice, and rubble cascaded through the Idaho panhandle to scour out the

scabland of eastern Washington before being funnelled down the narrow slot of the Columbia gorge.

Massive icebergs, rock, and sand, constricting the Columbia river channel near Rainier, Oregon, caused the torrent to back up into the Willamette Valley. The site of downtown Portland would have been under 400 feet of water with only isolated buttes projecting above the flood. From Portland to Eugene the valley was filled with the muddy waters of what is called Lake Allison. After a short interval, the water drained back into the Columbia River channel and on to the ocean at Astoria.

Ice-rafted boulders called erratics, some as large as a house, were transported on blocks of ice to be dropped along the Columbia gorge and throughout the Willamette Valley as flood water receded. Hundreds of these stones have been recognized and recorded where they litter the ground. The variety of granites, quartzites, gneiss, and slates that make up the erratics do not commonly occur in Oregon, and their composition readily points back to sources in Idaho and western Montana.

A regular succession of floods, each depositing a mantle of silt, marks the episodes. Sand and clay, carried in suspension, were left in layers up to 100 feet deep in places. In the southern Willamette Valley, these distinct silts have been used to trace the numerous flooding stages. It is presently estimated that at least 40 and probably many more glacial floods occurred, taking place in 100 to 200 year cycles.

The valley was also affected by glaciation that loosened debris, carrying it from the Cascades by streams. Moving almost like a delta, the glacio-fluvial sediments disrupted drainages and pushed the channel of the Willamette River off to the western side of the valley. Swamps and bogs, developed in the plugged drainage, preserve ice age elephants, giant sloths, and other large mammals that lived there.

THE SCENIC BEAUTY OF THE OLYMPIC PENINSULA with its jagged snow-capped ridges also owes its spectacular design to the last great glacial advance. Ice of the Puget lobe merged with valley glaciers on the north and east slopes of the peninsula, but the west and south sides were subjected to changes wrought by small alpine snow and ice fields. Glaciers in the lower valleys, fed by ice that mantled the central part of the range, reached the Pacific Ocean about 15,000 years ago. Along the coast from Point Grenville north to Hoh River, loosely consolidated silt and gravel, which locally cover the bedrock to depths of 100 feet, have been flushed out of the Olympics as glacial outwash.

Lakes Crescent and adjacent Sutherland southwest of Port Angeles occupy ice-scoured basins. At least 500 years ago ancient Lake Crescent was divided into the two bodies of water by landslides from Mount Storm King. Quileute Indian legends relate that the god of Mt. Storm King became angry during an Indian tribal battle and hurled a great chunk of rock down killing all the quarreling warriors. An interesting side effect of this Holocene partitioning of the lake waters is the presence of separate populations of native cutthroat trout in each lake that can readily be distinguished by color patterns and number of teeth.

Presently the majority of active glaciers within the continental United States are to be found among the high ridges of the Olympic and Cascade mountains. Because of heavy winter snowfall and cool summers, over 50 ice fields have accumulated in the Olympic Mountains, with the largest in the Mt. Olympus and Bailey Ranges. Of the six extensive glaciers, Hoh is the longest at over 3½ miles.

One by-product of Pleistocene glaciation was a global lowering of sea level by as much as 300 feet. Since the amount of water on earth is finite, huge polar ice caps, continental glaciers, and ice sheets covering many of these major mountain ranges deprived the ocean basins of significant amounts of water. During the period of maximum glacial extent, Pacific shorelines retreated well to the west of the present strand, exposing thousands of square miles of the upper shelf to erosion.

THE ORIGIN OF THE PUZZLING MIMA MOUNDS in Thurston County, Washington, has been the subject of much speculation since they were first sighted by the Wilkes scientific exploring expedition in 1845. At Mima Prairie, circular to elliptical mounds vary in shape, ranging from 1 to 7 feet in height and from 8 to 40 feet in diameter. As many as 20 to 25 of these features are crowded into an acre. The mounds are composed of soft black silt and pebbly sand lying atop a gravel base of glacial outwash. Known variously as hogwallows, prairie mounds, pimple mounds, pimpled plains, silt mounds, and yet other colorful descriptions, similar mound-like structures occur elsewhere in the United States, Africa, and South America.

A number of suggestions have been made to explain the formation of the mounds. The most familiar range from biological, involving imaginatively large rodents who dug and occupied the mounds, to those attributed to stream erosion. Water runoff, flood deposits around anchored trees or shrubs, or the formation of ice wedges in the ground under extremely cold conditions are thought to explain the development of some mounds. Many geologists seem to favor a periglacial origin related to the formation of permafrost. A recent study attributing mound formation to earthquakes notes that similar features on a small scale can be experimentally produced by shaking fine, loose sediment atop a rigid base.

EARTHQUAKE ACTIVITY in Puget Sound, the Willamette Valley, and on the Oregon and Washington coast in recent times has been only moderate to low. A clear picture of seismic events prior to the installation of sensitive recording instruments in the 1950s to 1960s is difficult to obtain since early accounts of earthquakes from pioneers, settlers, or even Indian legends are notoriously innacurate.

Historically the first cataloged earthquake in Puget Sound was in Thurston County on June 29, 1833. After this, the significant quakes in this region, which measure 5 or more in intensity on the Mercalli scale, took place in 1939, 1946, 1949, 1965, and 2001. At 7.1 on the scale, the 1949 quake, felt over a 150,000 square mile area, killed 8 people, injured many more, and caused over $25 million in damage to Seattle, Tacoma, and Olympia. Following the installation of seismic instruments at

Figure 11.25 Mima mounds south of Puget Sound (Washington Dept. of Transportation).

Seattle in 1949, precise records show an apparent trend toward a decreasing frequency in the rate of significant earthquakes within the past 25 years. Throughout this interval only four quakes around 6.0 magnitude took place in western Washington.

A powerful quake that jolted Puget Sound about 1000 years ago raised the south end of Bainbridge Island 20 feet, triggered landslides that swept trees into Lake Washington, and sent massive tsunami waves across the Sound. This event originated along the Seattle fault zone, running east-west under Seattle, Puget Sound, and Bainbridge Island. A broad Puget fault zone that follows a north-south direction between the Cascade Mountains and Coast Range intercepts the Seattle fault on the east, while a fault beneath Hood Canal truncates it on the

west. Earthquake activity as recent as February, 1995, can be attributed to the Seattle fault zone.

Much of Seattle is built atop loose glacial fill prone to landslides. The soil instability, combined with a dense population and reliance on a network of bridges and interconnected waterways, makes Seattle a candidate for severe losses. On February 28, 2001, at 10:45 A.M., Seattle experienced a large earthquake that caused between $1 and $2 billion in damages; however, since it originated at a depth of 32 miles, damage was less than had it taken place closer to the surface. It is thought that the fault which slipped was actually buried in the subducting Juan de Fuca tectonic slab. With an epicenter near Olympia, the quake reached 6.8 in magnitude. The state capitol sustained major cracks in the dome, and down-

Figure 11.26 The cause of the Mima mounds south of Olympia, Washington, has been a geologic puzzle for over 100 years (Washington Dept. of Natural Resources).

town Seattle, in particular the older part of the city, had severe structural wreckage to buildings with collapsing bricks and facades. The only recorded death was from a heart attack, but over 400 people were injured.

In the Willamette Valley, only three strong earthquakes were recorded from 1875 to 1993, and since record keeping began in Portland in 1882, significant earthquakes have taken place on the average of once every 5 years. Of the profound events in the northern part of the valley, two were in the immediate Portland area, and one was to the southeast where the epicenter was at Scotts Mills in Marion County. In 1993 the 5.6 event from the Mt. Angel fault, running from Scotts Mills toward Woodburn, came after dozens of smaller shocks had been recorded along this feature over several years. In spite of the apparent lack of powerful quakes, a close look at maps of the subsurface beneath Portland and surrounding areas reveals a network of crisscrossing faults that could become active.

Like Seattle, a combination of factors in the immediate Portland area suggests a high potential for a devastating earthquake. Only a few of the larger structures here, as elsewhere in the valley, have been constructed to withstand strong seismic movements, and many of the building foundations are on loose, poorly consolidated soils. It has been shown that rocks similar to those beneath Portland will shake like jello for a sustained period of time during and after a seismic event. Following heavy rains, soil liquifaction, landslides, and mudslides would be among the costly secondary effects. These factors, in conjunction with the population density of the northern part of the Willamette Valley, add up to a high risk for disaster.

Earthquakes in the coastal region of the Pacific Northwest closest to the active subduction zone have the greatest potential for catastrophic seismic movement. On the coast from British Columbia to southern Oregon, quakes originate in the shallow subsurface along the

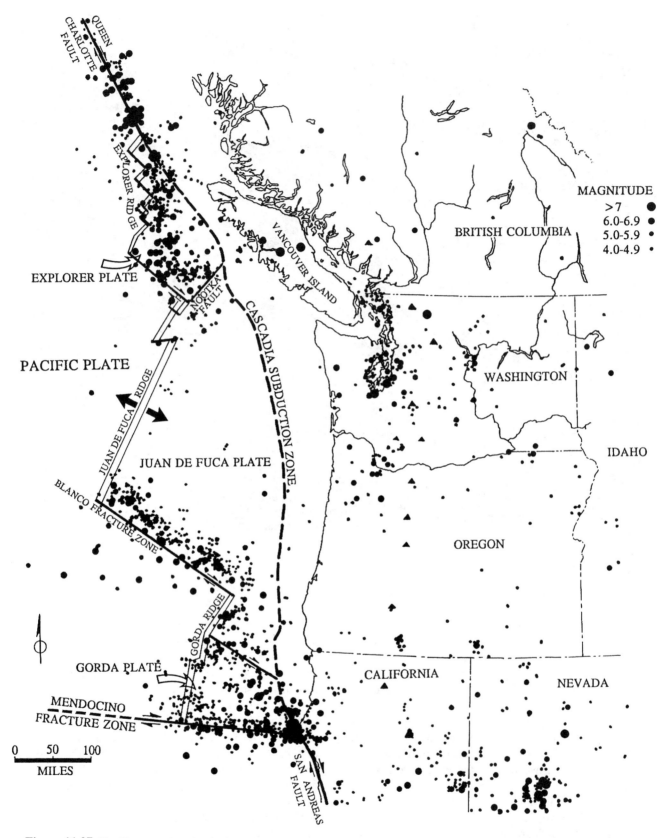

Figure 11.27 Significant earthquakes in the Pacific Northwest (after Crosson and Owens, 1987; Palmer, 1991; Walter 1986).

Figure 11.28 The earthquake of April 13, 1949, caused so much damage to Lafayette Elementary School in Seattle that the building had to be torn down and rebuilt (Washington Dept. of Natural Resources).

Figure 11.29 The 1996 floods in Oregon and Washington triggered many landslides and debris flows. Here the Royce flow just east of Multnomah Falls has moved this multi-ton boulder along its channel (photo courtesy Scott Burns).

Figure 11.30 After ocean storms removed the toe of an ancient landslide below The Capes housing development in Tillamook County, Oregon, the slide was reactivated in 1997, threatening these structures (photo courtesy Scott Burns).

Figure 11.31 "Stair-step" terraces at Cape Arago in southern Oregon are evidence of the rapidly rising coast (U.S. Forest Service).

boundary between the descending Juan de Fuca oceanic plate and the overriding North American continental plate. Offshore in northern California, the Gorda and Pacific plates meet near Cape Mendocino, the site of a number of historic events of magnitude 5.5 or higher.

Newly generated oceanic crust is typically warm and somewhat sticky, and the subduction process is rarely smooth. As the plates collide, the descending slab may catch or bind, building pressure that causes the upper continental crust to bulge upward. With a snapping motion, pressure is suddenly released when the uplifted area drops down to the accompaniment of strong earth movements. A similar geologic setting along the coast of Chile, where young crust is being subducted, has yielded earthquakes as high as 8.0 on the Richter scale.

On the Pacific Northwest coastline, the high incidence of sunken or buried Holocene marshes and bogs has been ascribed to this sudden seismic-related subsidence. Sediment records of the past 7000 years show multiple rapid subsidence events have taken place along the coasts of both Oregon and Washington. When the land drops abruptly from 2 to 8 feet, trees, along with other vegetation, have been covered by shallow marine

Figure 11.32 The community of Bandon, Oregon, sits atop a marine terrace that has been raised 60 feet. Offshore sea stacks are rocks of the Jurassic Otter Point Formation (Oregon Dept. of Geology and Mineral Industries).

sand and silt. In many cases the plants are preserved still upright. Multiple buried forests and mud-covered swamps delineate an ongoing record of past coastal seismic activity.

Both depressed coastal swamps and offshore submarine slides called turbidity currents have left unmistakable deposits that point to large-scale earth movement as well. This loose material flows rapidly downslope along the ocean floor where it spreads out over the abyssal plain. Measured in thickness of only inches, each thin turbidite flow layer grades from coarse sand at the base to fine-grained clay at the top. Ocean cores show a consistent record of submarine slides that may have been induced by strong earthquakes of magnitude 9.0 or more taking place on a 300 to 500 year cycle. Besides buried coastal swamps and turbidite layers in the deep ocean, evidence for seismic seawaves or tsunamis is also preserved in distinctive coastal wave deposits. Triggered by offshore earthquakes, enormous ocean waves, which reach the shoreline, may be up to 100 feet in height. As these high energy waves sweep through and scour the

shelves, estuaries, and beaches, they leave behind recognizable layers of sand within marsh and bog sediments that chronicle each event.

Tsunamis on the Oregon and Washington coastline may eventually be responsible for calamitous fatalities in situations where bridges or other escape routes from heavily populated sand spits are cut off before evacuation can take place. Although tsunamis from distant quakes can give coastal residents several hours to prepare, a near source may allow only a several-minute warning before a giant wave crashes ashore.

TERRACES ON THE COAST commonly occur as elevated step-like benches. A terrace begins as an erosionally flattened coastal apron just below high tide that is usually littered with a veneer of sand and gravel impregnated with the shells and skeletons of nearshore marine animals. Since the Pleistocene, marine terraces on the coastline from southern Oregon to Cape Flattery, Washington, have been progressively raised and tilted landward.

Figure 11.33 A view northward from Pt. Grenville, a headland of Eocene lavas capped with Pleistocene debris on the southcentral Washington coast, shows raised terraces of Quinault sands stretching to Cape Elizabeth in the background (courtesy W. Rau, Washington Dept. of Natural Resources).

Uplift and deformation of the coast and coast range are related to continued subduction of the Juan de Fuca Plate. As the descending slab is being thrust eastward, a growing accretionary wedge of sediments just below the outer margin of the North American Plate steadily raises the coast. Because subduction here proceeds from west to east, a tilting or seesaw effect elevates the coast to the west faster and higher than areas to the east. Cape Blanco, Oregon, closest to the subduction trench, is presently being elevated at the spectacular rate of 1 inch every 3 years, the fastest of any area in the northwest. North of the Cape, the elevation of coastal terraces diminishes from 150 feet at the Siuslaw River to about 60 feet at Tillamook Head.

In Washington, old terraces can be followed laterally from elevations of 40 feet at Willapa Bay and Point Grenville to 125 feet at Cape Elizabeth. North of Hoh Head, Pleistocene gravels, marking the raised terrace on the upper surface of Alexander Island, are about 120 feet above sea level.

Just as coast regions are being raised, large tracts of the Puget and Fraser lowlands and San Juan Islands are dropping relative to sea level. Seattle is presently sink-ing at the rate of about 1/10 inch per year, Vancouver, British Columbia, at 1/40 inch per year, and Victoria at a rate of 1/100 inch per year, reflecting an overall southeastward tilt to the land.

The oldest raised terraces are inland at higher elevations where many have been severely dissected by erosion making them difficult to trace. At Cape Blanco the youngest 80,000-year old terrace, at an elevation of 150 feet, is easily defined and covers much of the point, while the surface of the 1 million-year old Poverty Ridge terrace inland at 800 feet in elevation is only poorly delineated.

Determining the age of marine terraces requires precision, and newer chemical methods using fossils have been very successful. The ages of these ancient surfaces can be pinpointed particularly well by measuring the decay of radioactive elements in the skeletal calcite of corals and sea urchins buried in terrace sands as well as the steady alteration in amino acids preserved in mollusc shells.

OCEAN BEACHES AND SCENIC ROCKY SHORELINES have long been a major Pacific North-

Figure 11.34 Projecting far out into the Pacific Ocean, terraces at Cape Blanco, Oregon, are being raised about 1 inch every 3 years because of eastward tilting of the Coast Range (U.S. Forest Service).

west attraction. A variety of sandy stretches interspersed by quiet pocket coves and dramatic headlands supported by steep cliffs that plunge into the ocean provide unexpected vistas with each turn in the road. Sand spits, dunes, coves, and high coastlines are the product of ongoing geologic processes that have been active for tens of thousands of years. More durable rocks slowly emerge as cliffs and headlands, while less resistant or fractured strata evolve into inlets and coves. Sand transported and deposited by rivers and streams builds up as thin lengthy spits or is carried by longshore currents to beaches.

AT CAPE BLANCO, OREGON, a complex of resistant Jurassic age Otter Point conglomerates are capped by Eocene mudstones and Late Miocene sandstones of the Empire Formation. Jurassic strata also compose the myriad of sea stacks and smaller islands between Cape Blanco and Coos Head. Coos Head is armored by fossiliferous sandstones of the Empire Formation, while nearby Cape Arago, Gregory Point, and offshore stacks are remnants of the Eocene Coaledo sandstones, which once formed a continuous headland here.

The sand dunes at Florence are the longest in Oregon, stretching for 55 miles along the coast and averaging 2 miles in width. Declared a National Recreation Area in 1972, the dunes entrap a chain of small lakes, which fill low spots or former river channels. Some of the smaller lakes cover less than 10 acres, even though all reach depths of 25 to 35 feet. Just north of Florence, sheer cliffs at Heceta Head, Cape Perpetua, and Devils Churn are carved from the late Eocene Yachats Basalt and provide a barrier to further northward migration of the Florence dune field.

From Seal Rock to Tillamook Head, most of the prominent headlands are invasive basalt of the Miocene Columbia River lavas that flowed all the way from eastern Oregon and adjacent Washington and Idaho. On the coast, these basalts compose the 800-foot high south-facing cliffs eroded into Cape Lookout that dramatically juts 3/4 of a mile out to sea. Sea stacks, knobs, arches, and offshore islands like Elephant Rock are of the same lava. Rising 1136 feet above the ocean, Tillamook Head is a massive rocky complex of basalt on the northern coast that encompasses promontories, coves, and erosional

Figure 11.35 Bay Ocean peninsula in Tillamook County, Oregon, was the site of a community of 59 homes built in 1910. The town lasted only 30 years before beach erosion reclaimed the land by 1940 (Oregon Dept. of Geology and Mineral Industries).

Figure 11.36 The Tillamook lighthouse, built atop a Miocene basalt knob, has withstood the ocean waves for over 100 years (Oregon Dept. of Geology and Mineral Industries).

structures. The famous Tillamook lighthouse, built here in 1878, has been purchased by real estate entrepreneurs in Portland and modified into a columbarium.

Extensive sandy beaches and spits dominate both sides of the Columbia River at the boundary between Oregon and Washington. The wide beach at Clatsop Plains, Oregon, has grown and migrated steadily toward the ocean away from the older shoreline that lay to the east 1400 years ago. North of the river, Long Beach Spit, across the mouth of Willapa Bay, and the two spits adjacent to Grays Harbor derive their sand from the Columbia River.

Very recently, a marked trend toward coastal erosion has been attributed to the construction of dams high in the watershed of the Columbia River. Dams here restrict the supply of new sand to the lower reaches.

THE SOUTHERN WASHINGTON COAST, with smooth sweeping beaches, lacks the numerous scenic basalt headlands that characterize Oregon. North of Grays Harbor, coastal geology can be divided into harder sandstones and seacliffs of the Quinault Formation, the broken melange rocks of the Hoh Formation from Point Grenville to Hoh Head, and the chaotically jumbled blocks of Ozette melange sediments from there to Point of the Arches.

Along the coast from Point Grenville to Cape Elizabeth and Tunnel Island, marine sand, silt, and conglomerate of the Pliocene Quinault Formation form massive cliffs. Point Grenville, by contrast, is armored by resistant Eocene lava of the Crescent Formation. Landslides are prevalent in Quinault rocks, and enormous blocks of sand and gravel, some carrying trees, have tilted and slumped seaward at a number of localities.

Hogsback, Little Hogsback, Willoughby, and Split Rock islands are rugged offshore erosional remnants or stacks of the Hoh Formation, and the 2½-mile exposure of Hoh melange here is one of the most continuous in the region. To the north, the headlands at Browns Point, Starfish Point, and Hoh Head are the same heterogenous sandstone melange. Broken and sheared zones in the melange are easily eroded, and fresh cobbles derived from nearby cliffs litter the coves and beaches. The Hoh Formation appears offshore as flat-topped Destruction and Alexander islands, distinctive landmarks for early explorers and sailors. Alexander Island, one of the largest of the northern islands, is a narrow strip a quarter of a mile long and 500 feet wide.

From Hoh Head northward to Cape Flattery, the shoreline becomes more rugged, and Ozette melange rocks make up the promontories of Toleak Point, Strawberry Point, Teahwhit Head, and Cape Johnson. Abundant sea stacks and islets, projecting from the ocean, have been carved from the same shattered melange rocks. The most unusual of these are Giants Graveyard and Quillayute Needles near LaPush. Rising dramatically 85 feet above the water, Quillayute Needle tapers to a sharp spine 35 feet in diameter.

In northwest Washington, the irregularly-shaped Ozette Lake was impounded behind a mound of Pleistocene glacial gravel during a time of low sea level when the shoreline was further west. Cape Alava and

Figure 11.37 Willapa Bay on the southern Washington coast displays a dendritic pattern of tidal channels across shallow mudflats (Washington Dept. of Transportation).

Point of the Arches, a series of eroded remnants, are composed of older Mesozoic rocks, while Cape Flattery and Tatoosh Island, the most northerly reach of the Olympic Peninsula and the most westerly of the continental United States, are braced by an angular breccia layer within the Eocene Lyre Formation. The lighthouse on the island was installed in 1857 to guide ships in the straits. The northern tip of Cape Flattery is covered by

sand and silt deposited in the Tofino-Fuca depression, which parallels the Strait of Juan de Fuca.

THE OFFSHORE SHELF AND SLOPE of Oregon and Washington extend down to a depth of 9000 feet onto the abyssal plain. A subtle gradient change at 600 feet between the shelf and slope marks a profound transition from the smooth upper shelf with low banks to the deeper

Figure 11.38 Alexander Island, Washington, is the remnant of an old terrace that was uplifted as the coastline moved west. The base of the island is Hoh sandstone, the upper 20 feet are glacial deposits, and the white topping is from a large bird population (courtesy W. Rau, Washington Dept. of Natural Resources).

Figure 11.40 Quillayute Needles, near LaPush, Washington, have been carved from massive sandstone of the Hoh Formation (courtesy W. Rau, Washington Dept. of Natural Resources).

Figure 11.39 Giants Graveyard is a dense field of sea stacks, which are made up of broken Ozette melange rocks. The stacks extend along the northern Washington coast south from Teahwhit Head to Toleak Point (Washington Dept. of Natural Resources).

slope cut by fault traces, submarine canyons, and pressure-built ridges.

Typical of continental edges with an active volcanic archipelago, the width of the margin off Oregon and Washington is extremely narrow. Where it meets the abyssal plain, the base of the continental slope is only 35 miles from the strandline off Cape Blanco, but it is as much as 75 miles offshore along the Olympic Peninsula. The base delineates the extreme western edge of the accretionary wedge, a prism of sediments constructed just above the subduction zone between the Juan de Fuca and American plates. Although the conti-

nental shelf is mainly layered marine rocks that continue as the Coast Range and interior valleys, the slope is a jumbled mixture of abyssal sediments piled up into the accretionary wedge, which is covered by submarine landslides or turbidites that descended from the shelf above.

Topography of the continental shelf and slope are subdued but not featureless. On the Oregon shelf edge, faulting has produced Coquille Bank northwest of Cape Blanco, Heceta Bank at Florence, Perpetua Bank off Cape Perpetua, and Stonewall Bank near Waldport, all of which project up to 200 feet above the sea floor. Offshore on the Washington shelf, the Quinault Ridge extends from Willapa Bay to Lake Ozette. Only three small banks, two near the Strait of Juan de Fuca and Learmonth Bank at Dixon Entrance near Queen Charlotte Islands, are found off British Columbia.

Beneath the southern Washington continental shelf, Late Miocene melange sediments are being pierced by peculiar rising dome-like features called diapirs roughly 2 miles in diameter. Oligocene to Early Miocene melange rocks, which were scraped off the descending Juan de Fuca slab, form the centers of these circular faulted structures. While the exact origin of the diapirs is not clear, they relate to the ongoing subduction process that pushes materials from deep in the accretionary wedge upward toward the surface of the shelf.

At 5000 feet in depth, the continental slope may be divided into upper and lower portions. Between 600 and 5000 feet, the upper slope has a westward topographic plunge of 5 degrees and is cut by submarine canyons that include the Rogue and Astoria off Oregon, the Juan de Fuca and Quinault off Washington, along with the Barkley and Loudoun off southwest Vancouver Island. Canyons in the northern area of the province have been steadily filling with muds, silts, and sands for the past

Figure 11.41 The high rocky bluffs at Cape Flattery, the most northwesterly point of Washington, look out at Tatoosh Island, the Strait of Juan de Fuca, and the Pacific Ocean (Washington Dept. of Transportation).

5000 years, but the Quinault Canyon adjacent to the mouth of the Quinault River flushes itself of sediment about every 500 years, probably as a consequence of major earthquakes.

Between depths of 5000 feet and 9000 feet, the lower slope has a more gentle incline than the upper and a completely different internal structure. Advancing at about 1¾ inches per year, the oncoming Juan de Fuca Plate exerts a steady eastward pressure on this area. This relentless compression has imposed a series of long overlapping mud ridges and intervening basins parallel to the

shoreline. In addition to giving the lower slope a rumpled appearance, these narrow basins serve to trap and direct sediments collected from the submarine canyons above into the Cascadia and Astoria channels below.

From the base of the continental slope, the immense sheet-like Nitinat and Astoria submarine fans project in a southwesterly direction. Offshore from the mouth of the Columbia River and Willapa Canyon, the Astoria fan stretches for more than 100 miles across the abyssal plain and covers over 7000 square miles. The Nitinat encompasses an even larger area off the tip of the Olympic

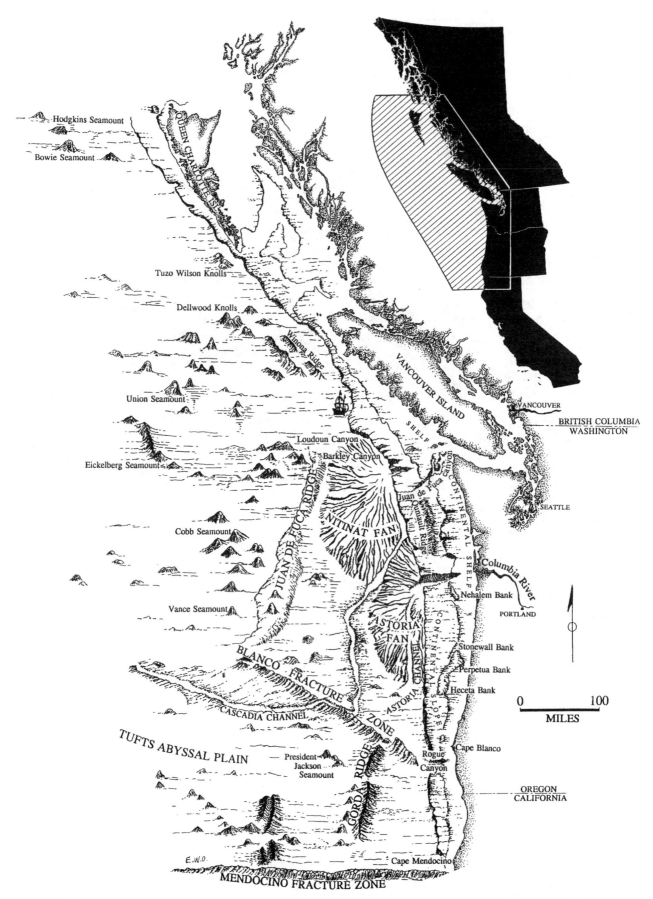

Figure 11.42 Topography of the offshore shelf, slope, and abyssal plain (after Barnard, 1978; Kulm and Fowler, 1974; Mammerickx and Smith, 1982).

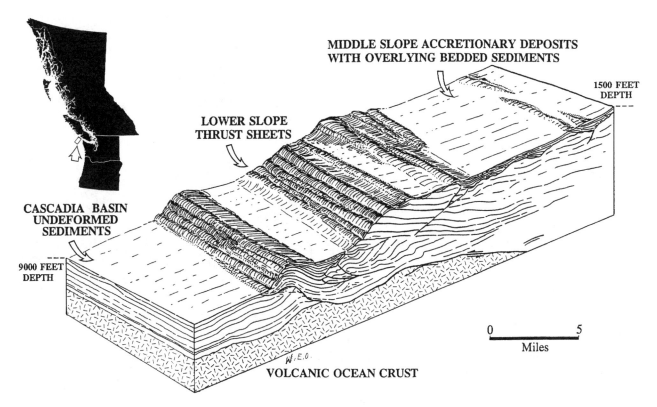

**MIDDLE SLOPE ACCRETIONARY DEPOSITS
WITH OVERLYING BEDDED SEDIMENTS**

1500 FEET
DEPTH

**LOWER SLOPE
THRUST SHEETS**

**CASCADIA BASIN
UNDEFORMED
SEDIMENTS**

9000 FEET
DEPTH

0   5
Miles

VOLCANIC OCEAN CRUST

Figure 11.43 Slope deposits and the buildup of sediments in an accretionary wedge behind the subduction zone between two plates. The bottom profile off Vancouver Island was taken during Leg 146, 1993, of the Ocean Drilling Program.

Peninsula. Stacked sequences of turbidite sands, silts, and clays within these submarine fans are 2 to 3 thousand feet thick at the base of the slope but thin outward toward the fan margins.

The abyssal plain is torn by zigzag lines of seamounts and submarine scarps that form the boundary between the diminutive east-bound Explorer, Juan de Fuca, and Gorda plates and the larger westward-moving Pacific Plate. Along this boundary, the Gorda and Juan de Fuca ridges are volcanic chains that generate ocean crust. At its point of origin along mid-ocean ridges, the abyssal plain is a remarkably rugged surface of submarine pillow basalts smoothed by a cover of clays and silts. Occasionally an isolated undersea volcano or seamount will rise well above the adjacent ocean plain and may even project as an island. Not infrequently, with continued plate movement, such seamounts will be once again carried below sea level.

Since the early 1980s a systematic examination of the Juan de Fuca ridge has been carried out by the U.S. and Canadian governments using manned submersibles and remote controlled submarines. In addition to deep marine communities of worms and molluscs near vents, fresh volcanic sheet flows and pillow basalt mounds mark the sites of recent eruptions along the ridge. By re-

visiting specific sites over a period of years, geologists have been able to map an ever changing sea floor of rugged volcanic topography.

Hundreds of seamounts or submarine volcanoes dot the abyssal plain adjacent to the Pacific Northwest, but four of the largest are Cobb, Union, Bowie, and Surveyor. Of these, Surveyor, roughly 150 miles west of Sitka, Alaska, is flat-topped due to erosion when it stood above sea level. Cobb seamount, west of the mouth of the Columbia River, rises over 8000 feet off the ocean floor to within 100 feet of the surface with pronounced ancient beach terraces at 250, 450, 550, and 600 feet below sea level. Bowie seamount, west of Queen Charlotte Islands, rises to within 100 feet of the surface from the abyssal plain at 9000 feet deep. This seamount, which forms one end of the Bowie-Kodiak chain, has a 550-foot deep terrace on its flanks that matches the one on Cobb seamount. Opposite Vancouver Island, Union seamount is a symmetrical volcanic cone that projects 9000 feet off the Pacific floor to within 880 feet of the surface. Although there is a distinct crater on one side, no terraces have been cut into its slopes. Exploration of Cobb and Bowie seamounts in 1976 using a submersible revealed a rich biologic array of invertebrates and fish clustered above these volcanic mountains.

MINERAL PRODUCTION

Mineral resources of the Coast Range province are meager. Prior to World War II, over 500 exploratory petroleum wells drilled throughout the Pacific Northwest Coast Range met with limited success. Situated near oil seeps, most of these wildcat wells were less than 5000 feet deep. After 1945, major companies began extensive explorations of the Willamette Valley, Coast Range, Olympic Peninsula, and offshore regions, but results were disappointing. All of the holes were plugged and abandoned except for the Mist gas field discovered in Columbia County, Oregon, in 1979. With 18 wells exploiting Cowlitz Formation rocks, the Mist field has produced almost 40 billion cubic feet of gas to date.

In 1899 one of the first exploratory wells was drilled on the coast near LaPush, Washington. Today a rusted boiler, drill pipe, and a steam engine lie covered with brush designating the spot. Bad luck plagued the operation from the beginning when the barge carrying the equipment from San Francisco broke up in the surf. Even after the heavy machinery was salvaged, difficulty in drilling and personnel problems brought the operation to a close within a year.

Through Herculean efforts a steam engine and other drilling machinery were skidded up the steep cliff at Hoh Head in 1913, but a test well here proved commercially unsatisfactory.

Gas and oil seeps with a strong petroleum odor are not uncommon in the broken, sheared melange rocks that line the northern Washington coast. A gas seep just north of the Quinault River is distinguished by a 50-foot high mound built by mud and natural gas extruded from the vent. Test wells drilled through highly fractured melange rocks nearby failed to locate significant accumulations of natural gas.

Low-grade Eocene coals have been strip-mined from sites in Pierce, Lewis, Kittitas, and King counties for over 80 years. At one time these were some of the most productive mines in both Washington or Oregon, but at present most of the sites are shut down or operated only periodically. One of the largest near Centralia supplied a power plant with millions of tons of subbituminous coal from the late 1960s.

A VARIETY OF MINERALS WAS BROUGHT TO THE SOUTHERN OREGON coast by Pleistocene rivers and streams that drained the interior from Cape Blanco to Cape Arago. Heavier grains accumulated in beach terraces, while the lighter quartz, feldspar, and mica were flushed offshore. Eroded from the Jurassic Galice Formation by the South and Middle forks of the Sixes River, gold, platinum, garnet, and other heavy

Figure 11.44 An old steam boiler, left over from a drilling operation, was used at one of the few natural gas seeps in Washington just north of the mouth of the Quinault River (courtesy W. Rau, Washington Dept. of Natural Resources).

minerals were usually mixed with dark black magnetite and chromite grains so the beach deposits were called "black sands."

In the middle 1800s, prospectors were initially interested in concentrations of gold sands near river entrances where they constructed long sluice boxes to extract the valuable metal. Once this source was exhausted, dilligent miners dug back into the black sands of the high terraces, but the loose unconsolidated layers made the tunnelling hazardous and recovery of gold difficult. In all, gold mining operations here were economically marginal and only lasted a few years.

WITHIN THE OLYMPIC PENINSULA, patches of red-colored sands are prevalent near Big and Little Hogsback and on Ruby Beach at the mouth of Cedar Creek in Jefferson County. This color is due to the abundant grains of almandite, a type of red garnet crystal, derived from fractured Hoh melange rocks. Like those on the Oregon coast, heavy minerals such as garnet, magnetite, hematite, and very small clear glassy crystals of zircon are winnowed out by waves and deposited in the nearshore. Most of the garnet crystals, originally with 12 faces, have been beach-worn into a spherical shape.

Figure 11.45 On the central Oregon coast, the lighthouse and irregular promontory at Yaquina Head are part of the Miocene Columbia River Basalt Group (Oregon Dept. of Geology and Mineral Industries).

ADDITIONAL READINGS

Adams, J., 1984

Adams, J., 1990

Algermissen, S.T., and Harding, S.T., 1986

Armentrout, J.M., 1987

Babcock, R.S., Suczek, C.A., and Engebretson, D.C., 1994

Baldwin, E.M., 1974

Barnard, W.D., 1978

Beeson, M.H., Perttu, R., and Perttu, J., 1979

Blunt, D.J., Easterbrook, D.J., and Rutter, N.W., 1987

Booth, D.C., and Goldstein, B., 1994

Buckovic, W.A., 1979

Chase, T.E., Menard, H.W., and Mammerickx, J., 1970

Christiansen, R.L., and Yeats, R.S., 1992

Crandall, D.R., 1965

Danner, W.R., 1955

Dengler, L., Carver, G., and McPherson, R., 1992

Easterbrook, D.J., 1979

Johnson, S.Y., Potter, C.J., and Armentrout, J.M., 1994

Kelsey, H.M., 1990

Komar, P.D., 1992

Lund, E.H., 1972

Mooney, W.D., and Weaver, C.S., 1989

Niem, A.R., and Snavely, P.D., 1991

Oles, K.F., et al., eds., 1980

Orr, E.L., and Orr, W.N., 1999

Palmer, S.P., 1991

Palmer, S.P., and Lingley, W.S., 1989

Rau, W.W., 1973

Rau, W.W., 1980

Riddihough, R., and Hyndman, R.D., 1989

Snavely, P.D., 1987

Snavely, P.D., and Wells, R.E., 1991

Stover, C., and Coffman, J., 1993

Tabor, R.W., 1975

Washburn, A.L., 1988

Wells, R.E., and Coe, R.S., 1985

Wells, R.E., Coe, R.S., and Heller, P.L., 1988

Wells, R.E., et al., 1984

Yeats, R.S., 1989

GLOSSARY

A

aa lava   A rough blocky, fragmented lava consisting of separate angular clinkers.

abyssal plain   A large flat region of the deep ocean from 2 to 4 miles below sea level. See subduction for illustration.

accretionary wedge   A prism of sediment that builds up on the outer margin of a moving tectonic plate. See subduction for illustration.

alkali   Water or sediment having a large amount of sodium and potassium carbonate.

alkaline   A basic substance with a pH of 7 or higher.

alluvial fan   See alluvial plain.

alluvial plain   A level land surface beneath a river or floodplain formed by the deposition of alluvium (clay, sand, and gravel). Sometimes fan-shaped.

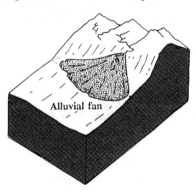

almandite   An iron-aluminum-rich member of the garnet mineral group. Typically deep red to purple.

alpine glaciers   See valley glaciers.

amalgamation (terranes)   When two or more small terranes merge and become joined before the larger mass is accreted or attached to a mainland.

amino acid   A group of organic compounds which are the primary components of proteins.

amphibolite   A coarse-grained metamorphic rock composed primarily of the minerals amphibole and plagioclase feldspar.

andesite   A dark-colored extruded volcanic rock composed of plagioclase feldspars and one or more mafic (iron- or magnesium-rich) minerals.

anticline   An elongated upward fold in rocks.

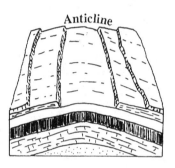

anticlinorium   A large geographic region displaying upwardly-domed or folded rocks.

aquifer   A layer of porous rock that holds and conducts a considerable amount of groundwater.

archipelago   A group or chain of islands in a seaway. See subduction for illustration.

arrastra   A device for crushing ore consisting of a heavy horizontal rotating wheel in a cylindrical pit.

artesian (spring)   A spring that flows by itself under hydrostatic pressure.

ash flow   Fine-grained fragments and high temperature incandescent gas clouds ejected from a vent and flowing down the side of a volcano before settling. See also ignimbrite.

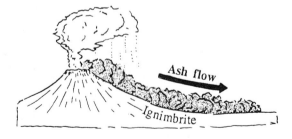

asthenosphere   A layer within the upper mantle of the earth below the lithosphere where the rocks are weak and easily deformed. About 60 to 200 miles below the surface. See mantle for illustration.

B

back-arc spreading The region of crustal spreading between a volcanic archipelago and the continental interior.

badlands A region marked by deep erosion to produce unusual shapes.

Badlands

bank An elevation from the seafloor to a shallow depth of less than 600 feet.

barchan dune A moving dune of sand that is crescent-shaped. The horns or points of the dune project downwind.

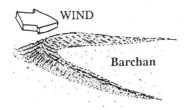

WIND

Barchan

basalt A dark-colored fine-grained extruded volcanic rock, rich in iron and magnesium, that is chiefly composed of the minerals plagioclase and pyroxene.

basin A circular or elliptical depressed area.

basin and range The regional topography of the Great Basin of the western United States characterized by raised faulted mountain blocks and alternating valleys.

Graben

PLAYA

batholith A plutonic mass covering at least 40 square miles intruded into the surrounding rock layers.

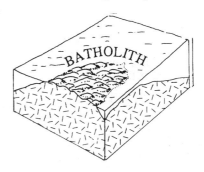

BATHOLITH

bathyal The ocean environments lying between a depth of 600 and 3000 feet.

bedrock The layer of solid rock that lies beneath soil or other loose surface material.

belemnite An extinct squid-like swimming mollusc with a bullet-shaped internal skeletal rod.

bioherm See reef.

bimodal volcanism Eruptions of rock and magma, which have markedly different compositions but which are from the same source or area, e.g., rhyolite and basalt.

bituminous coal Coal that produces 15 to 20% volatile matter when it is burned. The most common variety of coal is known as soft coal.

black sand Beach or stream sands rich in dark mineral grains. Usually the dark minerals are heavy, and the sand may be the source of gold and platinum.

black smoker A deep ocean floor hydrothermal spring exuding insoluble sulfide minerals.

blueschist A metamorphic rock produced by high pressure and low temperature in which blue minerals predominate.

bomb, volcanic A spindle-shaped blob of lava that was briefly airborne.

brackish Water that has a salinity between normal seawater and fresh water.

braided river A stream system so overloaded with sediment that there are many dividing and rejoining channels.

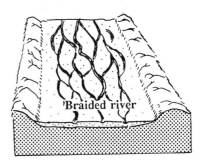

Braided river

breccia   Fragmented rock whose pieces are angular and unworn. Ordinarily derived from either earthquake (fault) or volcanic activity.

C

caldera   A large basin-shaped volcanic depression that formed when the central section of a volcano collapsed. Diameter is measured in miles.

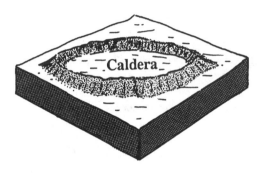

caldron   A caldera of immense size with a diameter measured in tens of miles.

carbonization   The accumulation or concentration of carbon by the slow underwater decay of organic matter.

cephalopod   A predatory swimming mollusc that includes octopus and squid.

chalicothere   An extinct mammal with heavily clawed feet related to horses.

chert   A dense finely crystalline sedimentary rock composed of silica.

chromite   A brown to black mineral that is the most important ore of chrome.

clastic   Broken-sedimentary rocks as sandstones, silts, and conglomerates.

coastal plain   A low, flat area of land adjacent to the ocean. See shelf, continental, for illustration.

colonnade   The lower portion of a lava flow with well-formed parallel shrinkage columns.

columnar jointing   Long joints in volcanic rock that split it into columns as the rock cools and contracts.

complex   An assemblage or association of rocks in which the structural relations are so intricate that the rocks cannot be easily differentiated or mapped.

concretion   A local hardened body enclosed in sedimentary rock that was precipitated around a central nucleus.

conglomerate   A sedimentary rock composed of rounded worn pebbles and cobbles.

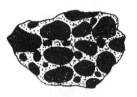

conodonts   Microfossils composed of calcium phosphate which are thought to be the teeth of tiny extinct fish-like organisms.

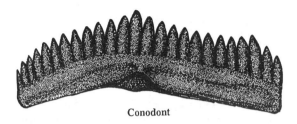

Conodont

continental glacier   A thick ice sheet covering large parts of a continent.

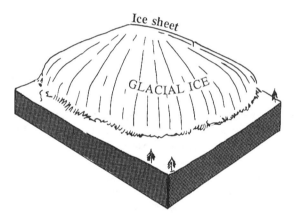

core complex   A mass of metamorphic or igneous rock, significantly older than the enclosing layers, which has been tectonically raised from deep in the crust to near the surface by faulting.

coulee   A dry trench-like intermittent stream bed or wash.

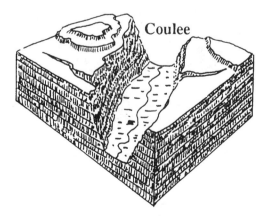

craton   The dome-like stable central interior core to a large continental mass.

crinoid   An attached echinoderm, somewhat cup-shaped, with feathery arms.

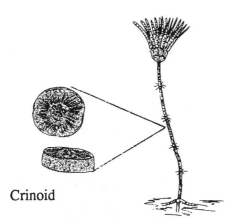

Crinoid

crust   The outermost shell of the earth above the mantle. See mantle for illustration.

D

dacite   A fine-grained extrusive (volcanic) rock that has a composition similar to andesite.

delta   A low flat tract of land at the mouth of a river composed of stream-deposited sediment, which is triangular in shape from an aerial view.

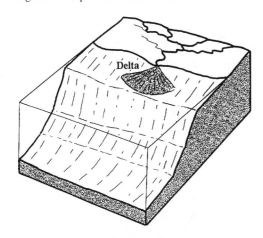

detritus   Loose material formed when rock decomposes.

diapir   A nonvolcanic forceful intrusion into sediments from below.

diatom   A microscopic single-celled plant with a shell of opaline silica.

diatomite   A rock formed of multiple shells of microscopic diatoms.

dike   A tabular igneous body that has been intruded across the surrounding rock.

dome (volcanic)   A volcanic extrusion of stiff viscous material forming a steep-sided topographic feature.

dredge (gold)   A floating machine for excavating and processing mineral-rich sediments from a river bed.

dripstone   Mineral deposits in a cave. See stalactite and stalagmite for illustration.

dunite   A peridotite rock in which the iron-magnesium-rich mineral is almost entirely olivine.

E

entablature   The upper portion of a lava flow that displays irregular, thin shrinkage columns.

erratic   A large rock which has been moved by glacial ice.

escarpment   See scarp.

estuary   An inlet from the sea such as the mouth of a river or bay.

exotic accreted terrane   A mass of rock that is different from the adjacent rock and that has been tectonically emplaced by faulting or crustal plate movement.

explosion pit   A small saucer-shaped crater produced by a single explosive volcanic eruption.

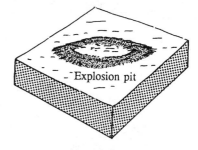

F

fault   A break in the earth's crust along which movement has taken place. Frequently occurs as a zone of numerous smaller faults. See normal fault.

fault trace   Intersection of faults with the ground surface (horizon); also called a fault line. See normal fault for illustration.

fauna   The entire animal population living in a given area, environment, or time span.

feldspar   A common rock-forming silicate mineral and one of the most abundant minerals in the earth's crust.

fissure eruption   A volcanic extrusion of material from a crack or rift in the crust that does not form a significant cone structure.

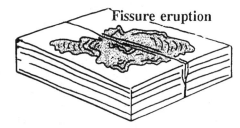

Fissure eruption

fjord   A deep steep-walled inlet to the sea that has been cut or deepened by a glacier.

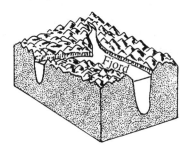

flood basalt   A plateau basalt. Successive flows of high temperature fluid basalt from fissure eruptions merge to form a continuous flat plateau.

floodplain   The flat area where a stream may overflow.

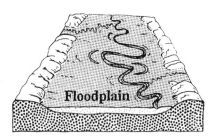

foraminifera   A microscopic animal (protozoa) with a shell of mineral material or foreign particles.

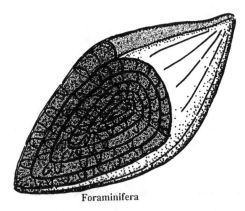

Foraminifera

fore-arc basin   A linear basin on the ocean side of a volcanic island chain that separates the archipelago from the subduction trench. See subduction for illustration.

formation   A body of identifiable rocks that can be traced with sufficient thickness and distribution to be represented on a geologic map.

G

gabbro   A dark, coarse-grained, intrusive rock with the same composition as basalt.

garnet   A silicate mineral common in metamorphic rocks.

gastropod   Molluscs commonly known as snails and slugs.

geophysics   The study of the interior of the earth.

geothermal resources   Energy utilized from naturally occurring steam and hot water within the earth's crust.

glacier   A mass of ice and snow that moves because of gravity.

gneiss   A banded metamorphic rock with distinct wavy layers of alternating granular and platy minerals.

graben   An elongated depressed block bordered by faults. See basin and range for illustration.

graded bedding   Rapid sedimentation leaving a distinctive texture of coarse grains followed by fine grains in a single layer.

granite   A coarse-grained intruded rock that has a high percentage of quartz and potassium feldspar.

greenschist   Metamorphic rocks, resulting from low pressure and temperature, with green minerals such as chlorite and epidote.

greywacke   A dirty gray sandstone. Its presence implies rapid deposition close to the source of sediments.

H

hanging valleys   Smaller glacial valleys feeding into a deep main glacier trough.

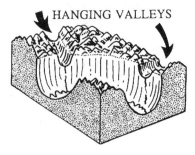

HANGING VALLEYS

headland   A high irregular promontory of the coast jutting out from the shore.

headwaters   Water near the origin of a stream system.

hematite   A common oxide ore of iron.

Holocene   The most recent 11,000 years of geologic time between the Pleistocene epoch and today.

hot spot   A stationary heat (volcanic) source in the earth's mantle below a moving crustal plate. The hot spot is expressed as a linear trail of successively older volcanic events leading in the direction of plate motion. Also called a mantle plume.

hot springs   Thermal springs where the water is warmer than the human body (98.6 degrees Fahrenheit).

hydrothermal   Related to heated water or mineralization from hot water.

I

ice sheet   See continental glacier.

ichthyosaur   A large extinct porpoise-like aquatic reptile, which lived during the Mesozoic era.

ignimbrite   A volcanic rock that forms from an ash flow, where a dense fiery incandescent cloud of ejected debris flows downslope then welds itself together as it settles. See ash flow.

interglacial period   The interval of significantly warmer climate between major glacial events.

intracanyon flow   Lava flows that follow stream pathways to plug the canyon or cover the bed and banks with a veneer of cooled volcanic rock.

intrusion   The process of emplacing magma into previously existing rock.

L

lacustrine   Lake environments or sediments.

lahar   A mudflow formed from a mixture of volcanic material and water.

lava lake   A liquid pool of basaltic lava trapped in a volcanic crater or depression. The lava may either harden to a smooth surface or drain away.

lignite   Brownish to black low-grade coal between peat and sub-bituminous coal in grade.

lineament   A conspicuous, large, straight feature visible on an aerial photograph that follows a structure, as a fault.

listric faulting   Tensional faults that display a high angle on the surface but which are nearly horizontal at depth. The faults are curved in a concave upward configuration.

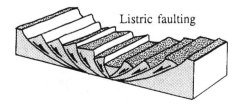

Listric faulting

lithosphere   The outer 60 miles of the earth between the crust and mantle where the rocks are harder and more brittle. See mantle for illustration.

lode   Deposits in which ores have been placed into rock fractures where they occur in veins.

loess   Nonstratified silt, clay, and dust, originating as glacial sediment, but redeposited by wind.

M

magma   A silica-rich molten fluid that appears on the surface as lava and which cools underground as a pluton.

magma chamber   A magma reservoir in the shallow part of the lithosphere.

mammoth   Several extinct species of elephant that were grazing animals.

mantle   The earth's interior below the lithosphere and above the outer core.

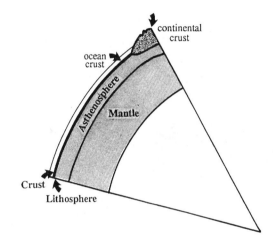

mantle plume   See hot spot.

marble   A metamorphic rock, composed mainly of calcium carbonate, that was derived from limestone.

massive sulfide   A hydrothermal deep ocean deposit of sulfide minerals associated with oceanic rifting.

meander   A sinuous curve and loop in a stream course.

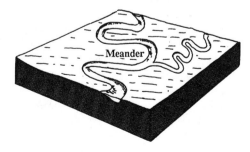

melange   A mixture or variety of rocks jumbled together by tectonic plate movement.

Mercali scale   The scale based on the destructive effects of earthquakes ranging from I to XII (total destruction).

metaliferous sediment   Oceanic sediment near a spreading ridge which is rich in a variety of metals. The metals were derived from both seawater and the underlying volcanic rocks.

metamorphism Alteration of rocks in the earth's crust due to heat, pressure, and chemically active fluids.

mica A group of silicate minerals with perfect cleavage in one direction, yielding thin sheets.

mollusc Invertebrates in the phylum Mollusca with a nonsegmented body, mantle, and shell. Typically includes clams, snails, and cephalopods.

moraine A general term for landforms composed of glacial till.

mudstone A sedimentary rock that contains a high proportion of clay and silt-size particles lacking the fine lamination of shale.

mylonite A zone of rocks that have been thoroughly sheared up by faulting or metamorphic processes.

N

nonmarine A fresh or nonsaline water environment.

normal fault Faulting due to tension where the hanging wall has dropped down relative to the foot wall.

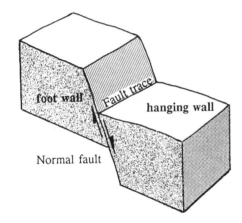

Normal fault

O

olistoliths An individual block or large exotic rock in an olistostrome.

olistostromes Chaotic rock mixtures of debris piled up in an underwater (submarine) slide.

olivine A silicate mineral rich in iron and magnesium but without aluminum.

ophiolite A sequence of ocean crust beginning with ultramafic rocks (gabbro) at the base grading upward to sheeted dikes, pillow lavas, and deep sea muds.

ore A mineral containing valuable metal or other constituents for which it is mined.

outwash plain A broad surface of sediment deposited at the margin of a glacier.

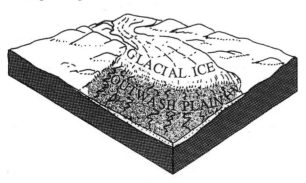

P

pahoehoe lava A smooth, glassy, billowy basaltic lava with a ropy appearance.

paleosol A buried prehistoric (fossil) soil layer (horizon).

parasitic cones A group of small volcanic cones and vents on the sides of a much larger volcano.

Parasitic cones

periglacial Features and processes at the margins of present and former glaciers.

phyllite A granular platy metamorphic rock with a texture between slate and schist.

pillow basalt Lava that has been extruded into water where it characteristically forms rounded blobs as it cools.

placer Ore deposits associated with stream gravels or sedimentary layers.

playa A basin in a dry region that may contain lake water intermittently. See basin and range for illustration.

pluton A large mass of intruded rock below the earth's surface. See batholith or dike for illustration.

pluvial lake A lake formed during heavy rains of the Pleistocene, which, at present, is much smaller than the original body of water.

potholes A bowl-shaped depression in a rock surface that has been carved by the whirling action of stones in a stream bed.

pumice A light-weight, frothy, glassy volcanic rock where the void space of bubbles exceeds the rock volume.

pyroclastic Broken rock material that has been shattered by a volcanic explosion.

Q

quartz Crystalline silica. The most important common rock-forming mineral.

quartzite Sandstone that has been altered by metamorphic processes, including heat and pressure.

R

radiolaria A microscopic single-celled animal (protozoa) with a delicate ornate skeleton of silica.

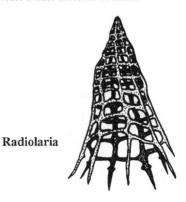

Radiolaria

rainshadow An area of low rainfall cut off from moisture-laden clouds by a topographic high.

reef A structure projecting from the sea floor that can be made of rocks, coral, etc. A reef that is exclusively organic is a bioherm.

retrenchment The reconstruction, usually by fluvial processes, of a previous valley or topographic low that has been filled or plugged by lava or sediments.

rhyolite The volcanic equivalent of granite; a quartz-rich grey or pink fine-grained rock.

Richter scale A measurement of earthquake magnitude from 1 through 9. Each point is 10 times higher than the previous number.

rift A narrow crevice or fissure in rock produced by splitting due to tension.

rip-rap A layer of selected large stones placed in order to slow erosion.

rock flour Unweathered fine rock in powder form that has been milled by glacial processes.

S

sand Rock and mineral fragments ranging in size between 1/16 to 2 millimeters in diameter.

scabland An irregular surface of basalt that has been scoured of its soil cover by floods.

scarp Escarpment; a sharp to flat steep cliff produced by faulting or erosion.

seamount A conical mound rising several thousand feet off the deep ocean floor. Almost invariably of volcanic origin.

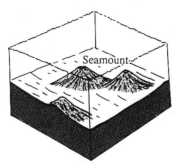

sea stack Small steep-sided promontories surrounded by water. Sea stacks have been separated from the rocky coast by erosion.

sediment starved A basin that is subsiding or sinking faster than it is being filled with debris.

seismic Having to do with earthquakes or earth vibrations.

serpentine A group of common minerals, which are somewhat greasy to the touch and green in color, derived from the metamorphic alteration of the silicate minerals olivine and pyroxene.

shear zone A sheet-like zone of broken rock with multiple parallel fractures.

shelf The flat submerged portion of the continental mass extending from the shoreline outward to a depth of 600 to 1000 feet.

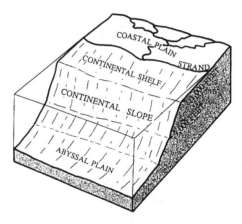

shield cone A volcano where the slopes are 10 degrees or less. Ordinarily of basaltic composition.

Shield cone

sill A tabular intrusion oriented parallel to the layers in the surrounding rock. See dike for illustration.

siltstone A sedimentary rock made up of particles smaller than very fine sand but larger than coarse clay in the size range between 1/256 and 1/16 millimeter.

slate A compact fine-grained metamorphic rock formed from clay and shale that has a characteristic sheet-like cleavage.

slope, continental   The ocean floor at the margin of a continental mass that lies below the shelf but above the continental rise. See shelf for illustration.

slumping   Mass movement of soil downslope along curved planes.

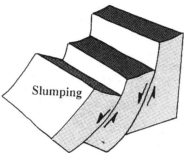

soil liquifaction   A phenomena related to mass movement where water-soaked soil or fine sediments mobilize and flow as a liquid.

spatter cone   A low basaltic volcanic cone built up of debris above a small vent.

spit   Sandspit, a linear ridge of sand extending from a headland along the coast.

spreading ridge   A shared plate boundary at which the crustal slabs are moving away from each other.

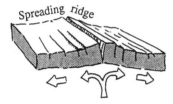

squeeze up   A small local extrusion of fluid lava from a fracture on the solidified surface of a larger flow.

stalactite (lava)   A volcanic feature where solidified drops of lava project from the roof of a lava cave.

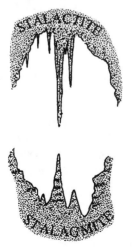

stalagmite   A cylindrical dripstone deposit that projects upward from the floor of a cave.

stamp mill   A device for crushing ores consisting of heavy steel rods, which rise then fall by the action of a rotating cam.

steptoe   An isolated protrusion of older bedrock such as a hill or mountain projecting like an island above a lava flow.

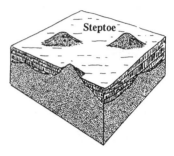

stitching plutons   Intrusive bodies that have invaded along faults to weld two rock masses together.

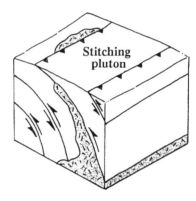

strait   A narrow waterway connecting two larger bodies of water.

strand   Beach and very shallow coastal area dominated by marine processes. See shelf for illustration.

stratigraphy   The study of stratified layered rocks.

stratocone   A volcano constructed of alternating layers of lava and broken volcanic fragments (pyroclastics).

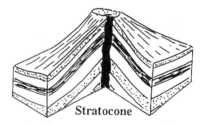

strike-slip fault   Faulting where the plane of the fault is nearly vertical and movement is parallel to the plane of the fault.

subduction   Plate collision boundary where one plate is overriding another.

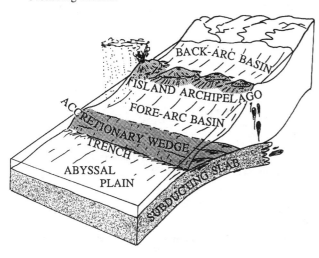

subduction trench   A distinct linear groove on the earth's crust marking the boundary between two colliding plates. See subduction for illustration.

submarine canyon   A steep V-shaped trench along or across a continental shelf, slope, and deep ocean floor.

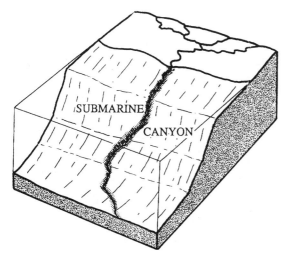

submarine fan   A marine accumulation of sediments similar to a delta but situated further basinward on the continental slope and rise. Of a much larger physical dimension than a delta.

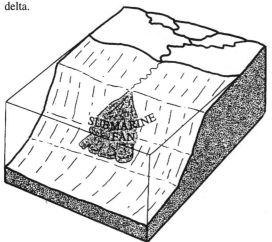

surficial deposits   Recently deposited loose sedimentary material found at or near the earth's surface.

T

tectonics   The study of the major structures of the earth such as mountains and ocean basins, which relate to the movement and interrelationship of large crustal plates.

tephra   Broken fragments of rock from a volcanic explosion.

terrace (marine)   A flat erosional surface extending seaward from the beach, which has been cut by waves and covered by a thin layer of marine sediment.

tethys   An east-west Mesozoic seaway from the Mediterranean to the Pacific, which was tropical in nature.

tidal channel   A major channel followed by tidal currents from offshore well into a tidal marsh or tide flat.

tidewater glacier   A valley glacier that reaches the sea, as in a fjord.

till   Loose sediment deposited by glaciers.

transcurrent fault   See strike-slip fault.

transgression   A shoreline that is elevated because the landmass is sinking or because sea level is rising.

tsunami   Enormous waves caused by submarine-landslides or fault movement.

tuff   A rock of compacted volcanic ash with particles smaller than 4 millimeters in diameter.

tuff ring   High circular wall of shattered debris around a volcanic vent resulting when ascending lava encountered groundwater near the surface with explosive results.

tundra   A level treeless plain characteristic of arctic and subarctic regions.

turbidite   The sediment left by a turbidity current. Typically coarse sand at its base grades upward to finer sizes.

turbidity current   A rapidly moving current of dense muddy fluids flowing by gravity along the bottom of an undersea slope.

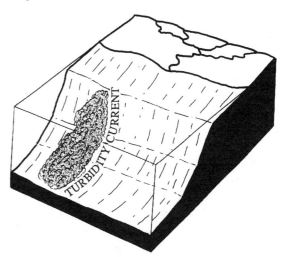

V

valley glacier　Glaciation in an alpine setting where ice masses occupy former stream channels.

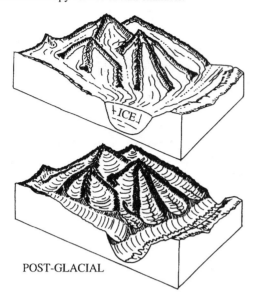

POST-GLACIAL

volcanic spine　An elongated pointed mass of solidified lava in the throat of a volcano that may form either by slow extrusion or weathering.

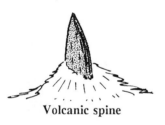

Volcanic spine

W

watershed　The total area drained by a stream system.

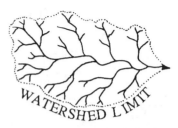

wetland　A poorly drained region of constant moisture and limited runoff.

Z

zircon　A silicate ore of zirconium, common in beach placers.

BIBLIOGRAPHY

Abegglen, D.E., Wallace, A.T., and Williams, R.E., 1970. The effects of drain wells on the ground-water quality of the Snake River Plain. Idaho Bur. of Mines and Geology, Pam. 148, 51p.

Adams, J., 1984. Active deformation of the Pacific Northwest continental margin. Tectonics, vol. 3, no. 4, pp. 449–472.

———— 1990. Paleoseismicity of the Cascadia subduction zone: evidence from turbidites off the Oregon-Washington margin. Tectonics, vol. 9, no. 4, pp. 569–583.

Albers, J.P., 1966. Economic deposits of the Klamath Mountains. In: Bailey, E.H., ed., Geology of northern California. Calif. Division of Mines and Geology, Bull. 190, pp. 51–62.

———— 1984. Geological and geophysical studies of chromite deposits in the Josephine peridotite, northwestern California and southwestern Oregon. U.S. Geol. Survey, Bull. 1546 A–D, pp. 1–5.

Algermissen, S.T., and Harding, S.T., 1986. The Puget Sound, Washington, earthquake of April 29, 1965. Wash. Div. Geol. and Earth Resources, Info. Circular 81, pp. 35–65.

Allen, J.E., 1979. The magnificent gateway: a layman's guide. Portland, Oregon, Timber Press, 144p.

———— Burns, M., and Sargent, S.C., 1986. Cataclysms on the Columbia. Portland, Oregon, Timber Press, 211p.

Alley, N.F. et al., 1986. Paleoclimatic implications of middle Wisconsin pollen and a paleosol from the Purcell trench, southcentral British Columbia. Canadian Jour. Earth Science, vol. 23, no. 8, pp. 1156–1168.

Allison, I.S., 1935. Glacial erratics in Willamette Valley, Geol. Soc. America, Bull. 46, pp. 615–632.

———— 1966. Fossil Lake, Oregon; its geology and fossil faunas. Oregon State Univ., Studies in Geology no. 6, 48p.

Allmendinger, R.W., 1982. Sequence of late Cenozoic deformation in the Blackfoot Mountains, southeastern Idaho. In: Bonnichsen, B., and Breckenridge, R.M., eds., Cenozoic geology of Idaho. Idaho Bur. of Mines and Geology, Bull. 26, pp. 505–516.

Alpha, Tau Rho, and Vallier, T.L., 1994. Physiography of the Seven Devils Mountains and adjacent Hells Canyon of the Snake River, Idaho and Oregon. In: Vallier, T.L., and Brooks, H.C., eds., Geology of the Blue Mountains region of Oregon, Idaho, and Washington; stratigraphy, physiography, and mineral resources of the Blue Mountains region. U.S. Geological Survey, Prof. Paper 1439, pp. 91–100.

Anders, M.H., 1990. Late Cenozoic evolution of Grand and Swan Valleys, Idaho. In: Roberts, S., ed., Geologic field tours of western Wyoming and parts of adjacent Idaho, Montana, and Utah. Geol. Survey of Wyoming, Public Info. Circular 29, pp. 14–25.

Anderson, A.L., 1961. Geology and mineral resources of the Lemhi Quadrangle, Lemhi County, Idaho Bur. of Mines and Geology, Pam. 24, 18p.

Anderson, J.L. et al., 1987. Distribution maps of stratigraphic units of the Columbia River Basalt Group. In: Schuster, J.E., ed., Selected papers on the geology of Washington. Wash. Division of Geology and Earth Resources, Bull. 77, pp. 183–195.

Armentrout, J.M., 1977. Cenozoic stratigraphy of southwestern Washington (field trip no.9). In: Brown, E.H., and Ellis, R.C., eds., Geological excursions in the Pacific Northwest. Geol. Soc. America, Annual Mtg., Seattle, pp. 227–264.

———— 1980. Field trip road log for the Cenozoic stratigraphy of Coos Bay and Cape Blanco; southwestern Oregon. In: Oles, K.F., et al., eds., Geologic field trips in western Oregon and southwestern Washington. Oregon Dept. Geol. and Mineral Indus., pp. 177–216.

———— 1987. Cenozoic stratigraphy, unconformity-bounded sequences, and tectonic history of southwestern Washington. In: Schuster, J.E., ed., Selected papers on the geology of Washington. Wash. Division of Geology and Earth Resources, Bull. 77, pp. 291–320.

———— and Suek, D.H., 1985. Hydrocarbon exploration in western Oregon and Washington. American Assoc. Petroleum Geologists Bull., vol. 69, no. 4, pp. 627–643.

———— Cole, M.R., and TerBest, H., eds., 1979. Cenozoic paleogeography of the Western United States. Soc. Econ. Paleontologists and Mineralogists, Pacific Sect., Pacific Coast Paleo. Symposium 3, 335p.

———— et al., 1983. Correlation of stratigraphic units in North America (COSUNA project): Northwest region. American Assoc. Petroleum Geologists, Chart.

———— ed., 1981. Pacific Northwest Cenozoic biostratigraphy: Geol. Soc. America, Special Paper 184. 172p.

Armstrong, J.E., 1977. Quaternary geology of the Fraser Lowland (Field trip no.6). In: Brown, E.H., and Ellis, R.C., eds., Geological excursions in the Pacific Northwest. Geol. Soc. America, Annual Mtg., Seattle, pp. 204–206.

———— 1981. Post-Vashon Wisconsin glaciation Fraser Lowland, British Columbia, Geological Survey Canada, Bull. 322, 34p.

———— et al., 1965. Late Pleistocene stratigraphy and chronology in southwestern British Columbia and northwestern Washington. Geol. Soc. America, Bull., vol. 76, pp. 321–330.

Armstrong, R.L., 1975. PreCambrian (1500 m.y. old) rocks of central Idaho—the Salmon River arch and its role in Cordilleran sedimentation and tectonics. American Jour. Science, vol. 275-A, pp. 437–467.

———— Leeman, W.P., and Malde, H.E., 1975. K-Ar dating, Quaternary and Neogene volcanic rocks of the Snake River Plain, Idaho. American Jour. Science, vol. 275, pp. 225–251.

Ash, S.R., 1991. A new Jurassic flora from the Wallowa terrane in Hells Canyon, Oregon and Idaho. Oregon Geology, vol. 53, pp. 27–33.

Asher, R.R., 1968. Geology and mineral resources of a portion of the Silver City region, Owyhee County, Idaho. Idaho Bur. of Mines and Geology, Pam. 138, 102p.

Atwater, B., 1994. Prehistoric earthquakes in western Washington. In: Lasmanis, R., and Cheney, E.S., convs., Regional geology of Washington state. Wash. Div. of Geology and Earth Resources, Bull. 80, pp. 219–222.

Atwater, T., 1989. Plate tectonic history of the northeast Pacific and western North America. In: Winterer, E.L., Hussong, D.M., and Decker, R.W., eds., The eastern Pacific Ocean and Hawaii. Geological Soc. America, The Geology of North America, vol. N, pp. 21–72.

Aune, Q.A., 1970. A trip to Castle Crags. Calif. Division of Mines and Geology, Mineral Information Service, vol. 23, no. 7, pp. 139–144.

Axelrod, D.I., 1944. The Alturas flora. Carnegie Inst., Wash., Contrib. to Paleontology, Publ. 553, pp. 225–306.

Babcock, R.S., Suczek, C.A., and Engebretson, D.C., 1994. The Crescent "terrane," Olympic Peninsula and southern Vancouver Island. In: Lasmanis, R., and Cheney, E.S., convs., Regional geology of Washington state. Wash. Div. Geology and Earth Resources Bull. 80, pp. 141–158.

———— et al., 1992. A rifted margin origin for the Crescent Basalts and related rocks in the northern Coast Range volcanic province, Washington and British Columbia. Jour. Geophys. Res. B: Solid Earth and Planets, vol. 97, no. 5, pp. 6799–6821.

Bailey, E.H., ed., 1966. Geology of northern California. Calif. Div. of Mines and Geology, Bull. 190, 508p.

Baird, D.M., 1964. Kootenay National Park; wild mountains and great valleys. Geol. Survey Canada, Misc. Rept. 9, 94p.

———— 1965. Glacier and Mount Revelstoke National parks; where rivers are born. Geol. Survey of Canada, Misc. Rept. 11, 104p.

Baker, V.R., 1987. Dry Falls of the channeled scabland, Washington. In: Hill, M.L., ed., Centennial Field Guide, Cordilleran Section, Geol. Soc. America, pp. 369–372.

———— and Nummedal, D., eds., 1978. The channeled scabland; a guide to the geomorphology of the Columbia Basin, Washington. National Aeronautics and Space Admin., Planetary Geology Field Conference Columbia Basin, June 5–8, 186p.

———— et al., 1987. Columbia and Snake River Plains. In: Graf, W.L., ed., Geomorphic systems of North America. Geol. Soc. America, Centennial Special Volume 2, pp. 403–468.

Baldwin, E.M., 1974. Eocene stratigraphy of southwestern Oregon. Oregon Dept. Geol. and Mineral Indus., Bull. 83, 40p.

———— 1976. Geology of Oregon. 3rd ed., Dubuque, Kendall-Hunt, 170p.

Ballard, S.M., 1924. Geology and gold resources of Boise basin, Boise County, Idaho. Idaho Bur. Mines and Geology, Bull. 9, 103p.

Ballard, W.W., Bluemle, J.P., and Gerhard, L.C., coord., 1983. Correlation of Stratigraphic Units of North America (COSUNA project): Northern Rockies/Williston Basin region. American Assoc. Petroleum Geologists, Chart.

Bally, A.W., and Palmer, A.R., 1989. The geology of North America; an overview. Geol. Soc. America. The Geology of North America, vol. A. 619p.

Bamber, E.W., et al., 1972. Biochronology: a standard of Phanerozoic time. In: Douglas, R.J.W., ed., Geology and economic minerals of Canada. Geological Survey of Canada. Economic Geology Report, no. 1, pp. 593–675.

Barksdale, J.D., 1975. Geology of the Methow valley, Okanogan County, Washington. Wash. Division of Geology and Resources, Bull. 68, 72p.

Barnard, W.D., 1978. The Washington continental slope: Quaternary tectonics and sedimentation. Marine Geology, vol. 27, pp. 79–114.

Beck, G.F., 1945. Ancient forest trees of the sagebrush area in central Washington. Jour. of Forestry, vol. 43, no. 5, pp. 334–338.

Beeson, M.H., Perttu, R., and Perttu, J., 1979. The origin of the Miocene basalts of coastal Oregon and Washington; an alternative hypothesis. Oregon Geology, vol. 41, no. 10, pp. 159–166.

———— Tolan T.L., and Anderson, J.L., 1989. The Columbia River Basalt Group in western Oregon; geologic structures and other factors that controlled flow emplacement patterns. In: Reidel, S., and Hooper, P.R., eds., Volcanism and tectonism in the Columbia River flood-basalt province. Geol. Soc. America, Special Paper 239, pp. 223–246.

Beget, J.E., 1981. Early Holocene glacier advance in the North Cascade Range, Washington. Geology, vol. 9, pp. 409–413.

Behrens, G.W., and Hansen, P.J., 1989. Geology and related construction problems of the Grand Coulee Dam. In: Joseph, N.J., et al., eds., Geologic guidebook for Washington and adjacent areas. Wash. Division of Geology and Earth Resources, Info. Circular 86, pp. 357–369.

Bell, W.A., 1956. Lower Cretaceous floras of western Canada. Geol. Survey of Canada, Memoir 285, 331p.

Bennett, E.H., n.d. Idaho's phosphate industry. Idaho Geol. Survey, GeoNote 8, [2p.].

———— n.d. Industrial minerals in Idaho. Idaho Geol. Survey, GeoNote 14 [2p.].

———— Siems, P.L., and Constantoplos, J.T., 1989. The geology and history of the Coeur d'Alene mining district, Idaho. In: Chamberlain, V.E., Breckenridge, R.M., and Bonnichsen, B., eds., Guidebook to the geology of northern and western Idaho and surrounding area. Idaho Geol. Survey, Bull. 28, pp. 137–156.

Berg, A.W., 1990. Formation of Mima Mounds; a seismic hypothesis. Geology (Boulder), vol. 18, no. 3, pp. 281–284.

Berry, E.W., 1929. A revision of the flora of the Latah Formation. U.S. Geol. Survey, Prof. Paper 154-H, pp. 225–264.

Bishop, C.C., and Davis, J.F., coord., 1984. Correlation of stratigraphic units of North America (COSUNA) project:

northern California region. American Assoc. of Petroleum Geologists, Chart.

Bjork, P.R., 1970. The carnivora of the Hagerman local fauna (late Pliocene) of southwestern Idaho. American Philos. Soc., Trans., n.s., vol. 60, pt.7, 54p.

Blake, M.C., et al., 1985. Tectonostratigraphic terranes in southwest Oregon. *In:* Howell, D.G., ed., Tectonostratigraphic terranes of the circum-Pacific. Council for Energy and Mineral Resources, Earth Science Series, vol. 1, pp. 147–157.

Blome, C.D., and Nestell, M.K., 1991. Evolution of a Permo-Triassic sedimentary melange, Grindstone terrane, east-central Oregon. Geol. Soc. America, Bull. 103, pp. 1280–1296.

—— et al., 1986. Geologic implications of radiolarian-bearing Paleozoic and Mesozoic rocks from the Blue Mountains province, eastern Oregon. *In:* Vallier, T.L., and Brooks, H.C., eds., Geology of the Blue Mountains region of Oregon, Idaho, and Washington. U.S. Geol. Survey, Prof. Paper 1435, pp. 79–93.

Blunt, D.J., Easterbrook, D.J., and Rutter, N.W., 1987. Chronology of Pleistocene sediments in the Puget Lowland, Washington. *In:* Schuster, J.E., ed., Selected papers on the geology of Washington. Wash. Div. of Geology and Earth Resources, Bull. 77, pp. 321–353.

Bond, J.G., 1963. Geology of the Clearwater embayment. Idaho Bureau of Mines and Geology, Pam. 128, 81p.

Bonnichsen, B., 1982. The Bruneau-Jarbidge eruptive center, southwestern Idaho. *In:* Bonnichsen, B., and Breckenridge, R.M., eds., Cenozoic geology of Idaho. Idaho Bur. of Mines and Geology, Bull. 26, pp. 237–254.

—— and Breckenridge, R.M., eds., Cenozoic geology of Idaho. Idaho Bureau of Mines and Geology, Bull. 26. 725p.

—— et al., 1998. Geologic field trip guide to the central and western Snake River Plain, Idaho, emphasizing the silicic volcanic rocks. *In:* Link, P.K., and Hackett, W.R., eds., Guidebook to the geology of central and southern Idaho. Idaho Geol. Survey, Bull. 27, pp. 247–281.

Booth, D.B., and Goldstein, B., 1994. Patterns and processes of landscape development by the Puget lobe ice sheet. *In:* Lasmanis, R., and Cheney, E.S., convs., Regional geology of Washington state. Wash. Div. Geology and Earth Resources, Bull. 80, pp. 207–218.

Bovis, M.J., 1987. The interior mountains and plateaus. *In:* Graf, W.L., ed., Geomorphic systems of North America. Geol. Soc. America, Centennial Special Volume 2, pp. 469–515.

Boyer, S.E., and Hossack, J.R., 1992. Structural features and emplacement of surficial gravity-slide sheets, northern Idaho-Wyoming thrust belt. *In:* Link, P.K., Kuntz, M.A., and Platt, L.B., eds., Regional geology of eastern Idaho and western Wyoming. Geol. Soc. America, Memoir 179, pp. 197–213.

Brandon, M.T., and Cowan, D.S., 1987. The Late Cretaceous San Juan thrust system, San Juan Islands, Washington. Geol. Soc. America, Special Paper 221, 81p.

—— Cowan, D.S, and Vance, J.A., 1988. The Late Cretaceous San Juan thrust system, San Juan Islands, Washington. Geol. Soc. America, Special Paper 221, 81p.

Breckenridge, R.M., 1989. Pleistocene ice dams and glacial Lake Missoula floods in northern Idaho and adjacent areas. *In:* Chamberlain, V.E., Breckenridge, R.M., and Bonnichsen, B., eds., 1989. Guidebook to the geology of northern and western Idaho and surrounding area. Idaho Geol. Survey, Bull. 28, 156p.

—— et al., 1988. Glacial geology of the Stanley Basin. *In:* Link, P.K., and Hackett, W.R., eds., Guidebook to the geology of central and southern Idaho. Idaho Geol. Survey, Bull. 27, pp. 209–221.

Bretz, J.H., 1923. The channeled scablands of the Columbia plateau. Jour. Geol., vol. 77, pp. 505–543.

—— 1932. The Grand Coulee. American Geographical Soc., Special Publ. no. 15, 86p.

Bright, R.C., 1982. Paleontology of the lacustrine member of the American Falls Lake beds, southeastern Idaho. *In:* Bonnichsen, B., and Breckenridge, R.M., eds., Cenozoic geology of Idaho. Idaho Bur. of Mines and Geology, Bull. 26, pp. 597–614.

Brooks, H.C., 1979. Plate tectonics and the geologic history of the Blue Mountains. Oregon Geology, vol. 41, pp. 71–80.

—— and Ramp, L., 1968. Gold and silver in Oregon. Oregon Dept. Geol. and Mineral Indus., Bull. 61, 337p.

—— and Vallier, T.L., 1978. Mesozoic rocks and tectonic evolution of eastern Oregon and western Idaho. *In:* Howell, D., and McDougall, K., eds., Mesozoic paleogeography of the western United States. Soc. Economic Paleontologists and Mineralogists, Pacific Coast Paleogeography Symposium 2, pp. 133–145.

Brown, A.S., 1968. Geology of the Queen Charlotte Islands. British Columbia. British Columbia, Dept. of Mines and Petroleum Resources, Bull. 54, 226p.

Brown, E.H., and Ellis, R.C., eds., 1977. Geological excursions in the Pacific Northwest; fieldtrips. Geol. Soc. America Annual Mtg., Seattle, 414p.

Buckovic, W.A., 1979. The Eocene deltaic system of west-central Washington. *In:* Armentrout, J.M., Cole, M.R., and TerBest, H., Cenozoic paleogeography of the western United States. Soc. Economic Paleontologists and Mineralogists, Pacific Coast Paleogeography Symposium 3, pp. 147–163.

Burchfiel, B.C., Cowan, D.S., and Davis, G.A., 1992. Tectonic overview of the Cordilleran orogen in the western United States. *In:* Burchfiel, B.C., Lipman, P.W., and Zoback, M.L., eds., The Cordilleran orogen; conterminous U.S. Geol. Soc. America, Geology of North America, vol. G-3, pp. 407–479.

—— Lipman, P.W., and Zoback, M.L., eds., 1992. The Cordilleran orogen; conterminous U.S. Geol. Soc. America, Geology of North America, vol. G-3, 724p.

Burnett, J.L., 1990. 1989 California mining review. California Geology, vol. 43, no. 10, pp. 210–224.

Busacca, A.J., and McDonald, E.V., 1994. Regional sedimentation of late Quaternary loess on the Columbia plateau: sediment source areas and loess distribution patterns. *In:* Lasmanis, R., and Cheney, E.S., convs., Regional geology of Washington state. Wash. Div. Geology and Earth Resources, Bull. 80, pp. 181–190.

Camp, V.E., 1981. Geologic studies of the Columbia plateau: Part II. Upper Miocene basalt distribution, reflecting source locations, tectonism, and drainage history in the Clearwater embayment, Idaho. Geol. Soc. America, Bull., Part I, vol. 92, pp. 669–678.

—— and Hooper, P.R., 1981. Geologic studies of the Columbia Plateau: I. Late Cenozoic evolution of the southeast part of the Columbia River basalt province. Geol. Soc. America, Bull., Part I, vol. 92, pp. 659–668.

—— et al., 1982. Columbia River basalt in Idaho: physical and chemical characteristics, flow distribution, and tectonic implications. *In:* Bonnichsen, B., and Breckenridge, R.M., eds., Cenozoic geology of Idaho. Idaho Bur. of Mines and Geology, Bull. 26, pp. 55–75.

Campbell, N.P., 1989. Structural and stratigraphic interpretation of rocks under the Yakima fold belt, Columbia Basin, based on recent surface mapping and well data. Geol. Soc. America, Special Paper 239, pp. 209–222.

Carson, R.J., and Pogue, K.R., 1996. Flood basalts and glacier floods: Roadside geology of parts of Walla Walla, Franklin, and Columbia Counties, Washington. Washington Div. Geology and Earth Resources, Info. Circular 90, 47p.

——— Tolan, T.L., and Reidel, S.P., 1987. Geology of the Vantage area, south-central Washington: an introduction to the Miocene flood basalts, Yakima fold belt, and the channeled scabland. In: Hill, M.L., ed., Centennial Field Guide, Geol. Soc. America, Cordilleran Sect., pp. 357–362.

Chamberlain, V.E., Breckenridge, R.M., and Bonnichsen, B., eds., 1989. Guidebook to the geology of northern and western Idaho and surrounding area. Idaho Geol. Survey, Bull. 28, 156p.

Chaney, R.W., Condit, C., and Axelrod, D.I., 1959. Miocene floras of the Columbia plateau. Carnegie Institute Wash., Publ. 617, 237p.

The channeled scablands of eastern Washington; the geologic story of the Spokane flood, 1982. U.S. Geol. Survey, 23p.

Charvet, et al., 1990. Tectono-magmatic evolution of Paleozoic and early Mesozoic rocks in the eastern Klamath Mountains, California, and the Blue Mountains, eastern Oregon-western Idaho. In: Harwood, D.S., and Miller, M.M., eds., Paleozoic and early Mesozoic paleogeographic relations; Sierra Nevada, Klamath Mountains, and related terranes. Geol. Soc. America, Special Paper 255, pp. 255–276.

Chase, T.E., Menard, H.W., and Mammerickx, J., 1970. Bathymetry of the north Pacific [map]. Inst. of Marine Resources, Univ. of Calif., San Diego, IMR Technical Report Series TR-9, Chart no. 4.

Cheney, E.S., 1980. Kettle dome and related structures of northeastern Washington. Geol. Soc. America, Memoir 153, pp. 463–483.

——— 1987. Major Cenozoic faults in the northern Puget Lowland of Washington. In: Schuster, J.E., ed., Selected papers on the geology of Washington, Wash. Div. Geology and Earth Resources, Bull. 77, pp. 149–168.

——— Rasmussen, M.G., and Miller, M.G., 1994. Major faults, stratigraphy, and identity of Quesnellia in Washington and adjacent British Columbia. In: Lasmanis, R., and Cheney, E.S., convs., Regional geology of Washington state. Wash. Div. Geology and Earth Resources, Bull. 80, pp. 49–72.

Chesterman, C.W., and Saucedo, G.J., 1984. Cenozoic volcanic stratigraphy of Shasta Valley. California Geology, vol. 37, no. 4, pp. 67–74.

Christiansen R.S., 1982. Late Cenozoic volcanism of the Island Park area, eastern Idaho. In: Bonnichsen, B., and Breckenridge, R.M., eds., Cenozoic geology of Idaho, Idaho Bur. Mines and Geology, Bull. 26, pp. 345–364.

——— and Yeats, R.L., 1992. Post-Laramide geology of the U.S. Cordilleran region. In: Burchfiel, B.C., Lipman, P.W., and Zoback, M.L., eds., The Cordilleran orogen: conterminous U.S. Geol. Soc. America, The Geology of North America, vol. G-3, pp. 261–406.

Church, B.N., 1985. Volcanology and structure of Tertiary outliers in south-central British Columbia. In: Templeman-Kluit, D.J., ed., Field guides to geology and mineral deposits in the southern Canadian Cordillera. Geol. Soc. America, Cordilleran Sect., Vancouver, B.C., 1985 Mtg., pp. 5-1 to 5–46.

Clague, J.J., 1981. Late Quaternary geology and geochronology of British Columbia. Part 2: summary and discussion of radiocarbon-dated Quaternary history. Geological Society of Canada, Paper 80–35, 41p.

——— 1989. Quaternary geology of the Canadian Cordillera. In: Fulton R.J., ed. Quaternary geology of Canada and Greenland. Geol. Soc. Canada No. 1, pp. 17–83.

——— and Bobrowsky, P.T., 1990. Holocene sea level change and crustal deformation, southwestern British, Columbia. In: Current Research, Part E., Geol. Survey of Canada, Paper 90-1E, pp. 245–250.

——— and Bobrowsky, P.T., 1994. Evidence for a large earthquake and tsunami 100–400 years ago on western Vancouver Island, British Columbia. Quaternary Research 41, pp. 176–184.

——— et al., 1987. Quaternary geology of the southern Canadian Cordillera. International Union for Quaternary Research, XIIth INQUA Congress, Field Excursion A-18, 67p.

Clark, W.B., 1970. Gold districts of California. Calif. Division of Mines and Geology, Bull. 193, 186p.

Cole, M.R., and Armentrout, J.A., 1979. Neogene paleogeography of the western United States. In: Armentrout, J.M., Cole, M.R., and TerBest, H., eds., Cenozoic paleogeography of the western United States. Soc. Economic Paleontologists and Mineralogists, Pacific Coast Paleogeography Symp. 3, pp. 297–323.

Coney, P.J., 1979. Tertiary evolution of Cordilleran metamorphic core complexes. In: Armentrout, J.M., Cole, M.R., and TerBest, H., eds., Cenozoic paleogeography of the western United States. Soc. Economic Paleontologists and Mineralogists, Pacific Coast Paleogeography Symp. 3, pp. 15–28.

Conley, C., 1982. Idaho for the curious; a guide. Cambridge, Idaho, Backeddy Books, 704p.

Coogan, J.C., 1992. Structural evolution of piggyback basins in the Wyoming-Idaho-Utah thrust belt. In: Link, P.K., Kuntz, M.A., and Platt, L.B., eds., Regional geology of eastern Idaho and western Wyoming, Geol. Soc. America, Memoir 179, pp. 55–81.

Coombs, H.A., 1936. The geology of Mount Rainier National Park. Univ. of Wash. Publ. in Geology, vol. 3, no. 2, pp. 131–212.

Cornelius, R.R., 1986. Geology and structure of the Rock Creek Butte area. California Geology, vol. 39, no. 9, pp. 195–201.

Cowan, D.S., 1994. Alternative hypotheses for the mid-Cretaceous paleogeography of the western Cordillera. GSA Today, vol. 4, no. 7, pp. 181, 184–186.

——— Brown, E.H., and Whetten, J.T., 1977. Geology of the southern San Juan Islands (Field trip no. 11). In: Brown, E.H., and Ellis, R.C., eds., Geological excursions in the Pacific Northwest. Geol. Soc. America, Annual Mtg., Seattle, pp. 309–338.

Craddock, J.P., 1992. Transgression during tectonic evolution of the Idaho-Wyoming fold-and-thrust belt. In: Link, P.K., Kuntz, M.A., and Platt, L.B., eds., Regional geology of eastern Idaho and western Wyoming. Geol. Soc. America, Memoir 179, pp. 125–140.

Crandall, D.R., 1965. The glacial history of western Washington and Oregon. In: Wright, H.E., and Frey, D., eds., The Quaternary history of the United States. Princeton, N.J., Princeton Univ. Press, pp. 341–353.

Crosson, R.S., and Owens, T.J., 1987. Slab geometry of the Cascadia subduction zone beneath Washington from earthquake hypocenters and teleseismic converted waves. Geophysical Res. Newsletter, vol. 14, pp. 824–827.

Cuvier, M., 1812. Recherches sur les ossemens fossiles de quadrupedes. Paris, Chez Deterville, vol. 2.

Dana, J.D., 1875. The geological story briefly told; an introduction to geology New York, American Book Co., 263p.

Danner, W.R., 1955. Geology of Olympic National Park. Seattle, Univ. of Washington, 68p.

Davis, E.E., and Riddihough, R.P., 1982. The Winona basin: structure and tectonics. Canadian Jour. Earth Science, vol. 19, pp. 767–788.

Davis, G.A., 1966. Metamorphic and granitic history of the Klamath Mountains. *In:* Bailey, E.H., ed., Geology of northern California. Calif. Div. Mines and Geology, Bull. 190, pp. 39–50.

———— 1979. Problems of intraplate extensional tectonics, western United States, with special emphasis on the Great Basin. *In:* Newman, G.W., and Goode, H.D., eds., Basin and Range Symposium and Great Basin field conference. Rocky Mountain Assoc. of Geologists and Utah Geol. Assoc., pp. 41–54.

Dawson, G.M. 1888. The mineral wealth of British Columbia. Montreal, Dawson Bros. Publ., Geol. and Natural History Soc. of Canada, 163p.

Dawson, K.M., et al., 1991. Regional metallogeny. *In:* Galbriese, H., and Yorath, C.J., eds., Geology of the Cordilleran orogen in Canada. Geol. Survey Canada, Geology of Canada no. 4, pp. 709–768.

Debiche, M.G., Cox, A., and Engebretson, D.C., 1987. The motion of allochthonous terranes across the north Pacific basin. Geol. Soc. America, Special Paper 207, 49p.

Dengler, L., Carver, G., and McPherson, R., 1992. Sources of north coast seismicity. California Geology, vol. 45, no. 2, pp. 40–53.

Derkey, R.E., 1994. Metallic mineral deposits. Washington Geology, vol. 22, no. 1, pp. 16–18.

Dicken, S.N., 1980. Pluvial Lake Modoc, Klamath County, Oregon, and Modoc and Siskiyou Counties, California. Oregon Geology, vol. 42, no. 11, pp. 179–187.

Dickinson, W.R., 1979. Mesozoic forearc basin in central Oregon. Geology, vol. 7, pp. 166–170.

———— and Thayer, T.P., 1978. Paleogeographic and paleotectonic implications of Mesozoic stratigraphy and structure in the John Day inlier of central Oregon. *In:* Howell, D.G., and McDougall, K.A., eds., Pacific Coast Paleogeography Symposium 2. Soc. Economic Paleontologists and Mineralogists, Pacific Sect., pp. 147–162.

Diller, J.S., 1914. Auriferous gravels in the Weaverville Quadrangle, California. U.S. Geol. Survey, Contributions to Economic Geology, Bull. 540, pp. 11–79.

———— et al., 1915. Guidebook of the western United States; Part D. The Shasta route and coast line. U.S. Geol. Survey, Bull. 614, 142p.

Dingler, C.M., and Breckenridge, R.M., 1982. Glacial reconnaissance of the Selway-Bitterroot Wilderness Area. *In:* Bonnichsen, B., and Breckenridge, R.M., eds., Cenozoic geology of Idaho. Idaho Bur. of Mines and Geology, Bull. 26, pp. 645–652.

Donnelly-Nolan, J.M., 1992. Medicine Lake volcano and Lava Beds National Monument. California Geology, vol. 45, no. 5, pp. 145–153.

Douglas, R.J.W., ed., 1972. Geology and economic minerals of Canada. Geological Survey of Canada, Economic Geology Report no. 1, 838p.

Doukas, M.P., and Swanson, D.A., 1987. Mount St. Helens, Washington, with emphasis on 1980–85 eruptive activity as viewed from Windy Ridge. *In:* Hill, M.L., ed., Centennial Field Guide, Geol. Soc. America, Cordilleran Sect., pp. 333–338.

Driedger, C.L., and Kennard, P.M., 1984. Ice volumes on Cascade volcanoes: Mount Rainier, Mount Hood, Three Sisters, and Mount Shasta. U.S. Geol. Survey, Open-file Report 84–581, 42p.

Duncan, R.A., and Kulm, L.D., 1989. Plate tectonic evolution of the Cascades arc-subduction complex. *In:* Winterer, E.L., Hussong, D.M., and Decker, R.W., eds., Geology of North America, vol. N: The eastern Pacific Ocean and Hawaii, Geol. Soc. America, pp. 413–438.

Dunne, G.C., and McDougall, K.A., eds., 1993. Mesozoic paleogeography of the western United States—II. Soc. Economic Paleontologists and Mineralogists, Pacific Sect., 494p.

Easterbrook D.J., 1979. The last glaciation of northwest Washington. *In:* Armentrout, J.M., Cole, M.R., and TerBest, H., eds., Cenozoic paleogeography of the western United States. Soc. Econ. Paleontologists and Mineralogists, Pacific Coast Paleogeography Symposium 3, pp. 177–189.

———— 1986. Stratigraphy and chronology of the Quaternary deposits of the Puget Lowland and Olympic Mountains of Washington and the Cascade Mountains of Washington and Oregon. *In:* Sibrava, V., Bowen, D.Q., and Richmond, G.M., eds., Quaternary glaciations in the northern hemisphere. Quaternary Sci. Reviews 5, pp. 145–158.

———— 1994. Chronology of pre-late Wisconsin Pleistocene sediments in the Puget lowland, Washington. *In:* Lasmanis, R., and Cheney, E.S., convs., Regional geology of Washington state. Wash. Div. Geology and Earth Resources, Bull. 80, pp. 191–206.

Eaton, G.P., 1979. Regional geophysics, Cenozoic tectonics, and geologic resources of the Basin and Range province and adjoining regions. *In:* Newman, G.W., and Goode, H.D., eds., Basin and Range Symposium and Great Basin field conference. Rocky Mountain Assoc. of Geologists and Utah Geol. Assoc., pp. 11–39.

Eisbacher, G.H., 1977. Vancouver geology; a short guide. Geological Assoc. of Canada, 51p.

Ekren, E.B., et al., 1982. Cenozoic stratigraphy of western Owyhee County, Idaho. *In:* Bonnichsen, B., and Breckenridge, R.M., eds., Cenozoic geology of Idaho. Idaho Bur. of Mines and Geology, Bull. 26, pp. 215–235.

Engebretson, D.C., Cox, A., and Gordon, R.C., 1985. Relative motions between oceanic and continental plates in the Pacific basin. Geol. Soc. America, Special Paper 206, 59p.

England, T.D.J., 1991. Late Cretaceous to Paleogene structural and stratigraphic evolution of Georgia Basin, southwestern British Columbia; implications for hydrocarbon potential. Washington Geology, vol. 19, no. 4, pp. 10–12.

Ettlinger, A.D., and Ray, G.E., 1988. Gold-enriched skarn deposits of British Columbia. B.C. Ministry of Energy, Mines, and Petroleum Res., Geological Fieldwork, 1987, Paper 1988-1, pp. 263–279.

Evans, J.E., and Johnson, S.Y., 1989. Paleogene strike-slip basins of central Washington: Swauk Formation and Chumstick Formation. *In:* Joseph, N.L., et al., eds., Geologic guidebook for Washington and adjacent areas. Washington Div. Geology and Earth Resources, Info. Circular 86, pp. 214–237.

Evans, J.R., 1963. Geology of some lava tubes, Shasta County. Calif. Division of Mines and Geology, Mineral Information service, vol. 16, no. 3, pp. 1–63.

Evenson, E.B., Cotter, J.F.P., and Clinch, J.M., 1982. Glaciation of the Pioneer Mountains: a proposed model for Idaho. *In:* Bonnichsen, B., and Breckenridge, R.M., eds., Cenozoic geology of Idaho. Idaho Bur. of Mines and Geology, Bull. 26, pp. 653–665.

Fecht, K.R., Reidel, S.P., and Tallman, A.M., 1987. Paleodrainage of the Columbia River system on the Columbia Plateau of Washington State—a summary. *In:* Schuster, J.E., ed., Selected papers on the geology of Washington. Wash. Division of Geology and Earth Resources, Bull. 77, pp. 219–248.

Ferrero, T., 1990. The Liberty gold mining district. California Geology, vol. 43, no. 6, pp. 123–133.

Field, M.E., et al., eds., 1980. Quaternary depositional environments of the Pacific Coast. Soc. Econ. Paleontologists and Mineralogists, Pacific Sect., Pacific Coast Paleogeography Symp.4, 355p.

Fields, R.W., et al., 1985. Cenozoic rocks of the intermontane basins of western Montana and eastern Idaho: a summary. *In:* Flores, R.M., and Kaplan, S.S., eds., Cenozoic paleogeography of the west-central United States. Soc. Economic Paleontologists and Mineralogists, Rocky Mtn. Sect., pp. 1–36.

Fiero, B., 1986. Geology of the Great Basin. Reno, Univ. Nevada Press, 197p.

Fiesinger, W.D., Perkins, W.D., and Puchy, B.J., 1982. Mineralogy and petrology of Tertiary-Quaternary volcanic rocks in Caribou County, Idaho. *In:* Bonnichsen, B., and Breckenridge, R.M., eds., Cenozoic geology of Idaho. Idaho Bur. of Mines and Geology, Bull. 26, pp. 465–488.

Files, J.T., 1970. Structure of the Shuswap complex in the Jordan River area, northwest of Revelstoke, British Columbia. *In:* Wheeler, J.O., ed., Structure of the southern Canadian Cordillera. Geol. Assoc. of Canada, Special Paper no. 6, pp. 87–98.

Fisher, F.S., and Johnson, K.M., 1995. Challis volcanic terrane. *In:* Fisher, F.S., and Johnson, K.M., eds., Geology and mineral resource assessment of the Challis 1° × 2° Quadrangle, Idaho. U.S. Geol. Survey, Prof. Paper 1525, pp. 41–47.

Fisher, R.V., and Rensberger, J.M., 1972. Physical stratigraphy of the John Day Formation, central Oregon. University of Calif., Publ. Geol. Sci., vol. 101, 45p.

Fitzgerald, J.F., 1982. Geology and basalt stratigraphy of the Weiser embayment, west central Idaho. *In:* Bonnichsen, B., and Breckenridge, R.M., eds., Cenozoic geology of Idaho. Idaho Bur. of Mines and Geology, Bull. 26, pp. 103–128.

Flores, R.M., and Kaplan, S.S., eds., 1985. Cenozoic paleogeography of the west-central United States. Soc. Economic Paleontologists and Mineralogists, 460p.

Flower, W.H., and Lydekker, R., 1811. An introduction to the study of mammals living and extinct. London, Black Publ., 763p.

Foley, D., and Street, L., 1988. Hydrothermal systems of the Wood River area, Idaho. *In:* Link, P.K., and Hackett, W.R., eds., Guidebook to the geology of central and southern Idaho. Idaho Bureau of Mines and Geology, Bull. 27, pp. 109–126.

Follo, M.F., 1994. Sedimentology and stratigraphy of the Martin Bridge Limestone and Hurwal Formation (upper Triassic to lower Jurassic) from the Wallowa terrane, Oregon. *In:* Vallier,

T.L., and Brooks, H.C., eds., Geology of the Blue Mountains region of Oregon, Idaho, and Washington; stratigraphy, physiography, and mineral resouces of the Blue Mountains region. U.S. Geol. Survey, Prof. Paper 1439, pp. 1–27.

Fox, K.F., 1983. Melanges and their bearing on late Mesozoic and Tertiary subduction and interplate translation at the west edge of the North American plate. U.S. Geol. Survey, Prof. Paper 1198, 40p.

———— and Rinehart, C.D., 1988. Okanogan gneiss dome; a metamorphic core complex in north-central Washington. Washington Geologic Newsletter, vol. 16, no. 1, pp. 3–12.

———— and Wilson, J.R., 1989. Kettle gneiss dome: a metamorphic core complex in north-central Washington. *In:* Joseph, N.L., et al., eds., Geologic guidebook for Washington and adjacent areas. Wash. Div. Geology and Earth Resources, Info. Circular 86, pp. 201–214.

———— Rinehart, C.D., and Engels, J.C., 1977. Plutonism and orogeny in north-central Washington—timing and regional context. U.S. Geol. Survey, Prof. Paper 989, 27p.

———— et al., 1976. Age of emplacement of Okanogan gneiss dome, north-central Washington. Geol. Soc. America, Bull., vol. 87, pp. 1217–1224.

Frizzell, V.A. et al., 1987. Late Mesozoic or early Tertiary melanges in the western Cascades of Washington. *In:* Schuster, J.E., ed., Selected papers on the geology of Washington. Wash. Div. Geology and Earth Resources, Bull. 77, pp. 129–148.

Fulton, R.J., 1986. Quaternary stratigraphy of Canada. *In:* Sibrava, V., Bowen, D.Q., and Richmond, G.M., eds., Quaternary glaciations in the northern hemisphere. New York, Pergamon Press, pp. 207–209.

———— ed., 1989. Quaternary geology of Canada and Greenland. Geol. Soc. America., Geology of North America, vol. K-1; *also* Geological Survey of Canada, Geology of Canada no. 1, 839p.

Fyles, J.T., 1970. Geologic setting of the lead-zinc deposits in the Kootenay Lake and Salmo areas of British Columbia. *In:* Weisenborn, A.E., ed., Lead-zinc deposits in the Kootenay Arc, northeastern Washington and adjacent British Columbia. Wash. Division of Mines and Geology, Bull. 61, pp. 41–53.

Gabrielse, H., and Brookfield, A.J., 1988. Correlation chart for the Canadian Cordillera (sheet 2). *In:* Gabrielse, H., and Yorath, C.J., eds., 1991. The Geology of the Cordilleran orogen in Canada. Geol. Soc. America, Geology of North America, vol. G-2, Chart.

———— and Yorath, C.J., eds., 1991. Geology of the Cordilleran orogen in Canada. Geol. Soc. America, Geology of North America, vol. G-2, 844p.

Garver, J.I., 1992. Provenance of Albian-Cenomanian rocks of the Methow and Tyaughton basins, southern British Columbia: a mid-Cretaceous link between North America and Insular terrane. Canadian Jour. Earth Science, vol. 29, pp. 1274–1295.

Gay, T.E., 1966. Economic mineral deposits of the Cascade Range, Modoc Plateau, and Great Basin region of northeastern California. *In:* Bailey, E.H., ed., Geology of northern California. Calif. Div. of Mines and Geology, Bull. 190, pp. 97–104.

Gaylord, D.R., Lundquist, J.H., and Webster, G.D., 1989. Stratigraphy and sedimentology of the Sweetwater Creek interbed, Lewiston basin, Idaho and Washington. *In:* Reidel, S.P., and Hooper, P.R., eds., Volcanism and tectonism in the Columbia River flood-basalt province, Geol. Soc. America, Special Paper 239, pp. 199–208.

Gehrels, G.E., 1992. Penrose conference: tectonic evolution of the Coast Mountains orogen. Geoscience Canada, vol. 19, no. 4, pp. 170–171.

Geology and economic minerals of Canada. 1947. 3rd ed., Ottawa Geologic Survey, Economic Geology Series, no. 1, 357p.

Gidley, J.W., 1930. A new Pliocene horse from Idaho. Jour. Mammalogy, vol. 11, no. 3, pp. 300–303.

Goldstrand, P.M., 1994. The Mesozoic geologic evolution of the northern Wallowa terrane, northeastern Oregon and western Idaho. *In:* Vallier, T.L., and Brooks, H.C., eds., Geology of the Blue Mountains region of Oregon, Idaho, and Washington; stratigraphy, physiography, and mineral resources of the Blue Mountains region. U.S. Geol. Survey, Prof. Paper 1439, pp. 29–53.

Graf, W.L., ed., 1987. Geomorphic systems of North America. Geol. Soc. America, Centennial Special Volume 2, 643p.

Grant, A.R., 1969. Chemical and physical controls for base metal deposition in the Cascade Range of Washington. Wash. Div. Mines and Geology, Bull. 58, 107p.

Greeley, R., 1982. The style of basaltic volcanism in the eastern Snake River Plain, Idaho. *In:* Bonnichsen, B., and Breckenridge, R.M., eds., Cenozoic geology of Idaho. Idaho Bur. Mines and Geology, Bull. 26, pp. 407–421.

——— and King, J.S., 1975. Geologic field guide to the Quaternary volcanics of the south-central Snake River Plain, Idaho. Field guide, Geol. Soc. America, Rocky Mtn. Sect., 28th Annual Mtg., Boise, 47p.

Green, N.L., 1990. Late Cenozoic volcanism in the Mount Garibaldi and Garibaldi Lake volcanic fields, Garibaldi volcanic belt, southwestern British Columbia. Geoscience Canada, vol. 7, no. 3, pp. 171–175.

Gresens, R.L., 1987. Early Cenozoic geology of central Washington state—I. Summary of sedimentary, igneous, and tectonic events. *In:* Schuster, J.E., ed., Selected papers on the geology of Washington. Wash. Div. Geology and Earth Resources, Bull. 77, pp. 169–177.

——— 1990. Late Cenozoic volcanism in the Mount Garibaldi Lake volcanic fields, Garibaldi volcanic belt, southwestern British Columbia. Geoscience Canada, vol. 17, no. 3, pp. 171–175.

——— et al., 1977. Tertiary stratigraphy of the central Cascades Mountains, Washington state (Field trip no. 3). *In:* Brown, E.H., and Ellis, R.C., eds., Geological excursions in the Pacific Northwest. Geol. Soc. America, Annual Mtg., Seattle, pp. 84–126.

Grimaldi, D., 1993. Forever in amber. Natural history, vol. 102, no. 6, pp. 59–61.

Gusey, D., and Brown, E.H., 1987. The Fidalgo ophiolite, Washington. *In:* Hill, M.L., ed., Centennial Field Guide, Geol. Soc. America, Cordilleran Sect., pp. 389–392.

Gustafson, E.P., 1978. The vertebrate faunas of the Pliocene Ringold Formation, south-central Washington. Univ. of Oregon, Mus. Natural Hist., Bull. 23, 62p.

Hacker, B.R., and Ernst, W.G., 1993. Jurassic orogeny in the Klamath Mountains: a geochronological analysis. *In:* Dunne, G.C., and McDougall, K.A., eds., Mesozoic paleogeography of the western United States—II. Soc. Economic Paleontologists and Mineralogists, Pacific Sect., pp. 37–60.

Hackett, W.R., and Morgan, L.A., 1988. Explosive basaltic and rhyolitic volcanism of the eastern Snake River Plain, Idaho. *In:*

Link, P.K., and Hackett, W.R., eds., 1988. Guidebook to the geology of central and southern Idaho. Idaho Geol. Survey, Bull. 27, pp. 283–301.

Haggart, J.W., 1989. New and revised ammonites from the upper Cretaceous Nanaimo Group of British Columbia and Washington state. Geol. Survey Canada, Bull. 396, pp. 181–221.

——— 1993. Latest Jurassic and Cretaceous paleogeography of the northern Insular belt, British Columbia. *In:* Dunne, G.C., and McDougall, K.A., eds., Mesozoic paleogeography of the western United States—II Soc. Economic Paleontologists and Mineralogists, Pacific Sect., pp. 463–476.

Hamilton, W., 1965. Geology and petrogenesis of the Island Park caldera of rhyolite and basalt, eastern Idaho. U.S. Geol. Survey, Prof. Paper 504-C, pp. C1–C37.

Hammond, P.E., 1983. Volcanic formations in the upper Western Cascade Group along the Klamath River near Copco Lake. California. California Geology, vol. 36, no. 5, pp. 99–109.

——— 1989. Guide to the geology of the Cascade Range; Portland, Oregon, to Seattle, Washington. American Geophysical Union, Field Trip Guidebook T306, 215p.

——— et al., 1977. Volcanic stratigraphy and structure of the southern Cascade Range, Washington (Field trip no.4). *In:* Brown, E.H., and Ellis, R.C., eds., Geological excursions in the Pacific Northwest. Geol. Soc. America, Annual Mtg., Seattle, pp. 127–169.

Harbaugh, J.W., 1975. Northern California, field guide. Dubuque, Iowa, Kendall-Hunt Pub. Co., 123p.

Harms, T.A., and Price, R.A., 1992. The Newport fault: Eocene listric normal faulting, mylonitization, and crustal extension in northeast Washington and northwest Idaho. Geol. Soc. America, Bull., vol. 104, pp. 745–761.

Harper, G.D., 1984. Middle to late Jurassic tectonic evolution of the Klamath Mountains, California-Oregon. Tectonics, vol. 3, no. 7, pp. 759–772.

——— Grady, K., and Wakabayashi, J., 1990. A structural study of a metamorphic sole beneath the Josephine ophiolite, western Klamath terrane, California-Oregon. *In:* Harwood, D.S., and Miller, M.M., eds., Paleozoic and early Mesozoic paleogeographic relations; Sierra Nevada, Klamath Mountains, and related terranes. Geol. Soc. America, Special Paper 255, pp. 379–396.

Harrington, C.R., 1975. Pleistocene muskoxen (*Symbos*) from Alberta and British Columbia. Canadian Jour. Earth Science, vol. 12, pp. 903–919.

——— Tipper, H.W., and Mott, R.J., 1974. Mammoth from Babine Lake, British Columbia. Canadian Jour. Earth Science, vol. 11, no. 2, pp. 285–303.

Harris, A.G., 1990. Geology of national parks. Dubuque, Kendall-Hunt, 652p.

Harris, S.L., 1988. Fire mountains of the west; the Cascade and Mono Lake volcanoes. Missoula, Mountain Press, 378p.

Harwood, D.S., and Miller, M.M., eds., 1990. Paleozoic and early Mesozoic paleogeographic relations; Sierra Nevada, Klamath Mountains, and related terranes. Geol. Soc. America, Special Paper 255, 422p.

Haugerud, R.A., 1989. Geology of the metamorphic core of the North Cascades. *In:* Joseph, N.L., et al., eds., Geologic guidebook for Washington and adjacent areas. Wash. Div. Geology and Earth Resources, Info. Circular 86, pp. 119–136.

Heller, P.L., Tabor, R.W., and Suczek, C.A., 1987. Paleogeographic evolution of the United States Pacific Northwest during Paleogene time. Canadian Jour. Earth Science, vol. 24, no. 8, pp. 1652–1667.

Hickson, C.J., 1990. The May 18, 1980, eruption of Mount St. Helens, Washington state: A synopsis of events and review of Phase I from an eyewitness perspective. Geoscience Canada, vol. 17, no. 3, pp. 127–130.

Hicock, S.R., Hobson, K., and Armstrong, J.E., 1982. Late Pleistocene proboscideans and early Fraser glacial sedimentation in eastern Fraser Lowland, British Columbia, Canadian Jour. Earth Science, vol. 19, pp. 899–906.

Hill, M.L., ed., 1987. Centennial field guide, vol. 1. Geol. Soc. America, Cordilleran Sect., 490p.

Hinds, N.E.A., 1952. Evolution of the California landscape, Calif. Dept. Natural Resources, Bull. 158, 240p.

Hintze, L.F., coord., 1985. Correlation of stratigraphic units in North America (COSUNA) project: Great Basin Region. American Assoc. of Petroleum Geologists, Chart.

Holder, R.W., Gaylord, D.R., and Golder, G.A., 1989. Plutonism, volcanism, and sedimentation associated with core complex and graben development in the central Okanogan highlands, Washington. *In:* Joseph, N.L., et al., eds., Geologic guidebook for Washington and adjacent areas. Wash. Div. Geology and Earth Resources, Info. Circular 86, pp. 189–200.

Hooper, P.R., and Conrey, R.M., 1989. A model for the tectonic setting of the Columbia River basalt eruptions. *In:* Reidel, S.P., and Hooper, eds., Volcanism and tectonism in the Columbia River flood-basalt province. Geol. Soc. America, Special Paper 239, pp. 239–306.

———— and Swanson, D.A., 1987. Evolution of the eastern part of the Columbia Plateau. *In:* Schuster, J.E., ed., Selected papers on the geology of Washington. Wash. Division of Geology and Earth Resources, Bull. 77, pp. 197–217.

———— and Swanson, D.A., 1990. The Columbia River basalt group and associated volcanic rocks of the Blue Mountains province. *In:* Walker, G.W., ed., Geology of the Blue Mountains region of Oregon, Idaho, and Washington: Cenozoic geology of the Blue Mountains region. U.S. Geol. Survey, Prof. Paper 1437, pp. 63–99.

Hotz, P.E., 1971. Geology of lode gold districts in the Klamath Mountains, California and Oregon. U.S. Geol. Survey, Bull. 1290, 91p.

———— 1971. Plutonic rocks of the Klamath Mountains, California and Oregon. U.S. Geol. Survey, Prof. Paper 684-B, pp. B1–B20.

Houser, B.B., 1992. Quaternary stratigraphy of an area northeast of American Falls reservoir, eastern Snake River Plain, Idaho. *In:* Link, P.K., and Platt, L.B., eds., Regional geology of eastern Idaho and western Wyoming. Geol. Soc. America, Memoir 179, pp. 269–288.

Howard, K.A., Shervais, J.W., and McKee, E.H., 1982. Canyon-filling lava dams on the Boise River, Idaho, and their significance for evaluating downcutting during the the last two million years. *In:* Bonnichsen, B., and Breckenridge, R.M., eds., Cenozoic geology of Idaho. Idaho Bur. of Mines and Geology, Bull. 26, pp. 629–641.

Howell, D.G., and McDougall, K.A., eds., 1978. Mesozoic paleogeography of the western United States. Soc. Economic Paleontologists and Mineralogists, 573p.

———— Alpha, T.R., and Joyce, J.M., 1980. Plate tectonics, physiography, and sedimentation of the northeast Pacific region. *In:* Field, M.E., et al., eds., Quaternary depositional environments of the Pacific Coast. Soc. Econ. Paleontologists and Mineralogists, Pacific Coast Paleogeography, Symposium 4, pp. 43–53.

———— ed., 1985. Tectonotstratigraphic terranes of the circum-Pacific region. Circum-Pacific Council for Energy and Mineral Resources, Earth Science Series, no. 1, 581p.

Hoy, T., 1989. Tectonics and mineralization in southeastern British Columbia: structure, stratigraphy, and mineral deposits of the Rossland Group, Nelson area. *In:* Joseph, N.L., et al., eds., Geologic guidebook for Washington and adjacent areas. Wash. Div. Geology and Earth Resources, Info. Circular 86, pp. 63–68.

Hughes, S.S., and Thackery, G.D., eds., 1999. Guidebook to the geology of eastern Idaho. Pocatello, Idaho Museum of Natural History, 342p.

———— et al., 1999. Mafic volcanism and environmental geology of the eastern Snake River plain, Idaho. *In:* Hughes, S.S., and Thackery, G.D., eds., Guidebook to the geology of eastern Idaho. Pocatello, Idaho Museum of Natural History, pp. 143–168.

Hunt, C.B., 1979. The Great Basin, an overview and hypotheses of its origin. *In:* Newman, G.W., and Goode, H.D., eds., Basin and Range symposium and Great Basin field conference, Rocky Mountain Assoc. of Geologists and Utah Geol. Assoc., pp. 1–9.

Hunting, M.T., 1956. Inventory of Washington minerals. Part II: Metallic minerals. Wash. Div. of Mines and Geology, Bull. 37, 428p.

Hutchinson, R.W., and Albers, J.P., 1992. Metallogenic evolution of the Cordilleran region of the western United States. *In:* Burchfiel, B.C., Lipman, P.W., and Zoback, M.L., eds., The Cordilleran orogen: conterminous U.S. Geol. Soc. America, The Geology of North America, vol. G-3, pp. 629–652.

Hyndman, D.W., 1980. Bitterroot dome-Sapphire tectonic block, an example of a plutonic-core gneiss-dome complex with its detached suprastructure: *In:* Crittenden, M.D., ed., Cordilleran metamorphic core complexes. Geol. Soc. America, Memoir 153, pp. 427–443.

Idaho's mineral industry; the first hundred years. Idaho Bur. of Mines and Geology, Bull. 18, 71p.

Imlay, R.W., 1980. Jurassic paleobiogeography of the conterminous United States in its continental setting. U.S. Geol. Survey, Prof. Paper 1062, 134p.

Irwin, W.P., 1966. Geology of the Klamath Mountains Province. *In:* Bailey, E.H., ed., Geology of northern California. Calif. Div. Mines and Geology, Bull. 190, pp. 19–38.

Jackson, L.E., et al., 1985. Slope hazards in the southern coast mountains of British Columbia. *In:* Templeman-Kluit, D.J., ed., Field guides to geology and mineral deposits in the southern Canadian Cordillera. Geological Survey Canada, Vancouver, B.C., pp. 4-1 to 4–34.

Janes, J.R., 1978. The great Canadian outback. Douglas & McIntyre, Vancouver, 144p.

Jayko, A.S., 1990. Stratigraphy and tectonics of Paleozoic arc-related rocks of the northernmost Sierra Nevada, California; the eastern Klamath and northern Sierra terranes. *In:* Harwood, D.S., and Miller, M.M., eds., Paleozoic and early Mesozoic paleogeographic relations; Sierra Nevada, Klamath Mountains, and related terranes. Geol. Soc. America, Special Paper 255, pp. 307–323.

————— and Blake, M.C., 1993. Northward displacements of forearc slivers in the Coast Ranges of California and southwest Oregon during the late Mesozoic and early Cenozoic. *In:* Dunne, G.C., and McDougall, K.A., eds., Mesozoic paleogeography of the western United States—II. Soc. Economic Paleontologists and Mineralogists, Pacific Sect., pp. 19–36.

Jenks, M.D., and Bonnichsen, B., 1989. Subaqueous basalt eruptions into Pliocene Lake Idaho, Snake River Plain, Idaho. *In:* Chamberlain, V.E., Breckenridge, R.M., and Bonnichsen, B., eds., Guidebook to the geology of northern and western Idaho and surrounding area. Idaho Geol. Survey, Bull. 28, pp. 17–34.

Johnson, K.M., et al., 1988. Cretaceous and Tertiary intrusive rocks of south-central Idaho. *In:* Link, P.K., and Hackett, W.R., eds., Guidebook to the geology of central and southern Idaho. Idaho Bur. of Mines and Geology, Bull. 27, pp. 55–86.

Johnson, L., and Monahan, D., 1979. General bathymetric chart of the oceans (GEBO). Ottawa, Canada. Hydrographic Service, 5th ed, Map no. 5–03.

Johnson, S.Y., 1991. Sedimentation and tectonic setting of the Chuckanut Formation, northwest Washington. Washington Geology, vol. 19, no. 4, pp. 12–13.

————— Potter, C.J., and Armentrout, J.M., 1994. Origin and evolution of the Seattle fault and Seattle basin, Washington. Geology, vol. 22, pp. 71–74.

Johnston, D.A., and Donnelly-Nolan, J., eds., 1981. Guides to some volcanic terranes in Washington, Idaho, Oregon, and northern California. U.S. Geol. Survey, Circular 838, 189p.

Jones, D.L., Silberling, N.J., and Hillhouse, J.W., 1977. Wrangellia—a displaced terrane in northwestern North America. Canadian Jour. Earth Science, vol. 14, pp. 2565–2577.

————— Silberling, N.J., and Hillhouse, J.W., 1978. Microplate tectonics of Alaska—significance for the Mesozoic history of the Pacific coast of North America. *In:* Howell, D.G., and McDougall, K.A., eds., Mesozoic paleogeography of the western United States—II. Soc. Economic Paleontologists and Mineralogists, Pacific Sect., pp. 71–84.

Joseph, N.L., 1987. Republic fossil locality open to public. Washington Geologic Newsletter, vol. 15, no. 4, p. 17.

————— 1991. Washington's mineral industry—1990. Washington Geology, vol. 19, no. 1, pp. 3–24.

————— et al., eds., 1989. Geologic guidebook for Washington and adjacent areas. Wash. Div. Geology and Earth Resources, Info. Circular 86, 369p.

Journeay, J.M., 1990. A progress report on the structural and tectonic framework of the southern Coast belt, British Columbia. *In:* Current Research, Part E., Geol. Survey Canada, Paper 90-1E, pp. 183–195.

Kaler, K.L., 1988. The Blue Lake rhinoceros. Wash. Geol. Newsletter, vol. 16, no. 4, pp. 3–8.

Kane, P., 1982. Pleistocene glaciation, Lassen Volcanic National Park. California Geology, vol. 35, no. 5, pp. 95–105.

Kaysing, B., 1984. Great hot springs of the west. Santa Barbara, Calif., Capra Press, 213p.

Keller, G., et al., 1984. Paleoclimatic evidence for Cenozoic migration of Alaskan terranes. Tectonics, vol. 3, no. 4, pp. 473–495.

Kellogg, K.S., 1992. Cretaceous thrusting and Neogene block rotation in the northern Portneuf Range region, southeastern Idaho. *In:* Link, P.K., Kuntz, M.A., and Platt, L.B., eds., Regional geology of eastern Idaho and western Wyoming. U.S. Geol. Survey, Memoir 179, pp. 95–113.

Kelsey, H.M., 1990. Late Quaternary deformation of marine terraces on the Cascadia subduction zone near Cape Blanco, Oregon. Tectonics, vol. 9, no. 5, pp. 983–1014.

Ketner, K.B., 1977. Late Paleozoic orogeny and sedimentation, southern California, Nevada, Idaho, and Montana. *In:* Steward, J.H., Stevens, C.H., and Fritsche, A.E., eds., Paleozoic paleogeography of the western United States, Soc. Economic Paleontologists and Mineralogists, pp. 363–369.

Kiilsgaard, T.H., Fisher, F.S., and Bennett, E.H., 1986. The trans-Challis fault system and associated precious-metal deposits. Economic Geology, vol. 81, pp. 721–724.

Kimmel, P.G., 1984. Stratigraphy, age, and tectonic setting of the Miocene-Pliocene lacustrine sediments of the western Snake River Plain, Oregon and Idaho. *In:* Bonnichsen, B., and Breckenridge, R.M., eds., Cenozoic geology of Idaho. Idaho Bur. of Mines and Geology, Bull. 26, pp. 559–578.

King, J.S., 1982. Selected volcanic features of the south-central Snake River Plain, Idaho. *In:* Bonnichsen, B., and Breckenridge, R.M., eds., Cenozoic geology of Idaho. Idaho Bur. of Mines and Geology, Bull. 26, pp. 439–451.

Kleinspehn, K.L., 1985. Cretaceous sedimentation and tectonics, Tyaughton-Methow basin, southwestern British Columbia. Canadian Jour. Earth Science, vol. 22, pp. 154–174.

Knowlton, F.H., 1926. Flora of the Latah Formation of Spokane, Washington, and Coeur d'Alene, Idaho. U.S. Geol. Survey, Prof. Paper 140-A, 55p.

Komar, P.D., 1992. Ocean processes and hazards along the Oregon coast. Oregon Geology, vol. 54, no. 1, pp. 3–19.

Koschmann, A.H., and Bergendahl, M.H., 1968. Principal gold-producing districts of the United States. U.S. Geol. Survey, Prof. Paper 610, 283p.

Kulm, L.D., and Fowler, G.A., 1974. Oregon continental margin structure and stratigraphy: a test of the imbricate thrust model. *In:* Burk, C.A., and Drake, M.J., eds., The geology of continental margins. New York, Springer-Verlag, pp. 261–283.

Kuntz, M.A., 1992. A model-based perspective of basaltic volcanism, eastern Snake River Plain, Idaho. *In:* Link, P.K., Kuntz, M.A., and Platt, L.B., eds., Regional geology of eastern Idaho and western Wyoming. Geol. Soc. America, Memoir 179, pp. 289–304.

————— Covington, H.R., and Schorr, L.J., 1992. An overview of basaltic volcanism of the eastern Snake River Plain, Idaho. *In:* Link, P.K., Kuntz, M.A., and Platt, L.B., eds., Regional geology of eastern Idaho and western Wyoming. Geol. Soc. America, Memoir 179, pp. 227–267.

————— et al., 1982. The Great Rift and the evolution of the Craters of the Moon lava field, Idaho. *In:* Bonnichsen, B., and Breckenridge, R.M., eds., Cenozoic geology of Idaho. Idaho Bur. of Mines and Geology, Bull. 26, pp. 423–437.

LaMotte, R.S., 1936. An upper Oligocene florule from Vancouver Island. *In:* Middle Cenozoic floras of western North America. Carnegie Inst. Wash., Publ. 455, pp. 51–56.

————— 1936. The upper Cedarville flora of northwestern Nevada and adjacent California. *In:* Middle Cenozoic floras of western North America. Carnegie Inst. Wash., Publ. 455, pp. 59–142.

Lasmanis, R., and Cheney, E.S., convs., 1994. Regional geology of Washington state. Wash. Div. Geology and Earth Resources, Bull. 80, 227p.

Lavine, A., 1994. Geology of Prisoners Rock and The Peninsula. California Geology, vol. 47, no. 4, pp. 95–103.

Leclair, A.D., Parrish, R.R., and Archibald, D.A., 1993. Evidence for Cretaceous deformation in the Kootenay Arc based on U-Pb and Ar 40/Ar 39 dating, southeastern British Columbia. Geol. Survey Canada, Paper 93-1A, pp. 207–220.

Leeman, W.P., 1982. Development of the Snake River Plain-Yellowstone plateau province, Idaho and Wyoming: an overview and petrologic model. *In:* Bonnichsen, B., and Breckenridge, R.M., eds., Cenozoic geology of Idaho. Idaho Bur. of Mines and Geology, Bull. 26, pp. 155–177.

———— 1982. Geology of the Magic Reservoir area, Snake River Plain, Idaho. *In:* Bonnichsen, B., and Breckenridge, R.M., eds., Cenozoic geology of Idaho. Idaho Bur. of Mines and Geology, Bull. 26, pp. 369–376.

Lewis, G.C., and Fosberg, M.A., 1982. Distribution and character of loess and loess soils in southeastern Idaho. *In:* Bonnichsen, B., and Breckenridge, R.M., eds., Cenozoic geology of Idaho. Idaho Bur. of Mines and Geology, Bull. 26, pp. 705–716.

Lewis, R.S., et al., 1987. Lithologic and chemical characteristics of the central and southeastern part of the southern lobe of the Idaho batholith. *In:* Vallier, T.L., and Brooks, H.C., eds., Geology of the Blue Mountains region of Oregon, Idaho, and Washington; the Idaho batholith and its border zone. U.S. Geol. Survey, Prof. Paper 1436, pp. 171–196.

Lewis, S.E., 1993. Insects of the Klondike Mountain Formation, Republic, Washington. Washington Geology, vol. 20, no. 3, pp. 15–19.

Lindgren, W., and Bancroft, H., 1914. The ore deposits of northeast Washington. U.S. Geol. Survey, Bull. 550, 215p.

Lindsley-Griffin, N., et al., 1993. Post-accretion history and paleogeography of a Cretaceous overlap sequence in northern California. *In:* Dunne, G.C., and McDougall, K.A., eds., Mesozoic paleogeography of the western United States—II. Soc. Econ. Paleontologists and Mineralogists, Pacific Sect., pp. 99–112.

Link, P.K., and Phoenix, E.C., 1996. Rocks, rails, and trails, 2d ed. Pocatello, Idaho Museum of Natural History, 193p.

———— Kuntz, M.A., and Platt, L.B., eds., 1992. Regional geology of eastern Idaho and western Wyoming. U.S. Geol. Survey, Memoir 179, 312p.

———— et al., 1988. Structural and stratigraphic transect of south-central Idaho: a field guide to the Lost River, White Knob, Pioneer, Boulder, and Smoky Mountains. *In:* Link, P.K., and Hackett, W.R., eds., Guidebook to the geology of central and southern Idaho. Idaho Geol. Survey, Bull. 27, pp. 5–42.

———— et al., 1999. Field guide to Pleistocene Lakes Thatcher and Bonneville and the Bonneville flood. *In:* Hughes, S.S., and Thackery, G.D., eds. Guidebook to the geology of eastern Idaho. Pocatello, Idaho Museum of Natural History, pp. 251–266.

Livingston, V.E., 1969. Geologic history and rocks and minerals of Washington. Wash. Division of Mines and Geology, Info. Circular No. 45, 42p.

Logan, R.L., and Schuster, R.L., 1991. Lakes divided: the origin of Lake Crescent and Lake Sutherland, Clallam County, Washington. Washington Geology vol. 19, no. 1, pp. 38–42.

Lund, E.H., 1972. Coastal landforms between Tillamook Bay and the Columbia River, Oregon. Ore Bin, vol. 34, no. 11, pp. 173–194.

———— 1972. Coastal landforms between Yachats and Newport, Oregon. Ore Bin, vol. 34, no. 5, pp. 73–91.

———— 1973. Landforms along the coast of southern Coos County, Oregon. Ore Bin, vol. 35, no. 12, pp. 189–210.

Mabey, D.R., 1982. Geophysics and tectonics of the Snake River Plain, Idaho. *In:* Bonnichsen, B., and Breckenridge, R.M., eds., Cenozoic geology of Idaho. Idaho Bur. of Mines and Geology, Bull. 26, pp. 139–153.

Mabey, M.A., et al., 1993. Earthquake hazard maps of the Portland Quadrangle, Multnomah and Washington Counties, Oregon, and Clark County, Washington. Oregon Dept. Geol. and Mineral Indus., GMS-79.

MacLeod, N.S., Walker, G.W., and McKee, E.H., 1976. Geothermal significance of eastward increase in age of upper Cenozoic rhyolitic domes in southeast Oregon. United Nations, Second Symposium on the development and use of geothermal resources. Proc., vol. 1, pp. 465–474.

Malde, H.E., and Cox, A., 1971. History of Snake River Canyon indicated by revised stratigraphy of Snake River Group near Hagerman and King Hill, Idaho. U.S. Geol. Survey, Prof. Paper 644-F, pp. F1–F21.

———— and Powers, H.A., 1962. Upper Cenozoic stratigraphy of western Snake River Plain, Idaho. Geol. Soc. America, Bull., vol. 73, pp. 1197–1220.

Maley, T., 1987. Exploring Idaho geology. Boise, Mineral Lands Publ., 232p.

Mammerickx, J., and Smith, S.M., 1982. General bathymetric chart of the oceans (GEBCO). Ottawa, Canada. Hydrographic Service, Map no. 5-07.

Mankinen, E.A., and Irwin, W.P., 1990. Review of paleomagnetic data from the Klamath Mountains, Blue Mountains, and Sierra Nevada; implications for paleogeographic reconstructions. *In:* Harwood, D.S., and Miller M.M., eds., Paleozoic and early Mesozoic paleogeographic relations; Sierra Nevada, Klamath Mountains, and related terranes. Geol. Soc. America, Special Paper 255, pp. 397–409.

Markman, H.C., 1952. Fossil mammals. Colorado, Denver Museum Natural History, Museum Pictorial no. 4, 62p.

Marsh, S.P., Kropschot, S.J., and Dickinson, R.G., 1984. Wilderness mineral potential; assessment of mineral-resource potential in U.S. Forest Service lands studied, 1964–1984. U.S. Geol. Survey, Prof. Paper 1300, 1183p.

Massey, N.W.D., 1986. Metchosin igneous complex, southern Vancouver Island: ophiolite stratigraphy developed in an emergent island setting. Geology, vol. 14, pp. 602–605.

Mathewes, R.W., 1989. The Queen Charlotte Islands refugium: a paleoecological perspective. *In:* Fulton, R.J., ed., Quaternary geology of Canada and Greenland. Geol. Soc. America, The geology of North America, vol. K-1. *Also:* Geological Survey of Canada, Geology of Canada no. 1, pp. 486–491.

———— and Clague, J.C., 1994. Detection of large prehistoric earthquakes in the Pacific Northwest by microfossil analysis. Science, vol. 264, pp. 688–691.

Mathews, W.H., 1951. The Table, a flat-topped volcano in southern British Columbia. American Jour. Science, vol. 249, pp. 830–841.

———— 1952. Mount Garibaldi, a supra-glacial Pleistocene volcano in Southwestern British Columbia. American Jour. Science, vol. 250, pp. 81–103.

———— 1975. Garibaldi geology; a popular guide. Geol. Assoc. Canada, Cordilleran Sect., 48p.

———— 1987. Garibaldi area, southwestern British Columbia; volcanoes versus glacier ice. *In:* Hill, M.L., ed., Centennial Field Guide. Geol. Soc. America, Cordilleran Sect., pp. 403–407.

Matthews, R.A., and Burnett, J.L., eds., 1969. Geologic guide to the Lassen Peak, Burney Falls, and Lake Shasta area, California. Geol. Soc. of Sacramento, Annual field trip guidebook, 1969, 193p.

McCurry, M., Hackett, W. R., and Hayden, K., 1999. Cedar Butte and cogenetic Quaternary rhyolitic domes of the eastern Snake River Plain. *In:* Hughes, S.S., and Thackray, G.D., eds., Guidebook to the geology of eastern Idaho. Pocatello, Idaho Museum of Natural History, pp. 169–179.

McDonald, G.A., 1966. Geology of the Cascade Range and Modoc Plateau. *In:* Bailey, E.H., ed., Geology of northern California, Calif. Div. Mines and Geology, Bull. 190, pp. 65–96.

———— 1968. Geology of the Modoc Plateau. Calif. Division of Mines and Geology, Mineral Information Service, vol. 21, no. 6, pp. 92–93.

McFaddan, M.D., Measures, E.A., and Isaacson, P.E., 1988. Early Paleozoic continental margin development, central Idaho. *In:* Link, P.K., and Hackett, W.R., eds., Guidebook to the geology of central and southern Idaho. Idaho Bur. of Mines and Geology, Bull. 27, pp. 129–152.

McGinitie, H.D., 1937. Flora of the Weaverville beds of Trinity County, California. *In:* Eocene flora of western America. Carnegie Inst. Washington, Contrib. to Paleontology 465, pp. 85–151.

McGroder, M.F., and Miller, R.B., 1989. Geology of the eastern North Cascades. *In:* Joseph, N.L., et al., eds. Geologic guidebook for Washington and adjacent areas. Wash. Div. Geology and Earth Resources, Info. Circular 86, pp. 97–118.

McInelly, G.W., and Kelsey, H.M., 1990. Late Quaternary tectonic deformation in the Cape Arago-Bandon region of coastal Oregon as deduced from wave-cut platforms. Jour. Geophys. Research, vol. 95, no. B5, pp. 6699–6713.

McIntyre, D.H., Ekren, E.B., and Hardyman, R.F., 1982. Stratigraphic and structural framework of the Challis volcanics in the eastern half of the Challis 1 × 2 Quadrangle, Idaho. *In:* Bonnichsen, B., and Breckenridge, R.M., eds., Cenozoic geology of Idaho. Idaho Bur. of Mines and Geology, Bull. 26, pp. 3–22.

McKee, Bates, 1972. Cascadia; the geologic evolution of the Pacific Northwest. New York, McGraw-Hill, 394p.

McLearn, F.H., 1972. Ammonoids of the lower Cretaceous sandstone member of the Haida Formation, Skidgate Inlet, Queen Charlotte Islands, western British Columbia. Canada Geol. Survey, Bull. 188, 78p.

McLeod, J.D., and Welhan, J.A., [n.d.]. Snake River Plain aquifer: Idaho's hidden gem. Idaho Geol. Survey, GeoNote 15, [p. 2].

McNutt, S., 1989. Medicine Lake Highland, September 1988 earthquake swarm, Siskiyou County. California Geology, vol. 42, no. 3, pp. 51–52.

McPhee, J., 1981. Basin and Range. New York, Farrar, Straus, and Giroux, 215p.

———— 1993. Assembling California. New York, Farrar, Straus, and Giroux, 303p.

McTaggart, K.C., 1970. Tectonic history of the northern Cascade Mountains. Geological Assoc. of Canada, Special Paper 6, pp. 137–148.

Mertzman, S.A., 1981. Pre-Holocene silicic volcanism on the northern and western margins of the Medicine Lake Highland, California. *In:* Johnston, D.A., and Donnelly-Nolan, J., eds., Guides to some volcanic terranes in Washington, Idaho, Oregon, and northern California. U.S. Geol. Survey, Circular 838, pp. 163–176.

Middleton, L.T., Porter, M.L., and Kimmel, P.G., 1985. Depositional settings of the Chalk Hills and Glenns Ferry formations west of Bruneau, Idaho. *In:* Flores, R.M., and Kaplan, S.S., eds., Cenozoic paleogeography of west-central United States. Soc. Economic Paleontologists and Mineralogists, Rocky Mountain Sect., pp. 37–53.

Miller, C.D., 1990. Volcanic hazards in the Pacific Northwest. Geoscience Canada, vol. 17, no. 3, pp. 183–179.

Miller, E.L., et al., 1992. Late Paleozoic paleogeographic and tectonic evolution of the western U.S. Cordillera. *In:* Burchfiel, B.C., Lipman, P.W., and Zoback, M.L., eds., The Cordilleran orogen: conterminous U.S. Geol. Soc. America, The Geology of North America, vol. G-3, pp. 57–106.

Miller, F.K., 1994. The Windermere Group and late Proterozoic tectonics in northeastern Washington and northern Idaho. *In:* Lasmanis, R., and Cheney, E.S., convs., Regional geology of Washington state. Wash. Div. Geology and Earth Resources, Bull. 80, pp. 1–19.

Miller, M.M., 1987. Dispersed remnants of a northeast Pacific fringing arc: upper Paleozoic terranes of Permian McCloud faunal affinity western U.S. Tectonics, vol. 6, no. 6, pp. 807–830.

———— and Harwood, D.S., 1990. Paleogeographic setting of upper Paleozoic rocks in the northern Sierra and eastern Klamath terranes, northern California. *In:* Harwood, D.S., and Miller, M.M., eds., Paleozoic and early Mesozoic paleogeographic relations: Sierra Nevada, Klamath Mountains, and related terranes. Geol. Soc. America, Special Paper 255, pp. 175–192.

———— and Wright, J.E., 1987. Paleogeographic implications of Permian Tethyan fossils from the Klamath Mountains, California. Geology, vol. 15, pp. 266–269.

Miller, P.J.B., 1975. Guidebook for the geology and scenery of the Snake River on the Idaho-Oregon border from Brownlee Dam to Hells Canyon Dam. Idaho Bureau of Mines and Geology, Info. Circular 28, 20p.

Miller, P.R., and Orr, W.N., 1986. The Scotts Mills Formation: mid-Tertiary geologic history and paleogeography of the central Western Cascade range, Oregon. Oregon Geology, vol. 48, no. 12, pp. 139–151.

———— and Orr, W.N., 1988. Mid-Tertiary transgressive rocky coast sedimentation: central Western Cascade range, Oregon. Jour. Sed. Pet., vol. 58, no. 6, pp. 959–968.

Miller, R.B., 1985. The ophiolitic Ingalls complex, north-central Cascade Mountains, Washington. Geol. Soc. America, Bull., vol. 96, pp. 27–42.

———— et al., 1993. Tectonostratigraphic evolution of Mesozoic rocks in the southern and central Washington Cascades. *In:* Dunne, G., and McDougall, K., eds., Mesozoic paleogeography of the Western United States—II. Soc. Economic Paleontologists and Mineralogists, Pacific Sect., pp. 81–98.

———— et al., 1994. Tectonostratigraphic framework of the northeastern Cascades. *In:* Lasmanis, R., and Cheney, E.S., convs., Regional geology of Washington state. Wash. Div. of Geology and Earth Resources. Bull. 80, pp. 73–92.

Milne, W.G., et al., 1978. Seismicity of western Canada. Canadian Jour. Earth Science, vol. 15, pp. 1170–1193.

Misch, P., 1977. Bedrock geology of the North Cascades (Field trip no.1). *In:* Brown, E.H., and Ellis, R.C., eds., Geological excursions in the Pacific Northwest. Geol. Soc. Amer., Annual Mtg., Seattle, pp. 1–62.

Mitchell, E., 1966. Faunal succession of extinct North Pacific marine mammals. Norsk Hvalfangst-Tidende, no. 3, pp. 47–60.

Monger, J.W.H., 1985. Structural evolution of the southwestern Intermontane Belt, Ashcroft and Hope map areas, British Columbia. *In:* Current Research, Part A. Geological Survey of Canada, Paper 85-1A, pp. 349–358.

———— 1991. Late Mesozoic to Recent evolution of the Georgia Strait-Puget Sound region, British Columbia and Washington. Washington Geology, vol. 19, no. 4, pp. 3–9.

———— 1993. Georgia Basin project—geology of Vancouver map area, British Columbia. *In:* Current Research, Part A. Geological Survey of Canada, Paper 93-1A, pp. 149–157.

———— Price, R.A., and Templeman-Kluit, D.J., 1982. Tectonic accretion and the origin of the two major metamorphic plutonic welts in the Canadian Cordillera. Geology, vol. 10, pp. 70–75.

———— et al., 1994. Jurassic-Cretaceous basins along the Canadian Coast Belt: their bearing on pre-mid-Cretaceous sinistral displacements. Geology, vol. 22, pp. 175–178.

Moon, M., 1977. Ogopogo; the Okanagan mystery. Vancouver, B.C., J.J. Douglas Publ., 195p.

Mooney, W.D., and Weaver, C.S., 1989. Regional crustal structure and tectonics of the Pacific coastal states; California, Oregon, and Washington. *In:* Pakiser, L.C., and Mooney, W.D., eds., Geophysical framework of the continental United States. Geol. Soc. America, Memoir 172, pp. 129–161.

Moore, G.W., 1994. Geologic catastrophes in the Pacific Northwest. Oregon Geology, vol. 56, no. 1, pp. 3–6.

Morgan, L.A., 1992. Stratigraphic relations and paleomagnetic and geochemical correlations of ignimbrites of the Heise volcanic field, eastern Snake River Plain, eastern Idaho and western Wyoming. *In:* Link, P.L., Kuntz, M.A., and Platt, L.B., eds., Regional geology of eastern Idaho and western Wyoming, Geol. Soc. America, Memoir 179, pp. 215–226.

Morris, E.M., and Wardlaw, B.R., 1986. Conodont ages for limestones of eastern Oregon and their implication for pre-Tertiary melange terranes. *In:* Vallier, T.L., and Brooks, H.C., eds., Geology of the Blue Mountains region of Oregon, Idaho, and Washington. U.S. Geol. Survey, Prof. Paper 1435, pp. 59–63.

Moye, F.J., 1987. Republic graben, Washington. *In:* Hill, M.L., ed., Centennial Field Guide. Geol. Soc. America, Cordilleran Sect., pp. 399–402.

———— et al., 1988. Regional geologic setting and volcanic stratigraphy of the Challis volcanic field, central Idaho. *In:* Link, P.K., and Hackett, W.R., eds., Guidebook to the geology of central and southern Idaho. Idaho Bur. of Mines and Geology, Bull. 27, pp. 87–97.

Muhs, D.R., et al., 1987. Pacific coast and mountain system. *In:* Graf, W.L., ed., Geomorphic systems of North America. Geol. Soc. America, Centennial Special Volume 2, pp. 517–581.

Murchey, B.L., and Blake, M.C., 1993. Evidence for subduction of a major ocean plate along the California margin during the middle to early late Jurassic. *In:* Dunne, G.C., and McDougall, K.A., eds., Mesozoic paleogeography of the western United States. Soc. Economic Paleontologists and Mineralogists, Pacific Sect., pp. 1–18.

Mustard, P.S., 1991. Stratigraphy and sedimentology of the Georgia Basin, British Columbia and Washington state. Washington Geology, vol. 19, no. 4, pp. 7–9.

Mustoe, G.E., 1993. Eocene bird tracks from the Chuckanut Formation, northwest Washington. Canadian Jour. Earth Science, vol. 30, pp. 1205–1208.

Newman, G.W., and Goode, H.D., eds., 1979. Basin and Range symposium and Great Basin field conference. Rocky Mountain Assoc. of Geologists and Utah Geol. Assoc., 662p.

Newton, C.R., 1986. Late Triassic bivalves of the Martin Bridge Limestone, Hells Canyon, Oregon: taphonomy, paleoecology, paleozoogeography. *In:* Vallier, T.L., and Brooks, H.C., eds., Geology of the Blue Mountains region of Oregon, Idaho, and Washington. U.S. Geol. Survey, Prof. Paper 1435, pp. 7–17.

Niem, A.R., and Snavely, P.D., 1991. Geology and preliminary hydrocarbon evaluation of the Tertiary Juan de Fuca basin, Olympic Peninsula, northwest Washington. Washington Geology, vol. 19, no. 4, pp. 27–34.

Nilsen, T.H., 1977. Paleogeography of Mississippian turbidites in south-central Idaho. *In:* Stewart, J.H., Stevens, C.H., and Fritsche, A.E., eds., Paleozoic paleogeography of the western United States. Soc. Economic Paleontologists and Mineralogists, Pacific Sect., Pacific Coast Paleogeography Symposium 1, pp. 275–299.

———— 1984. Stratigraphy, sedimentology, and tectonic framework of the upper Cretaceous Hornbrook Formation, Oregon and California. *In:* Nilsen, T.H., ed., Geology of the upper Cretaceous Hornbrook Formation, Oregon and California. Soc. Economic Paleontologists and Mineralogists, Pacific Sect., Field trip guidebook, vol. 42, pp. 51–88.

Norris, R.M., and Webb, R.W., 1990. Geology of California. New York, Wiley, 541p.

O'Connor, J.E., 1993. Hydrology, hydraulics, and geomorphology of the Bonneville flood. U.S. Geol. Survey, Special Paper 274, 83p.

———— and Baker, V.R., 1992. Magnitudes and implications of peak discharges from glacial Lake Missoula. Geol. Soc. America, Bull., vol. 104, pp. 267–279.

Oles, K.F., et al., eds., 1980. Geologic field trips in western Oregon and southwestern Washington. Oregon, Dept. Geol. and Mineral Indus., Bull. 101, 232p.

Oregon. Dept. of Geology and Mineral Industries, 1991. Cascade Range volcanoes are just dormant. Oregon Geology, vol. 53, no. 4, 88p.

Orr, E.L., and Orr, W.N., 1999. Geology of Oregon. 5th ed. Dubuque, Kendall-Hunt, 254p.

———— and Orr, W.N., 1999. Oregon fossils. Dubuque, Kendall-Hunt, 381p.

Orr, K.E., and Cheney, E.S., 1987. Kettle and Okanogan domes, northeastern Washington and southern British Columbia. *In:* Schuster, J.E., ed., Selected papers on the geology of Washington. Wash. Div. of Geology and Earth Resources, Bull. 77, pp. 55–71.

Orr, W.N., 1986. A Norian (late Triassic) ichthyosaur from the Martin Bridge Limestone, Wallowa Mountains, Oregon. U.S. Geological Survey, Prof. Paper 1435, pp. 41–47.

Osborn, H.F., 1929. The titanotheres of ancient Wyoming, Dakota, and Nebraska. U.S. Geol. Survey, Monograph 55, vol. 2.

Pacht, J.A., 1984. Petrologic evolution and paleogeography of the late Cretaceous Nanaimo basin, Washington and British Columbia: implications for Cretaceous tectonics. Geol. Soc. America, Bull., vol. 95, pp. 766–778.

———— 1987. Sedimentology of the upper Cretaceous Nanaimo Basin, British Columbia. *In:* Hill, M.L., ed., Centennial Field Guide, Geological Soc. America, Cordilleran Sect., pp. 419–423.

Palmer, S.P., 1991. Modified Mercalli intensity VI and greater earthquakes in Washington state, 1928–1990. Washington Geology, vol. 19, no. 2, pp. 3–7.

———— and Lingley, W.S., 1989. An assessment of the oil and gas potential of the Washington outer continental shelf. University of Washington, Washington Sea Grant Program, 83p.

Pardee, J.T., and Bryan, K., 1926. Geology of the Latah Formation in relation to the lavas of the Columbia plateau near Spokane, Washington. U.S. Geol. Survey, Prof. Paper 140-A, 16p.

Parrish, R.R., 1983. Cenozoic thermal evolution and tectonics of the Coast Mountains of British Columbia. 1. Fission track dating, apparent uplift rates, and patterns of uplift. Tectonics, vol. 2, no. 6, pp. 601–631.

Patty, E.N., 1921. The metal mines of Washington. Wash. Geol. Survey, Bull.23, 366p.

Pearce, S., Schlieder, G., and Evenson, E.B., 1988. Glacial deposits of the Big Wood River valley. *In:* Link, P.K., and Hackett, W.R., eds., Central and southern Idaho. Idaho Geol. Survey, Bull. 27, pp. 203–207.

Pease, R., 1965. Modoc County and its regional setting in geologic time. University of California, Berkeley, Univ. Calif. Publ. in Geography, vol. 17, 304p.

Peck, D.L., 1964. Geology of the central and northern parts of the Western Cascade Range in Oregon. U.S. Geol. Survey, Prof. Paper 449, 56p.

Pelto, M.S., 1992. Current behavior of glaciers on the North Cascades and effect on regional water supplies. Washington Geology, vol. 21, no. 2, pp. 3–10.

Pessagno, E.A., and Blome, C.D., 1986. Faunal affinities and tectonogenesis of Mesozoic rocks in the Blue Mountains province of eastern Oregon and western Idaho. *In:* Vallier, T.L., and Brooks, H.C., eds., Geology of the Blue Mountains region of Oregon, Idaho, and Washington. U.S. Geol. Survey, Prof. Paper 1435, pp. 65–78.

Pierce, K.L., and Morgan, L.A., 1992. The track of the Yellowstone hot spot: volcanism, faulting, and uplift. *In:* Link, P.K., Kuntz, M.A., and Platt, L.B., eds., Regional geology of eastern Idaho and western Wyoming. Geol. Soc. America, Memoir 179, pp. 1–53.

———— and Scott, W.E., 1982. Pleistocene episodes of alluvial-gravel deposition, southeastern Idaho. *In:* Bonnichsen, B., and Breckenridge, R.M., eds., Cenozoic geology of Idaho. Idaho Bur. Mines and Geology, Bull. 26, pp. 685–702.

Piety, L.A., Sullivan, J.T., and Anders, M.H., 1992. Segmentation and paleoseismicity of the Grand Valley fault, southeastern Idaho and western Wyoming. *In:* Link, P.K., Kuntz, M.A., and Platt, L.B., eds., Regional geology of eastern Idaho and western Wyoming. Geol. Soc. America, Memoir 179, pp. 155–182.

Poole, F.G., and Sandberg, C.A., 1977. Mississippian paleogeography and tectonics of the western United States. *In:* Stewart, J.H., Stevens, C.H., and Fritsche, A.E., eds., Paleozoic paleogeography of the western United States. Soc. Economic Paleontologists and Mineralogists, Pacific Sect., Pacific Coast Paleogeography Symposium 1, pp. 67–85.

Porter, S.C., 1976. Pleistocene glaciation in the southern part of the North Cascade Range, Washington. Geol. Soc. America, Bull., vol. 87, pp. 61–75.

Potter, A.W., Hotz, P.E., and Rohr, D.M., 1977. Stratigraphy and inferred tectonic framework of lower Paleozoic rocks in the eastern Klamath Mountains, northern California. *In:* Stewart, J.H., Stevens, C.H., and Fritsche, A.E., eds., Paleozoic paleogeography of the western United States. Soc. Economic Paleontologists and Mineralogists, Pacific Sect., Pacific Coast Paleogeography Symposium 1, pp. 421–440.

———— et al., eds., 1990. Early Paleozoic stratigraphic, paleogeographic, and biogeographic relations of the eastern Klamath belt, northern California. *In:* Harwood, D.S., and Miller, M.M., eds., Paleozoic and early Mesozoic paleogeographic relations, Sierra Nevada, Klamath Mountains, and related terranes. Geol. Soc. America., Special Paper 255, pp. 57–74.

Powers, H.A., and Wilcox, R.C., 1964. Volcanic ash from Mount Mazama (Crater Lake) and from Glacier Peak. Science, vol. 144, pp. 1334–1336.

Prakash, U., and Barghoorn, E.S., 1961. Miocene fossil woods from the Columbia basalts of central Washington, II. Jour. of the Arnold Arboretum, vol. 42, pp. 347–358.

Price, R.A., 1994. Cordilleran tectonics and the evolution of the western Canada sedimentary basin. *In:* Mossop, G., and Shetsin, I., eds., Geological atlas of the western Canada sedimentary Basin, Alberta Research Council and Canadian Soc. Petroleum Geologists, pp. 13–24.

———— and Carmichael, D.M., 1986. Geometric test for late Cretaceous-Paleogene intracontinental transform faulting in the Canadian Cordillera. Geology, vol. 14, pp. 468–471.

———— Monger, J.W.H., and Roddick, J.A., 1985. Cordilleran cross-section, Calgary to Vancouver. *In:* Templeman-Kluit, D.J., ed., Field guides to geology and mineral deposits in the southern Canadian Cordillera. Geol. Soc. America, Cordilleran Sect., Vancouver, B.C., May, 1985, pp. 3-1 to 3-85.

Priest, G.R., and Vogt, B.F., 1983. Geology and geothermal resources of the central Oregon Cascade Range. Oregon Dept. Geol. and Mineral Indus., Special Paper 15, 123p.

Rau, W.W., 1973. Geology of the Washington coast between Point Grenville and the Hoh River. Wash. Division Geology and Earth Resources, Bull. 66, 58p.

———— 1980. Washington coastal geology between the Hoh and Quillayute River. Wash. Div. of Geology and Earth Resources, Bull. 72, 57p.

———— 1987. Melange rocks of Washington's Olympic coast. *In:* Hill, M.L., ed., Centennial field guide, Geol. Soc. America, Cordilleran Sect., pp. 373–376.

Ray, G.E., Ettlinger, A.D., and Meinert, L.D., 1989. Gold skarns: their distribution, characteristics, and problems in classification. *In:* Geological Fieldwork, Geological Survey of Canada, Paper 1990–1, pp. 237–246.

Read, P.B., 1990. Mount Meager complex, Garibaldi belt, southwestern British Columbia. Geoscience Canada, vol. 17, no. 3, pp. 167–170.

Reesor, J.E., 1970. Some aspects of structural evolution and regional setting in part of the Shuswap metamorphic complex. Geol. Assoc. Canada, Special Paper No. 6, pp. 73–86.

Rehrig, W.A., Reynolds, S.J., and Armstrong, R.L., 1987. A tectonic and geochronologic overview of the Priest River crystalline complex, northeastern Washington and northern Idaho. In: Schuster, J.E., ed., Selected papers on the geology of Washington. Wash. Div. Geology and Earth Resources, Bull. 77, pp. 1–14.

Reidel, S.P., and Hooper, P.R., eds, 1989. Volcanism and tectonism in the Columbia River flood-basalt province. Geol. Soc. America, Special Paper 239, 386p.

——— Hooper, P.R., and Price, S.M., 1987. Columbia River Basalt Group, Joseph and Grande Ronde canyons, Washington. In: Hill, M.L., ed., Centennial field guide, Geological Soc. America, Cordilleran Sect., pp. 351–356.

——— et al., 1994. Late Cenozoic structure and stratigraphy of south-central Washington. In: Lasmanis, R., and Cheney, E.S., convs., Geology of Washington state. Wash. Div. Geology and Earth Resources, Bull. 80, pp. 159–180.

Rhodes, P.T., 1987. Historic glacier fluctuations at Mount Shasta, Siskiyou County. California Geology, vol. 40, no. 9, pp. 205–211.

——— Harms, T.A., and Hyndman, D.W., 1989. Geology of the southern Priest River complex and the Newport fault. In: Joseph, N.L., et al., eds., Geologic guidebook for Washington and adjacent areas. Wash. Div. Geology and Earth Resources, Info. Circular 86, pp. 165–186.

Richards, M.A., Duncan, R.A., and Courtillot, V.E., 1989. Flood basalts and hot spot tracks: plume heads and tails. Science, vol. 246, pp. 103–107.

Richmond G.M., 1986. Stratigraphy and correlation of glacial deposits of the Rocky Mountains, the Colorado plateau, and the ranges of the Great Basin. In: Sibrava, V., Bowen, D.Q., and Richmond, G.M., eds., Quaternary glaciations in the northern hemisphere. New York, Pergamon Press, pp. 99–127.

——— 1986, Tentative correlation of deposits of the Cordilleran ice-sheet in the northern Rocky Mountains. In: Sibrava, V., Bowen, D.Q., and Richmond, G.M., eds., Quaternary glaciations in the northern hemisphere. New York, Pergamon Press, pp. 129–144.

Riddihough, R.P., and Hyndman, R.D., 1989. Queen Charlotte Islands margin. Geol. Soc. America, The Geology of North America, vol. N, pp. 403–411.

——— and Hyndman, R.D., 1991. Modern plate tectonic regime of the continental margin of western Canada. In: Gabrielse, H., and Yorath, C.J., eds., Geology of the Cordilleran orogen in Canada, Geological Survey of Canada, Geology of Canada, no. 4, pp. 437–455.

——— Currie, R.G., and Hyndman, R.D., 1980. The Dellwood knolls and their role in triple junction tectonics off northern Vancouver Island. Canadian Jour. Earth Science, vol. 17, pp. 577–593.

Ringe, D., 1970. Sub-loess basalt topography in the Palouse Hills, southeastern Washington, Geological Soc. America, Bull. 81, pp. 3049–3060.

Roberts, S., and Fountain, D., eds., 1982. Tobacco Root Geological Society, 1980 Field conference guidebook. Spokane Tobacco Root Geological Society, 93p.

Robinson, P.T., Walker, G.W., and McKee, E.H., 1990. Eocene(?), Oligocene, and lower Miocene rocks of the Blue Mountains region. In: Walker, G.W., ed., Geology of the Blue Mountains region of Oregon, Idaho, and Washington: Cenozoic geology of the Blue Mountains region. U.S. Geol. Survey, Prof. Paper 1437, pp. 29–61.

Rodgers, D.W., and Janecke, S.U., 1992. Tertiary paleogeologic maps of the western Idaho-Wyoming-Montana thrust belt. In: Link, P.K., Kuntz, M.A., and Platt, L.B., eds., Regional geology of eastern Idaho and western Wyoming. Geol. Soc. America, Memoir 179, pp. 83–94.

Romer, A.S., 1971. Vertebrate paleontology. University of Chicago Press, 661p.

Ross, C.A., and Ross, J.R.P., 1983. Late Paleozoic accreted terranes of western North America. In: Stevens, C.H., ed., Pre-Jurassic rocks in western North American suspect terranes. Soc. Economic Paleontogists and Mineralogists, Pacific Sect., Mtg., Sacramento, pp. 7–22.

Ross, C.P., 1961. Geology of the southern part of the Lemhi Range, Idaho. U.S. Geol. Survey, Bull 1081-F, pp. 189–255.

Ross, S.H., 1971. Geothermal potential of Idaho. Idaho Bur. of Mines and Geology, Pamphlet 150, 72p.

——— and Savage, C.N., 1967. Idaho earth science; geology, fossils, climate, water, and soils. Idaho Bur. of Mines and Geology, Earth Science Series no. 1, 271p.

Ruppel, E.T., 1982. Cenozoic block uplifts in east-central Idaho and southwest Montana. U.S. Geol. Survey, Prof. Paper 1224, 24p.

Russell, R.H., ed., 1975. Geology and water resources of the San Juan Islands, San Juan County, Wash. Wash. Dept. of Ecology, 171p.

Ryder, J.M., and Clague, J.J., 1986. British Columbia, Quaternary stratigraphy and history; Cordilleran ice sheet. In: Fulton, R.J., ed., Quaternary geology of Canada and Greenland, Geological Survey of Canada, Geology of Canada, no. 1, pp. 48–58.

Rytuba, J.J., Vander Meulen, D.B., and Barlock, V.E., 1991. Tectonic and stratigraphic controls on epithermal precious metal mineralization in the northern part of the Basin and Range, Oregon, Idaho, and Nevada. In: Buffa, R.H., and Coyner, A.R., eds., Geology and ore deposits of the Great Basin. Field Trip Guidebook Compendium, Geol. Soc. Nevada, Reno, pp. 635–695.

——— et al., 1990. Field guide to hot-spring gold deposits in Lake Owyhee volcanic field, eastern Oregon. Geol. Soc. Nevada and U.S. Geol. Survey, Spring field trip guidebook; field trip no. 10, 15p.

Saleeby, J.B., and Harper, G.D., 1993. Tectonic relations between the Galice Formation and the Condrey Mountain schist, Klamath Mountains, northern California. In: Dunne, G.C., and McDougall, K.A., eds., Mesozoic paleogeography of the western United States—II. Soc. Economic Paleontologists and Mineralogists, Pacific Section, pp. 61–80.

Savage, C.N., 1958. Geology and mineral resources of Ada and Canyon Counties. Idaho Bur. of Mines and Geology, County Report 3, 94p.

——— 1965. Geologic history of Pend Oreille Lake region in northern Idaho. Idaho Bur. Mines and Geology, Pam. 134, 18p.

Schuster, J.E., ed., 1987. Selected papers on the geology of Washington. Wash. Div. Geology and Earth Resources, Bull. 77, 396p.

Scott, W.B., 1977. A history of land mammals in the western hemisphere. New York, MacMillan, 693p.

Scott, W.E., 1967. Geology and mineral resources of Bonner County. Idaho, Bureau of Mines and Geology, County Report No. 6, 131p.

——— et al. 1982. Revised Quaternary stratigraphy and chronology in the American Falls area, southeastern Idaho. *In:* Bonnichsen, B., and Breckenridge, R.M., eds., Cenozoic geology of Idaho. Idaho Bureau of Mines and Geology, Bull. 26, pp. 581–595.

Sears, J.W., and Price, R.A., 1978. The Siberian connection; a case for Precambrian separation of the North American and Siberian cratons. Geology, vol. 6, pp. 267–270.

Shotwell, J.A., 1970. Pliocene mammals of southeast Oregon and adjacent Idaho. Univ. of Oregon. Museum Natl. Hist., Bull. 17, 103p.

Sibrava, V., Bowen, D.Q., and Richmond, G.M., eds., 1986. Quaternary glaciations in the northern hemisphere. New York, Pergamon Press, 514p.

Silberling, N.J., et al., 1992. Lithotectonic terrane map of the North American Cordillera. U.S. Geological Survey, Misc. Investigation Series, Map I-2176, 2 sheets.

Silberman, M.L., and Danielson, J., 1993. Gold-bearing quartz veins in the Klamath Mountains in the Redding 1×2 Degree quadrangle, northern California. California Geology, vol. 46, no. 2, pp. 35–44.

Simony, P.S., et al., 1980. Structural and metamorphic evolution of northeast flank of Shuswap complex, southern Canoe River area, British Columbia. Geol. Soc. America, Memoir 153, pp. 445–461.

Skipp, B., and Link, P.K., 1992. Middle and late Proterozoic tectonics in the southern Beaverhead Mountains, Idaho and Montana: a preliminary report. *In:* Link, P.K., Kuntz, M.A., and Platt., L.B., eds., Regional geology of eastern Idaho and western Wyoming. Geol. Soc. America, Memoir 179, pp. 141–154.

——— Sando, W.J., and Hall, W.E., 1979. The Mississippian and Pennsylvanian (Carboniferous) systems in the United States— Idaho. U.S. Geol. Survey, Prof. Paper 1110-A, pp. AA1–AA42.

Smiley, C.J., 1963. The Ellensburg flora of Washington. Univ. Calif. Publ. Geol. Science, vol. 35, Part 3, pp. 159–240.

——— 1989. The Miocene Clarkia fossil area of northern Idaho. *In:* Chamberlain, V.E., Breckenridge, R.M., and Bonnich sen, B., eds., Guidebook to the geology of northern and western Idaho and surrounding area. Idaho Geol. Survey, Bull. 28, pp. 35–48.

——— and Rember, W.C., 1979. Guidebook and road log to the St. Maries River (Clarkia) fossil area of northern Idaho. Idaho Bur. of Mines and Geology, Info. Circular 33, 18p.

——— Shah, S.M.I., and Jones, R.W., 1975. Guidebook for the later Tertiary stratigraphy and paleobotany of the Weiser area, Idaho. Idaho Bur. of Mines and Geology, Info. Circular 28, 12p.

Smith, A.G., Hurley, A.M., and Briden, S.C., 1981. Phanerozoic paleocontinental world maps. Cambridge, England, Cambridge Univ. Press, 102p.

Smith, G.A., 1991. A field guide to depositional processes and facies geometry of Neogene continental volcaniclastic rocks, Deschutes basin, central Oregon. Oregon Geology, vol. 53, no. 1, pp. 3–20.

——— Bjornstad, B.N., and Fecht, K.R., 1989. Neogene terrestrial sedimentation on and adjacent to the Columbia plateau; Washington, Oregon, and Idaho. *In:* Reidel, S.P., and Hooper, P.R., eds., Volcanism and tectonism in the Columbia River flood-basalt province. Geol. Soc. America, Special Paper 239, pp. 187–198.

Smith, G.R., et al., 1982. Fish biostratigraphy of late Miocene to Pleistocene sediments of the western Snake River Plain, Idaho. *In:* Bonnichsen, B., and Breckenridge, R.M., eds., Cenozoic geology of Idaho. Idaho Bur. of Mines and Geology, Bull. 26, pp. 519–541.

Smith, R.L., and Luedke, R.G., 1984. Potentially active volcanic lineaments and loci in western conterminous United States. *In:* U.S. Natl. Research Council, Geophysics Study Committee, Explosive volcanism: inception, evolution, and hazards. Studies in geophysics, National Acad. Press, Washington, D.C., pp. 47–66.

Snavely, P.D., 1987. Tertiary geologic framework, neotectonics, and petroleum potential of the Oregon-Washington continental margin. *In:* Scholl, D.W., Grantz, A., and Vedder, J.G., comps. and eds., Geology and resources potential of the continental margin of western North America and adjacent ocean basins Beaufort Sea to Baja, California. Circum-Pacific Council for Energy and Mineral Res., Earth Science Series, vol. 6, pp. 305–335.

——— and Wells, R.E., 1991. Cenozoic evolution of the continental margin of Oregon and Washington. U.S. Geol. Survey, Open-file Report 91-441-B, 34p.

——— et al., 1980. Geology of the west-central part of the Oregon Coast Range. *In:* Oles, K.F., et al., eds., Geologic field trips in western Oregon and southwestern Washington. Oregon. Dept. Geol. and Mineral Indus., Bull. 101, pp. 39–76.

Snyder, C.T., Hardman, G., and Zdenek, F.F., 1964. Pleistocene lakes of the Great Basin. U.S. Geol. Survey, Misc. Inv. Map I-416.

Souther, J.G., 1991. Volcanic regimes. *In:* Gabrielse, H., and Yorath, C.J., eds., Geology of the Cordilleran orogen in Canada. Geol. Survey of Canada, Geology of Canada, no. 4, pp. 459–490.

Staatz, M.H., et al., 1972. Geology and mineral resources of the northern part of the North Cascades National Park, Washington. U.S. Geol. Survey, Bull. 1359, 132p.

Stanley, G.D., 1986. Late Triassic coelenterate faunas of western Idaho and northeast Oregon: implications for biostratigraphy and paleoecology. U.S. Geol. Survey, Prof. Paper 1435, pp. 23–39.

——— 1986. Travels of an ancient reef. Natural History, vol. 87, no. 11, pp. 36–42.

——— and Beauvais, L., 1990. Middle Jurassic corals from the Wallowa terrane, west-central Idaho. Jour. Paleo., vol. 64, no. 3, pp. 352–362.

——— and McRoberts, C.A., 1993. A coral reef in the Telkwa Range, British Columbia: the earliest Jurassic example. Canadian Jour. Earth Science vol. 30, pp. 819–831.

Stearns, H.T., 1928. A guide to the Craters of the Moon National Monument, Idaho. Idaho Bur. of Mines and Geology, Bull. 13, 57p.

Stephens, G.C., 1988. History of gold mining on the Yankee Fork River, Custer County. *In:* Link, P.K., and Hackett, W.R., eds. Guidebook to the geology of central and southern Idaho. Idaho Bureau of Mines and Geology, Bull. 27, pp. 223–226.

——— 1988. Holocene volcanism at Craters of the Moon National Monument. *In:* Link, P.K., and Hackett, W.R., eds.,

Guidebook to the geology of central and southern Idaho. Idaho Geol. Survey, Bull. 27, pp. 241–244.

Stevens, C.H., 1977. Permian depositional provinces and tectonics, western United States. *In:* Stewart, J.H., Stevens, C.H., and Fritsche, A.E., eds., Paleozoic paleogeography of the western United States. Soc. Economic Paleontologists and Mineralogists, Pacific Sect., Mtg. Sacramento, Pacific Coast Paleogeography Symposium 1, pp. 113–135.

——— ed., 1983. Pre-Jurassic rocks in western North American suspect terranes. Society Economic Paleontologists and Mineralogists, Pacific Sect., Mtg. Sacramento, 141p.

——— Yancey, T.E., and Hanger, R.A., 1990. Significance of the provincial signature of early Permian faunas of the eastern Klamath terrane. *In:* Harwood, D.S., and Miller, M.M., eds., Paleozoic and early Mesozoic paleogeographic relations; Sierra Nevada, Klamath Mountains, and related terranes. Geol. Society America, Special Paper 255, pp. 201–218.

Stewart, J.H., and Suczek, C.A., 1977. Cambrian and latest PreCambrian paleogeography and tectonics in the western United States. *In:* Stewart, J.H., Stevens, C.H., and Fritsche, A.E., eds., Paleozoic paleogeography of the western United States. Soc. Economic Paleontologists and Mineralogists, Pacific Sect., Paleogeography Symposium 1, pp. 1–18.

——— Stevens, C.H., and Fritsche, A.E., eds., 1977. Paleozoic paleogeography of the western United States. Society Economic Paleontologists and Mineralogists, Pacific Sect., Paleogeography Symposium 1, 502p.

Stockwell, C.H., ed., 1957. Geoleogy and economic minerals of Canada. Ottawa, Dept. of Mines and Technical Surveys, 514p.

Stover, C., and Coffman, J., 1993. Seismicity of the U.S., 1568–1989 (revised). U.S. Geol. Survey, Prof. Paper 1527, 418p.

Stradling, D.F., and Kiver, E.P., 1989. The northern Columbia plateau from the air. *In:* Joseph, N.L. et al., eds., Geologic guidebook for Washington and adjacent areas. Wash. Division of Geology and Earth Resources, Info. Circular 86, pp. 349–356.

Suek, D.H., and Knaup, W.W., 1979. Paleozoic carbonate buildups in the Basin and Range province. *In:* Newman, G.W., and Goode, H.D., eds., Basin and Range symposium and Great Basin field conference. Rocky Mountain Assoc. of Geologists and Utah Geologic Assoc., pp. 245–257.

Swanson, D.A., and Wright, T.L., 1981. Guide to geologic field trip between Lewiston, Idaho, and Kimberly, Oregon, emphasizing the Columbia River Basalt Group. *In:* Johnston, D.A., and Donnelly-Nolan, J., eds., Guides to some volcanic terranes in Washington, Idaho, Oregon, and northern California. U.S. Geol. Survey, Circular 838, pp. 1–14.

Tabor, R.W., 1987. Geology of Olympic National Park. Seattle, Pacific Northwest Parks and Forests Assoc., 144p.

——— 1987. A Tertiary accreted terrane: oceanic basalt and sedimentary rocks in the Olympic Mountains, Washington. *In:* Hill, M.L., ed., Centennial field guide, Geol. Soc. America, Cordilleran Sect., pp. 377–382.

——— and Cady, W.M., 1978. The structure of the Olympic Mountains, Washington—analysis of a subduction zone. U.S. Geol. Survey, Prof. Paper 1033, 38p.

——— Zartman, R.E., and Frizzell, V.A., 1987. Possible tectonostratigraphic terranes in the North Cascades crystalline core, Washington. *In:* Schuster, J.E., ed., Selected papers on the geology of Washington. Wash. Div. Geology and Earth Resources, Bull. 77, pp. 107–127.

——— et al., 1989. Accreted terranes of the North Cascades Range, Washington. American Geophys. Union, Field trip guidebook T307, 62p.

Taylor, D.G., et al., 1984. Jurassic ammonite biogeography of western North America; the tectonic implications. *In:* Westermann, G.E.G., ed., Jurassic-lower Cretaceous biochronology and biogeography of North America. Geol. Assoc. of Canada, Special Paper 27, pp. 121–124.

Taylor, E.M., 1990. Volcanic history and tectonic development of the central High Cascade range, Oregon. Jour. Geophys. Res., vol. 95, no. B12, pp. 19, 611–19, 622.

——— 1981. Central High Cascade roadside geology, Bend, Sisters, McKenzie Pass, and Santiam Pass, Oregon. *In:* Johnson, D.A., and Donnelly-Nolan, J., eds., Guides to some volcanic terranes in Washington, Idaho, Oregon, and northern California. U.S. Geol. Survey, Circular 838, pp. 55–58.

Templeman-Kluit, D.J., ed., 1985. Field guides to geology and mineral deposits in the southern Canadian Cordillera. Geological Survey of Canada, Cordilleran Sect. Mtg., Vancouver, B.C., May, 1985, various pagings.

Tidwell, W.D., 1975. Common fossil plants of western North America. Provo, Brigham Young University Press, 197p.

Tolan, T.L., and Reidel, S.P., 1989. Revisions to the estimates of the areal extent and volume of the Columbia River Basalt group. *In:* Reidel, S.P., and Hooper, P.R., eds., Volcanism and tectonism in the Columbia River flood-basalt province. Geol. Soc. America, Special Paper 239, pp. 1–19.

Trehu, A.M., et al., 1994. Crustal architecture of the Cascadia forearc. Science, vol. 266, pp. 237–243.

U.S. Bureau of Mines, [various dates]. Minerals yearbook. [various pagings].

U.S. National Oceanic and Atmospheric Administration, 1974. North Pacific Ocean. NOS Seamap Series, no. 12042-12B.

U.S. National Park Service, 1986. North Cascades; a guide to the North Cascades. Guidebook 131, 112p.

Vallier, T.L., and Brooks, H.C., eds., 1987. Geology of the Blue Mountains region of Oregon, Idaho, and Washington: the Idaho batholith and its border zone. U.S. Geol. Survey, Prof. Paper 1436, 196p.

——— and Brooks, H.C., eds., 1994. Geology of the Blue Mountains region of Oregon, Idaho, and Washington: stratigraphy, physiography, and mineral resources of the Blue Mountains region. U.S. Geol. Survey, Prof. Paper 1439, 198p.

——— and Brooks, H.C., eds., 1995. Geology of the Blue Mountains region of Oregon, Idaho, and Washington; petrology and tectonic evolution of pre-Tertiary rocks. U.S. Geol. Survey, Prof. Paper 1438, 540p.

——— Brooks, H.C., and Thayer, T.P., 1977. Paleozoic rocks of eastern Oregon and western Idaho. *In:* Stewart, J.H., Stevens, C.H. and Fritsche, A.E., eds. Paleozoic paleogeography of the western United States. Soc. Economic Paleontologists and Mineralogists, Pacific Coast Paleogeography Symposium 1, pp. 455–466.

Vance, J.A., et al., 1987. Early and middle Cenozoic stratigraphy of the Mount Rainier-Tieton River area, southern Washington Cascades. *In:* Schuster, J.E., ed., Selected papers on the geology of Washington. Wash. Div. Geology and Earth Resources, Bull. 77, pp. 269–290.

Vennum, W., 1994. Geology of Castle Crags, Shasta and Siskiyou counties. California Geology, vol. 47, no. 2, pp. 31–38.

Von Huene, R., 1989. Continental margins around the Gulf of Alaska. *In:* Winterer, E.L., Hussong, D.M., and Decker, R.W., eds. The eastern Pacific Ocean and Hawaii. Geol. Soc. America, vol. N, pp. 383–401.

Wagner, D.L., 1988. Geology of Del Norte and Siskiyou counties and adjacent portions of Humboldt, Shasta, and Trinity counties. California Geology, vol. 41, no. 12, pp. 267–275.

Wagner, F.J.E., 1958. Paleoecology of the marine Pleistocene faunas of southwestern British Columbia. Geol. Survey Canada, Bull. 52, 67p.

Wahlstrom, R., and Rogers, G.C., 1992. Relocation of earthquakes west of Vancouver Island, British Columbia, 1965–1983. Canadian Jour. Earth Science, vol. 29, pp. 953–961.

Waitt, R.B., 1987. Evidence for dozens of stupendous floods from glacial Lake Missoula in eastern Washington, Idaho, and Montana. *In:* Hill, M.L., ed., Centennial Field Guidebook, Geol. Soc. America, Cordilleran Sect., pp. 345–350.

Walker, G.W., 1979. Revisions to the Cenozoic stratigraphy of Harney Basin, southeastern Oregon. U.S. Geol. Survey, Bull. 1475, 34p.

———— 1990. Miocene and younger rocks of the Blue Mountains region, exclusive of the Columbia River Basalt Group and associated mafic lava flows. U.S. Geol. Survey, Prof. Paper 1437, pp. 101–118.

———— and McCloud, N.S., 1991. Geologic map of Oregon. U.S. Geological Survey.

———— and Nolf, B., 1981. High Lava Plains, Brothers Fault zone to Harney Basin, Oregon. *In:* Johnston, D.A., and Donnelly-Nolan, J., eds., Guides to some volcanic terranes in Washington, Idaho, Oregon, and northern California. U.S. Geol. Survey, Circular 838, pp. 105–111.

———— and Robinson, P.T., 1990. Cenozoic tectonism and volcanism of the Blue Mountains region. *In:* Walker, G.W., ed., 1990. Geology of the Blue Mountains region of Oregon, Idaho, and Washington: Cenozoic geology of the Blue Mountains region. U.S. Geol. Survey, Prof. Paper 1437, pp. 119–134.

———— and Robinson, P.T., 1990. Paleocene(?), Eocene, and Oligocene(?) rocks of the Blue Mountains region. *In:* Walker, G.W., ed., 1990. Geology of the Blue Mountains region of Oregon, Idaho, and Washington: Cenozoic geology of the Blue Mountains region. U.S. Geol. Survey, Prof. Paper 1437, pp. 13–27.

———— ed., 1990. Geology of the Blue Mountains region of Oregon, Idaho, and Washington: Cenozoic geology of the Blue Mountains region. *In:* Walker, G.W., ed., 1990. Geology of the Blue Mountains region of Oregon, Idaho, and Washington: Cenozoic geology of the Blue Mountains region. U.S. Geol. Survey, Prof. Paper 1437, 135p.

Walter, S.R., 1986. Intermediate-focus earthquakes associated with Gorda plate subduction in northern California. Seismological Soc. America, Bull., vol. 75, pp. 237–249.

Ward, P.D., 1976. Upper Cretaceous ammonites (Santonian-Campanian) from Orcas Island, Washington. Jour. Paleo., vol. 50, no.3, pp. 454–461.

Waring, G., 1964. Thermal springs of the United States and other countries of the world—a summary. U.S. Geol. Survey, Prof. Paper 492, 383p.

Washburn, A.L., 1988. Mima mounds; an evaluation of proposed origins with special reference to the Puget Lowlands. Wash.

Div. of Geology and Earth Resources, Report of Investigations 29, 53p.

Waters, A.C., 1992. Captain Jack's stronghold; the geologic events that created a natural fortress, Siskiyou County California Geology, vol. 45, no. 5, pp. 135–144.

———— Donnelly-Nolan, J.M., and Rogers, B.W., 1990. Selected caves and lava-tube systems in and near Lava Beds National Monument, California. U.S. Geol. Survey, Bull. 1673, 102p.

Watkinson, A.J., and Ellis, M.A., 1987. Recent structural analysis of the Kootenay arc in northeast Washington. *In:* Schuster, J.E., ed., Selected papers on the geology of Washington. Wash. Div. Geology and Earth Resources, Bull. 77, pp. 41–53.

Weaver, C.E., 1912. Geology and ore deposits of the Index mining district. Wash. Geol. Survey, Bull. 7, 93p.

Webster, G.D., and Nunez, L., 1982. Geology of the steptoes and palouse hills of eastern Washington, a roadlog of the area south of Spokane, Washington. *In:* Roberts, S., and Fountain, D., eds., Field conference guidebook, Spokane Washington, 1980. Tobacco Root Geological Society, pp. 45–57.

Wells, M.W., 1963. Rush to Idaho. Idaho Bur. of Mines and Geology, Bull. 19, 46p.

———— 1983. Gold camps & silver cities. 2d ed., Idaho Bur. of Mines and Geology, Bull. 22, 165p.

Wells, R.E., and Coe, R.S., 1985. Paleomagnetism and geology of Eocene volcanic rocks of southwest Washington, implications for mechanisms of tectonic rotation. Jour. Geophys. Res., vol. 90, no. 32, pp. 1925–1947.

———— and Heller, P.L., 1988. The relative contribution of accretion, shear, and extension to Cenozoic tectonic rotation in the Pacific Northwest, Geol. Soc. America, Bull., vol. 100, pp. 325–338.

———— et al., 1984. Cenozoic plate motions and the volcano-tectonic evolution of western Oregon and Washington. Tectonics, vol. 3, no. 2, pp. 275–294.

Wheeler, G., 1982. Problems in the regional stratigraphy of the Strawberry volcanics. Oregon Geology, vol. 44, no. 1, pp. 3–7.

Wheeler, J.O., ed., 1970. Structure of the southern Canadian Cordillera. Geol. Assoc. Canada, Special Paper No. 6, 166p.

———— and McFeely, P., comps., 1991. Tectonic assemblage map of the Canadian Cordillera and adjacent parts of the United States of America. Geol. Survey Canada, Map. 1712-A.

Whetten, J.T., et al., 1980. Allochthonous Jurassic ophiolite in northwest Washington. Geol. Soc. America, Bull., Part 1, vol. 91, pp. 359–368.

White, D.L., and Vallier, T.L., 1994. Geologic evolution of the Pittsburg Landing area, Snake River Canyon, Oregon and Idaho. *In:* Vallier, T.L., and Brooks, H.C., eds., Geology of the Blue Mountains region of Oregon, Idaho, and Washington: Stratigraphy, physiography, and mineral resources of the Blue Mountains region. U.S. Geol. Survey, Prof. Paper 1439, pp. 55–73.

White, J.D.L., 1994. Intra-arc basin deposits within the Wallowa terrane, Pittsburg Landing area, Oregon and Idaho. *In:* Vallier, T.L., and Brooks, H.C., eds., Geology of the Blue Mountains region of Oregon, Idaho, and Washington; stratigraphy, physiography, and mineral resources of the Blue Mountains region. U.S. Geol. Survey, Prof. Paper 1439, pp. 75–89.

———— et al., 1992. Middle Jurassic strata link Wallowa, Olds Ferry, and Izee terranes in the accreted Blue Mountains island arc, northeastern Oregon. Geology, vol. 20, pp. 729–732.

Whitney, D.L., and McGroder, M.F., 1989. Cretaceous crustal section through the proposed Insular-Intermontane suture, North Cascades, Washington. Geology, vol. 17, pp. 555–558.

Williams, H., 1976. The ancient volcanoes of Oregon. Oregon State System of Higher Education, Eugene, Oregon. Condon Lectures, 6th ed., 70p.

Williams, I.A., 1991. Geologic history of the Columbia Gorge as interpreted from the historic Columbia River scenic highway. 3rd ed., rev. Portland, Oregon Historical Soc. Press, 137p.

Wills, C.J., 1991. Active faults north of Lassen Volcanic National Park, northern California. California Geology, vol. 44, no. 3, pp. 51–58.

Wise, W.S., 1970. Cenozoic volcanism in the Cascade Mountains of southern Washington. Wash. Division of Mines and Geology, Bull. 60, 45p.

Wolfe, J.A., and Wehr, W.C., 1991. Significance of the Eocene fossil plants at Republic, Washington. Washington Geology, vol. 19, no. 3, pp. 18–24.

Womer, M.B., Greeley, R., and King, J.S., 1982. Phreatic eruptions of the eastern Snake River Plain of Idaho. *In:* Bonnichsen, B., and Breckenridge, R.M., eds., Cenozoic geology of Idaho. Idaho Bur. of Mines and Geology, Bull. 26, pp. 453–464.

Wood, C.A., and Kienle, J., eds., 1990. Volcanoes of North America; United States and Canada. New York, Cambridge Univ. Press, 353p.

Woods, M.C., 1988. Ice age geomorphology in the Klamath Mountains. California Geology, vol. 41, no. 12, pp. 273–275.

Worl, R.G., et al., 1995. Geology and mineral resources of the Hailey 1° × 2° Quadrangle and the western part of the Idaho Falls 1° × 2° Quadrangle, Idaho. U.S. Geological Survey, Bull. 2064, Ch. A–R.

Wright, H.E., and Frey, D.G., eds., 1965. The Quaternary history of the United States. Princeton Univ. Press, Princeton, N.J., 723p.

Wust, S.L., and Link, P.K., 1988. Field guide to the Pioneer Mountains core complex, south-central Idaho. *In:* Link, P.K., and Hackett, W.R., eds., Guidebook to the geology of central and southern Idaho. Idaho Geol. Survey, Bull. 27, pp. 43–54.

Yeats, R.S., 1989. Current assessment of earthquake hazard in Oregon. Oregon Geology, vol. 51, no. 4, pp. 90–92.

———— et al., 1977. Structure, stratigraphy, plutonism, and volcanism of the central Cascades, Washington (Field trip no. 10). *In:* Brown, E.H., and Ellis, R.C., eds., Geological excursions in the Pacific Northwest. Geol. Soc. America, Annual Mtg., Seattle, pp. 265–308.

Yorath, C.J., 1980. The Apollo structure in Tofino basin, Canadian Pacific continental shelf. Canadian Jour. Earth Science, vol. 17, pp. 758–775.

———— 1990. Where terranes collide. Victoria B.C., Orca Books Publisher, 250p.

———— and Chase, R.L., 1981. Tectonic history of the Queen Charlotte Islands and adjacent areas—a model. Canadian Jour. Earth Science, vol. 18, pp. 1717–1739.

———— and Hyndman, R.D., 1983. Subsidence and thermal history of Queen Charlotte Basin. Canadian Jour. Earth Science, vol. 20, no. 1, pp. 135–159.

Ziegler, A.M., et al., 1979. Paleozoic paleogeography. Ann. Rev. Earth Planetary Science. vol. 7, pp. 473–502.

INDEX

Items in **bold** indicate an illustration.